Next-Generation Algae

Scrivener Publishing
100 Cummings Center, Suite 541J
Beverly, MA 01915-6106

Publishers at Scrivener
Martin Scrivener (martin@scrivenerpublishing.com)
Phillip Carmical (pcarmical@scrivenerpublishing.com)

Next-Generation Algae

Volume I: Applications in Agriculture, Food and Environment

Edited by
Charles Oluwaseun Adetunji
Julius Kola Oloke
Naveen Dwivedi
Sabeela Beevi Ummalyma
Shubha Dwivedi
Daniel Ingo Hefft
and
Juliana Bunmi Adetunji

Scrivener Publishing

WILEY

This edition first published 2023 by John Wiley & Sons, Inc., 111 River Street, Hoboken, NJ 07030, USA and Scrivener Publishing LLC, 100 Cummings Center, Suite 541J, Beverly, MA 01915, USA

For more information about Scrivener publications please visit www.scrivenerpublishing.com.

Wiley Global Headquarters
111 River Street, Hoboken, NJ 07030, USA

For details of our global editorial offices, customer services, and more information about Wiley products visit us at www.wiley.com.

Library of Congress Cataloging-in-Publication Data

ISBN 978-1-119-85727-3

Cover image: Pixabay.Com
Cover design by Russell Richardson

Set in size of 11pt and Minion Pro by Manila Typesetting Company, Makati, Philippines

Printed in the USA

10 9 8 7 6 5 4 3 2 1

Contents

Preface

The global population is projected to reach 9 billion by the year 2050. It is imperative to begin preparing for the challenges that come with accommodating this rapidly growing population. One of the most significant challenges will be to ensure that we can provide adequate food and nutritious diets to this growing population, as well as a healthy environment. The orthodox agricultural practice has depended heavily on non-renewable inputs such as pesticides, fertilizers, herbicides, and insecticides. Due to the rapid pace of industrialization and the continuous growth of human population, we are witnessing an alarming increase in environmental pollution caused by various anthropogenic and industrial activities. This has resulted in extensive damage to our environment. The majority of these pollutants are derived from inappropriate utilization and discharge of industrial effluents, fertilizers, pesticides, smelting and mining of ores, as well as the release of automobile exhaust and effluent from storage batteries. Additionally, the release of metalloids, heavy metals, petrochemicals, and petroleum hydrocarbons also contribute significantly to the problem. However, their introduction has led to increased agricultural production and significant advancements for humankind, but the application of agrochemicals has also resulted in numerous environmental and health hazards. Algae have been identified as a sustainable biotechnological resource that could help in resolving several of these problems.

This book provides up-to-date and cutting-edge information on the application of algae in producing sustainable solutions to various challenges that arise from an increase in agricultural production, as well as its utilization in bioremediation of industrial wastewater. Moreover, this book provides detailed information about the recent advancements in smart microalgae wastewater treatment using Internet of Things (IoT) and edge computing applications. Other topics covered include the use of microalgae in various applications (with past, present and future projections); the use of algae to remove arsenic; algae's role in plastic biodegradation, heavy metal bioremediation, and toxicity removal from industrial wastewater; the application of DNA transfer techniques in algae; the use of algae as food and in the production of food, ascorbic acid, health food, supplements, and food surrogates; relevant biostimulants and biofertilizers that could be derived from cyanobacterials and their role in sustainable agriculture; and algae's application in the effective production of biofuels and bioenergy.

This book is aimed at a diverse audience, including global leaders, industrialists, individuals in the food industry, agriculturists, the fishery sector, animal husbandry practitioners, investors, innovators, farmers, policy makers, extension workers, educators, researchers, and those in other interdisciplinary fields of science. It also serves as an educational resource manual and project guide for undergraduate and postgraduate students, as well as for educational institutions that seek to carry out research in the field of algae. Additionally, this book unites experts in relevant fields to describe the successful application of algae and its derivatives for bioremediation of extremely polluted environments, especially in water, air and soil. This book is highly recommended to a diverse community of professionals, scientists, environmentalists, industrialists, researchers, students in higher education, innovators, and policy makers who have an interest in bioremediation technologies and sustainable development.

I want to express my deepest appreciation to all the contributors who have dedicated their time and efforts to make this book a success. Furthermore, I want thank my coeditors for their effort and dedication during this project. Moreover, I wish to gratefully acknowledge the suggestions, help, and support of Martin Scrivener and others from Scrivener Publishing.

Charles Oluwaseun Adetunji
(Ph.D, AAS affiliate MNYA; MBSN; MNSM, MNBGN)
Dean Faculty of Science, Edo State University, Uzairue, Nigeria
March 2023

1

Smart Microalgae Wastewater Treatment: IoT and Edge Computing Applications with LCA and Technoeconomic Analysis

Mohd. Zafar[1]*, Avnish Pareek[1], Taqi Ahmed Khan[1], Ramkumar Lakshminarayanan[2] and Naveen Dwivedi[3]

[1]Department of Applied Biotechnology, College of Applied Sciences & Pharmacy, University of Technology and Applied Sciences - Sur, Sultanate of Oman
[2]Department of Information Technology, College of Computing & Information Sciences, University of Technology and Applied Sciences - Sur, Sultanate of Oman
[3]Department of Biotechnology, S. D. College of Engineering and Technology, Muzaffarnagar, India

Abstract

The application of microalgae in applied biotechnological studies for different biomaterials, such as biodiesel, bioethanol, and other high-value bioproducts, has been gaining attention in recent years. Large-scale integrated microalgae-wastewater treatment facilities have emerged as a promising technology. Technoeconomic and life cycle analyses of integrated algae technology in municipal wastewater treatment plants (WWTPs) can reveal its potential as a viable market technology. Thus, integrated microalgae WWTPs is seen as a promising field and is getting attention from the scientific community due to its multifold benefits in terms of nitrogen and phosphorous removal with reduction of organic load, accumulation of heavy metals, and simultaneous production of value-added biomaterials.

This chapter was designed to provide concise details on recent advancements in biological and technological approaches, LCA studies, and IoT and edge computing-based modeling and monitoring of integrated microalgae WWTPs with a technoeconomic feasibility analysis for its assessment as a promising market technology. It is noteworthy that stakeholders have an interest in integrated microalgae WWTPs, but are looking for a standardized process, including design,

**Corresponding author*: mohdz.sur@cas.edu.om

Charles Oluwaseun Adetunji, Julius Kola Oloke, Naveen Dwivedi, Sabeela Beevi Ummalyma, Shubha Dwivedi, Daniel Ingo Hefft and Juliana Bunmi Adetunji (eds.) Next-Generation Algae: Volume I: Applications in Agriculture, Food and Environment, (1–48) © 2023 Scrivener Publishing LLC

data availability, and management aspects, along with a legislative framework that makes it simple to implement.

***Keywords*:** Microalgae biorefinery, wastewater treatment, life cycle assessment, emergy analysis, IoT and edge computing

1.1 Introduction

It is noteworthy that global warming is considered a major issue for many countries around the world. Due to the recent pace of industrialization and urbanization, the emission of different greenhouse gases (GHGs), such as carbon dioxide (CO_2), has led to climate change. Thus, the agreement between world nations known as the Kyoto Protocol was enforced in 1997 to ensure the specific reduction of GHGs by countries. Among the different GHGs, CO_2 is considered to be the largest contributor to the greenhouse effect, and CO_2 mitigation strategies will directly affect the total GHGs emissions. In order to remove the excess atmospheric CO_2 emission, the following methods have been adopted worldwide: (i) Physicochemical processes, including solvent scrubbing, adsorption, absorption, cryogenics and membranes, (ii) Ocean storage of CO_2 and (iii) Biological transformation and mitigation of CO_2 to organic matter using a biological system [1].

Globally, about a 40% water deficit is predicted by 2030, along with several unavoidable challenges associated with societal and economic development in view of current perspectives on the increasing demand for water and lack of water reclamation technology [2]. The conventional wastewater treatment processes, viz. aerobic activated sludge-based process, nitrification-denitrification, and phosphorous removal, are facing challenges to meet the stringent nutrients discharge standards and a large amount of wastewater effluent is still being discharged with nutrients contents, resulting in eutrophication in the aquatic environment [2]. In addition, there are several other disadvantages, such as the high energy consumption, carbon emission, additional sludge discharge, and instability associated with these conventional processes, which can hinder the sustainability-based low carbon, low energy consumption, and resource recycling associated wastewater treatment [1].

Thus, the microalgae-related wastewater treatment (MBWT) process has been gaining attention in recent years and is considered as one of the most promising advanced technologies for sustainable wastewater treatment and efficient nutrient recovery from wastewater. The feasibility of microalgae-related treatment of wastewater generated from different sources, such as municipal, agricultural, and industrial, is being exploited as a tertiary wastewater treatment by many researchers because of its advantages as a highly efficient process for nutrient removal [3–7]. A symbiotic relationship between microalgae and the bacterial population of wastewater was reported by Oswald *et al.*, who observed the efficiency of microalgae in the enhancement of hazardous compounds removal with protection of the bacterial population [8]. Under symbiotic relationship, microalgae utilize CO_2 (produced through aerobic metabolism of bacteria) through the process of photosynthesis and generated O_2 could be utilized by the heterotrophic bacterial population for the assimilation of waste organic compounds. This created the idea of utilizing microalgae for wastewater treatment for the removal of excess nutrients of wastewater effluent and to reduce the risk of the eutrophication threat to natural water bodies. Furthermore, the microalgal biomass produced in wastewater treatment could be considered for "value-added product from waste" as a feedstock in further biorefinery processes [9–11] (Figure 1.1).

In this chapter, recent advancements with respect to diversity of microalgae, process system, internet of things (IoT) and edge computing-based process monitoring and control, and life cycle assessment (LCA)-based techno-economic feasibility analysis of microalgae-based wastewater treatment process are discussed. The details of psychrophilic, thermophilic and acidophilic microalgae and their roles in high-tech, low-cost, and environmentally friendly wastewater treatment process are discussed. Also, the different process systems associated with CO_2 bio-fixation with simultaneous

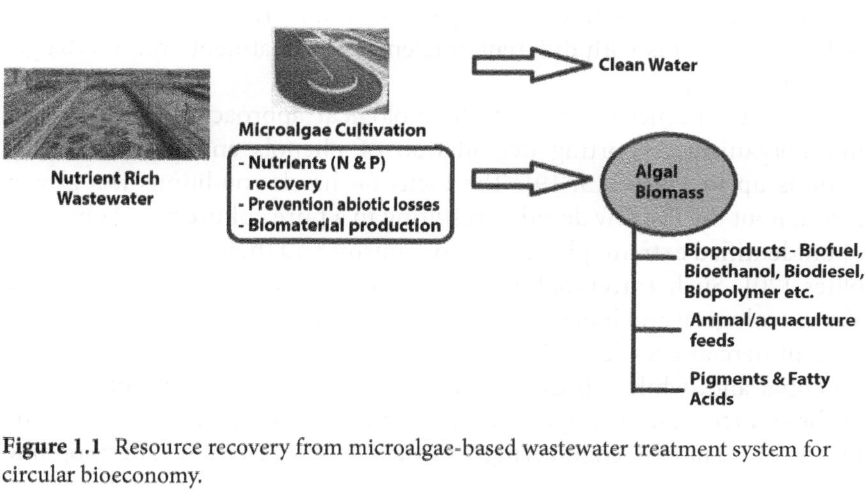

Figure 1.1 Resource recovery from microalgae-based wastewater treatment system for circular bioeconomy.

wastewater treatment are discussed. In addition, the application of emerging technologies, such as IoT automation, to microalgae-related technologies and machine learning approaches for data acquisition, monitoring and analysis of microalgae-based wastewater treatment system is discussed in view of the establishment of an integrated microalgae-wastewater treatment-based biorefinery and bioeconomy. Finally, the evaluation of microalgae-based carbon capture technology associated with wastewater is provided in terms of life cycle assessment, emergy analysis, and material flow analysis.

1.2 Importance and Potential of Extremophilic Microalgae-Based Wastewater Treatment (WWT) Plant

The essential importance of water to life on Earth is threatened by water pollution, which is a significant environmental concern [12]. Water contamination may be caused by anthropogenic or natural activity. The most important causes of human-made water pollution are harmful products from manufacturing processing and effluent making from businesses such as petrochemical plants and pulp and paper mills [13]. The hazardous and carcinogenic organic pollution emitted by crude oil, pharma, petrochemical and coal industries is recognized as being phenol and phenol compounds [14, 15]. Several research studies have examined the biological removal by microalgae of carbonate, nitrogen, and phosphorus through wastewater fluids. Different microalgal species are used in diverse types of wastewaters, including municipal, farming, brewery, refineries and industrial effluents with different efficiencies of treatment and microalgal growth [16–18].

There are numerous benefits to biological approaches, with certain microorganisms reporting degradation of phenols and phenolic compounds up to 1 g/L [15, 19]. The focus on harsh conditions has grown throughout the last few decades, resulting in a pure culture being obtained of unidentified extremophilic microorganisms and their associated metabolites [20]. Such extremophilic bacteria can provide crucial knowledge about ecological and biochemical responses and can lead to biotechnology or commercial uses [21, 22]. Extreme thermophiles currently have great potential and, while utilizing a contemporary understanding of genetics of these microbes, their application in renewable feedstock production by means of metabolic engineering will further increase [23]. Thermophilic microalgae are also used to find enough enzymes that then are integrated into plant genomes to increase their output and resistance to production [15]. Micro-algae separation and selection allow high quantities of biomass and important chemicals such as lipids to be produced in an industrial way [22, 24, 25]. The capacity to extract ammonium from wastewater at temperatures of 40-42 °C and light intensities of 2,500 µmol m^2/s for 5 h in a day was studied using a green microalga *Chlorella sorokiniana* isolated from a wastewater stabilization pond at La Paz, Baja California Sur, Mexico [15, 26]. Thermophilic microalgae may obviously be utilized as a gene pool to identify thermostable enzymes which can be employed in dry locations for improved stability and culture in such settings [27]. Thermophilic microalgae are becoming increasingly more important since they can live at high CO_2 levels. This characteristic makes them attractive candidates for CO_2 emissions from industrial flue gases and adds a step towards global warming reduction. Thermophilic microalgae are efficiently employed to bioremediate harmful industrial effluents and wastewater regardless of origin [15].

1.3 Status of Microalgae-Based WWT Plants

1.3.1 Conditions and Requirements (Abiotic and Biotic Requirements, Nutrients Requirement)

Wastewater remediation is required for preventing pollution and contamination of freshwater bodies as well as for effective reuse of the treated wastewater for sustainability. An ever-increasing population, reduction of freshwater availability, expanding industrialization, and growing human development index (HDI) has increased the demand for wastewater recycling and its sustainable utilization to help manage the precious potable water resources globally in the 21st century [28].

Wastewater is treated conventionally using four types of treatment methods based on the technology used or the category of inflow water. The different treatment plant types are sewage treatment plants (STPs), effluent treatment plants (ETPs), activated sludge plants (ASPs), and common and combined effluent treatment plants (CETPs). Most of the resultant treated water is used for non-potable applications after secondary treatments itself because of technological and/or logistical limitations [29, 30] and non-mandatory status of the tertiary treatment. However, this type of treated water often does not meet the minimum quality standards of water reuse and once released into water bodies it rapidly brings down the dissolved oxygen (DO) and causes pH fluctuation, resulting in the creation of dead aquatic zones and an increase in the overall toxicity [31, 32]. Moreover, these conventional wastewater treatment plants (CWWTPs) are energy intensive and require high operational and maintenance cost [33, 34]. In such a scenario, where the conventional systems are already posing challenges, an ever-increasing population will further stress the global wastewater treatment and reuse scenario as the nutrient load of nitrogen and phosphorus will increase, which will further call for a mandatory tertiary treatment [35–38]. Studies have shown that microalgae are excellent candidates for nitrogen and phosphorus removal and are better than other classes of microorganisms. Being photosynthetic and highly adaptive in their environment, microalgae are also considered the best candidates for tertiary treatment systems. The autotrophic nature of these organisms reduces the system's energy footprint and atmospheric carbon sequestering along with N and PO_4^- removal, which is an added bonus to the environment [39–44].

Wastewaters are complex systems, their treatment is not as straightforward as often understood in terms of biochemical oxygen demand (BOD), chemical oxygen demand (COD) and sludge. Their temporal and spatial characteristics depend on their source, geophysical conditions, factors such as temperature and pH, effluent and nutrient load, physical and chemical impurities, biotic load and flow regime, treatment system size, treatment protocol, transformation products and treatment technology, etc. Besides the composition of the wastewater, wastewater treatment at a national/regional level also depends upon the environmental policy, water resource availability, water withdrawn and water stress [45]. Nevertheless, the present discussion is focused on microalgae-based wastewater treatment plants and only the factors that directly affect these plants will be discussed in this section. The following table shows some of the recent works on biotic and abiotic factors of microalgae-based WWT. This will help to develop more clarity on biotic-abiotic factors and growth conditions for microalgae as well as its potential as a wastewater treatment candidate. From Table 1.1 it can be clearly understood that microalgae are a good candidate for nitrogen and phosphorus removal under all different system configurations. They are even effective in untreated wastewaters and can be employed along with conventional treatment methods. It can be further observed from the literature cited in the table that the best results are obtained with natural consortia instead of using a single isolated species [39]. In addition to the use of natural consortia, a combination with aerobic bacteria seems to give better results as has been suggested in many studies in the literature cited in this table. It is also proved that aerobic

Table 1.1 Microalgae-based WWT abiotic and biotic requirements, nutrients requirement.

Abiotic factor	Biotic factor	Organism used	Treatment level/ sampling	Treatment system	Findings	Reference
Lighting, pH Temperature CO_2 Total Nitrogen (TN) Total Phosphorus (TP)	Species of microalgae and bacterial consortia, cell density, cell size and biovolume *Franceia amphitricha* *Scenedesmus* sp. *Chlorella* sp. *Chlorellaceae* *Chlamydomonas* sp. *Desmodesmus* sp.	New isolated species	Anaerobic digested (AD) effluent sample	600-L horizontal tubular photo-bioreactors	- 99% removal of (TN) and Total Phosphorus (TP).	[40]

(*Continued*)

Table 1.1 Microalgae-based WWT abiotic and biotic requirements, nutrients requirement. (*Continued*)

Abiotic factor	Biotic factor	Organism used	Treatment level/ sampling	Treatment system	Findings	Reference
Irradiance Temperature Orbital Shaking	3 fluoroquinolones- ofloxacin, ciprofloxacin, norfloxacin; 3 macrolides - azithromycin, erythromycin, clarithromycin; and 3 antibiotics - trimethoprim, pipemidic acid, sulfapyridine	*Chlamydomonas reinhardtii* (UTEX ID 2243), *Chlorella sorokiniana* (UTEX ID 1663), *Dunaliella tertiolecta* (UTEX ID LB999) and *Pseudokirchneriella subcapitata* (UTEX ID 1648)	Direct toilet water	1200L Photobioreactor and a suspect screening methodology for assessment of the transformation products (TPs) generated from 9 antibiotics	- Microalgae ability for macrolide biotransformation. - 40 different TPs were identified.	[53]
COD, TSS, Total nitrogen (TN), Total Phosphorus (TP) pH, Temp Total solar irradiance	Interaction of several species	*Leptolyngbya* sp. *Synechococcus* sp. *Chlorella* sp. *Parachlorella* sp. *Dictyosphaerium* sp. *Scenedesmus* sp. *Desmodesmus* sp. *Pediastrum* sp. *Zooplankton Daphnia* sp.	Untreated influent wastewater	High-rate algal ponds (HRAPs)	- Microalgae biodiversity plays critically essential role in high productivity of HRAPs treating municipal wastewater.	[54]

(*Continued*)

Table 1.1 Microalgae-based WWT abiotic and biotic requirements, nutrients requirement. (*Continued*)

Abiotic factor	Biotic factor	Organism used	Treatment level/ sampling	Treatment system	Findings	Reference
CO_x, NO_x, SO_x, pH, Light, temperature, wind (m/s), precipitation (mm), relative humidity (%) DO	Consortium of local freshwater green algae	*Chlorella* sp., *Scenedesmus dimorphus*, *Scenedesmus quadricauda*, and *Desmodesmus armatus*, *Coelastrum microporum*	Municipal untreated wastewater and CO_2 from CHP plant	Raceway pond systems	- In wastewater treatment process, the interaction between bacteria and microalgae plays a crucial role.	[55]
nitrogen (N), phosphorus (P), magnesium (Mg), carbonate (CO_3) and gamma radiation	-	*Chlorella vulgaris*	Synthetic media	Lab-scale setup	- Biomass increase with high N and P and low Mg and CO_3, Lipid accumulation increase with low N and P and high Mg and CO_3. - Gamma radiation has negative effect on biomass and lipid accumulation.	[56]

(*Continued*)

Table 1.1 Microalgae-based WWT abiotic and biotic requirements, nutrients requirement. (*Continued*)

Abiotic factor	Biotic factor	Organism used	Treatment level/ sampling	Treatment system	Findings	Reference
Autotrophic and heterotrophic growth conditions	-	*Auxenochlorella protothecoides*	Synthetic media	Lab-scale setup	- Hub genes defined	[57]
Ammonium urea, and Nitrate as nitrogen source	Algal consortium	*Tetraselmis* sp. (UTEX LB 2767), *Raphidocelis subcapitata* (UTEX 1648), *Chlamydomonas reinhardtii* (UTEX 2243), and *Scenedesmus obliquus* (UTEX 393) *Navicula* sp.	Wastewater as a feedstock	Lab-scale setup	- In heterogeneous nitrogen environments, functional diversity increases with species complementarity and productivity	[58]

(*Continued*)

Table 1.1 Microalgae-based WWT abiotic and biotic requirements, nutrients requirement. (*Continued*)

Abiotic factor	Biotic factor	Organism used	Treatment level/ sampling	Treatment system	Findings	Reference
Ammonium as Nitrogen source	Bacteria derived from the AD effluents interactions with the *Chlorella* species	*Chlorella vulgaris* (KCTC AG10002) and *Chlorella protothecoides* (UTEX 1806)	AD effluents from four different lab-scale anaerobic digesters	Lab-scale setup	- A viable way to treat and value-add the wastewater effluents by *Chlorella* cultured on AD effluents	[59]
pH, EC, TS, TDS, TSS, DO, COD, Ammonia, Nitrate, Phosphate	Varying concentrations of same algal species at different HRT	*Chlorella vulgaris*	Raw domestic wastewater	Lab-scale setup	- Addition of microalgae to CWWTs can be a solution for pollution control	[32]
pH Nitrogen and phosphorus	Microalgae consortia	Different naturally occurring sewage algal species	Comparative study on wastewater and artificial media	Lab-scale setup	- Microalgae consortia has effectively removed phosphate and nitrogen with real wastewater instead of from synthetic media	[44]

bacteria support microalgal photosynthetic rates by reducing the micro-environments around the microalga and thereby help faster, better, energy smart and sustainable treatment of wastewater; whereas the conventional wastewater treatment is both oxygen and energy intensive, and thus less environmentally friendly and less sustainable [46]. Moreover, from Table 1.1, it can be further observed that if the microalgae are autotrophic there are fewer requirements on the surrounding media and the biomass produced can be further utilized or valorized, unlike the CWWTs [30, 42, 47, 48]. Microalgae has proven to be good in most of the wastewater treatment studies, except for complex wastes like phenols [49] and hydrocarbons [40, 50–52].

1.3.2 Microalgae-Based WWT System – Photobioreactor System in Suspension and Immobilized Model

Microalgae culture systems are vast. In wastewater treatment, local consortia of microalgae is preferably cultured in open raceway ponds or high-rate algal ponds (HRAPs). However, algae cultivation is done in a photobioreactor (PBR) either for culture valorization, biomaterial production or for high lipid production as well as to study the finer nuances of R&D on a specific species or an improved strain [60–62]. Nevertheless, the use of a photobioreactor for treatment of wastewater could undermine the overall cost and energy efficiency [63].

Microalgae is adventitious over filamentous as well as macroalgae in terms of its feasibility of culture in suspension as well as in immobilized forms [64]. With the advancement of information technology, control and feedback loops, automation, etc., PBR has gone from lab scale to pilot scale in the last two decades. Although giving a complete overview of the two decades of PBR algal cultivation is difficult and beyond the scope here, a few suspensions and immobilized algal culture studies are presented in Table 1.2.

1.3.3 Evaluation of Treatment Performance

Performance evaluation (PE) of a system is important for optimization of a process and is extensively applied in wastewater treatment processes. It is reported that the PE data do not provide suitable operational information for the optimization of individual units involved in a WWTP; however, they are important indicators for the overall performance of the system [78]. A good system performance can significantly reduce the operation

Table 1.2 Microalgae growth systems – suspension and immobilization in PBRs.

Aim	Culture type	PBR scale	Organism	Output	Reference
Anaerobic food processing wastewater for biodiesel production and wastewater purification	Suspension	Pilot scale	*Chlorella pyrenoidosa*	- Effective pollutants purification	[65]
Primarily treated pulp and paper wastewater	Suspension	Membrane photobioreactor (MPBR)	*Chlorella vulgaris*	- Moderate purification	[66]
Testing and comparison of 2-system MFC-PBR with a control MFC	Suspension	MFC-PBR (photobioreactor)	*Chlorella* sp.	- ηCOD values up to 99%	[67]

(Continued)

Table 1.2 Microalgae growth systems – suspension and immobilization in PBRs. (*Continued*)

Aim	Culture type	PBR scale	Organism	Output	Reference
Study on hydrodynamic conditions using computational fluid dynamics (CFD)	Suspension	Hybrid tubular photobioreactor	Mixed filamentous and smaller microalgae	- Importance of CFD simulations for scale-up in production of microalgae	[68]
Advanced pH control	Suspension	Raceway and thin-layer open photobioreactors	*Scenedesmus*	- With lower CO_2 consumption, improvement in system performance	[69]
Phosphate and nitrate recovery from wastewater	Immobilized	Design and operation of twin-layer photobioreactor for culturing green alga *Halochlorella rubescens* on vertical sheet-like surfaces	*Halochlorella rubescens*	- 70–99% removal of Nitrogen and Phosphorus	[70]

(*Continued*)

Table 1.2 Microalgae growth systems – suspension and immobilization in PBRs. (*Continued*)

Aim	Culture type	PBR scale	Organism	Output	Reference
Treatment of dairy effluents with high organic load	Immobilized	2-stage treatment –the first one consisting of a 1L PBR with immobilized *Chlorella pyrenoidosa*, whereas later includes two column sand bed filtration	*Chlorella pyrenoidosa*	- Within 96 hour of 2-stage purification process, complete removal of NH_4^+-N and 98% removal of PO_4^{3-}-P	[71]
Treatment of effluents from aquaculture	Immobilized	Synthetic textile used as a support medium for immobilized/packed bed bioreactor	*Picochlorum* sp.	- C and N removal rates up to 95%	[72]
Treatment of untreated palm oil mill effluent (POME)	Immobilized	3L capacity flat bioreactor	*Chlorella* sp.	- Removal of total nitrogen ranged between 11 to 62.46% along with COD removal between 23 to 63.1% using beads made from 8% Na-alginate concentration	[73]

(*Continued*)

Table 1.2 Microalgae growth systems – suspension and immobilization in PBRs. (*Continued*)

Aim	Culture type	PBR scale	Organism	Output	Reference
Removal of heavy metal ion (Copper (Cu^{2+})	Suspended and Immobilized	30-L photobioreactor	Oven-dried mixed microalgae of *Chlorella sorokiniana, Monoraphidium* sp. and *Scenedesmus obliquus* bound in Na-Alginate is used as biosorbent	- 96.4% removal efficiency	[74]
Treatment of ADE with highly concentrated organic matter	Suspended and Immobilized	Two-sequencing batch PBRs to compare suspension/ immobilization effect	*Microcystis aeruginosa*	- Microalgae immobilization is better than suspension for the ADE treatment	[75]

(*Continued*)

Table 1.2 Microalgae growth systems – suspension and immobilization in PBRs. (*Continued*)

Aim	Culture type	PBR scale	Organism	Output	Reference
Optimization of PBR with respect to light and CO_2 for algal biomass	Immobilized	Twin-layer photobioreactors (TL-PBRs), a type of porous substrate bioreactor (PSBR)	*Halochlorella rubescens*	- Surface productivity of 31.2 $g/m^2/d$ of dry biomass obtained using a combination of 1023 μmol photons per m^2/s and 3% of CO_2	[76]
Scale-up feasibility studies for production of Astaxanthin	Suspended and Immobilized	Small-scale angled twin-layer porous substrate photobioreactor (TL-PSBR)	*Haematococcus pluvialis*	- 6.5 g/m^{-2} of optimal initial biomass density	[77]

Table 1.3 Performance evaluation of WWTPs.

Source/plant	Method/technique for PE	Result/conclusion	Reference
Wastewater treatment plant with extended aeration sludge process	BOD, COD, TSS & PO_4	- Performance of WWTP, w.r.t. to various physicochemical properties was evaluated along with effluent discharge characteristics in a water body (Yamuna River).	[81]
Constructed wetlands	Analytic hierarchy process (AHP) entropy weight method Preference ranking organization method	- 48% organic matter removal by a vertical-flow wetland process, and 31.2% of NH_3-N, and 32.4% of TN removals by an integrated-flow wetland process.	[82]
Extended aeration plant and Trickling filter plant	BOD and COD estimation before and after treatment	- BOD removal of 79.5% and 90.7% was reported through trickling filter, and trickling filter with extended aeration processes, respectively. - The removal efficiency of COD was 60% and 86% through trickling filter, and trickling filter with extended aeration processes, respectively.	[84]

(*Continued*)

Table 1.3 Performance evaluation of WWTPs. (*Continued*)

Source/plant	Method/technique for PE	Result/conclusion	Reference
Discharge water treatment plant	Physicochemical and biological parameters	- Data verified against atomic adsorption spectroscopy, bacteriological analysis, photometer and flame photometer, and turbidity meter.	[85]
Sewage treatment plant	pH, BOD, COD, TSS	- The treated effluents met the discharge standards.	[86]
WWTPs of several metropolitan municipalities	Stepwise weight assessment ratio analysis (SWARA) method Output-oriented data envelopment analysis (DEA)	- Improvement in total, technical, and scale efficiencies was shown in multiple metropolitan municipalities.	[87]
Industrial WWTP	STOAT software used for modeling and PE	- Removal efficiency of WWTP: BOD, 90%; COD, 93.02%, and TSS, 96.12%.	[88]
Wastewater treatment plant in Souss-Massa region	Physicochemical and microbiological studies	- Removal of impurities between 97.5% and 100%.	[89]
Sewage water treatment plant	Evaluation of physicochemical indicators and fecal coliform prevalence	- WWTP performance was reported in accordance with the prescribed general limits.	[90]

(*Continued*)

Table 1.3 Performance evaluation of WWTPs. (*Continued*)

Source/plant	Method/technique for PE	Result/conclusion	Reference
Mashhad wastewater treatment plant	Optimized NN model using genetic algorithm	- The most important factors affecting the performance of Mashhad treatment plant were inlet flow rate, TCODin/ TBODin ratio, temperature and load of organic matter.	[91]
Membrane bioreactor WWTP	Influent and effluent sample analysis	- The average BOD and COD removal efficiencies were reported as 97.6% and 96.5%, respectively.	[92]
Tabriz WWTP	Support vector machine (SVM) and ANN model	- Efficient results using ensemble methods in predicting the performance of Tabriz WWTP.	[93]
Municipal WWTPs	Multi-criteria decision-making technique for order of preference by similarity to ideal solution	- In environmental monitoring systems, a field base approach, w.r.t. suitability of the weight allocation method and fuzzy approach is proposed.	[94]

and maintenance cost of the running plants. Furthermore, performance modeling and cost evaluation of processes are essential for designing, constructing, and predicting future economic requirements. The future economic requirements may have the labor requirement, project construction, consistence maintenance, material and energy requirements, and amortization costs of a WWTPs [79, 80]. Nonetheless, since wastewater treatment plants are associated with pollution control and the environment, it is obligatory for these plants to comply with the local/global regulatory authority [81]. In this case, PE becomes very important for all aspects, viz. technological, management, economic, environmental, social, and compliance, of running a WWTP [82, 83]. Table 1.3 shows some of the recent studies on PE of WWTPs.

1.4 IoT and Edge Computing-Based Monitoring and Modeling of Integrated Microalgae-Based WWT Plant

In recent years, environmental IoT sensors have been receiving attention as an important tool for monitoring and modeling of the environmental processes, including wastewater treatment. The IoT-based technology is being extensively used to connect everyday objects with sensors for network-based cost-effective data collection and transfer. It is noteworthy that IoT-based smart sensors and devices can be used efficiently in a monitoring system to send alerts to prevent accidents and also reduce the workload by reducing the physical monitoring of infrastructure. In addition, the cloud computing technology facilitates the cost-effective data transfer to server and processing units without latency in processing. Thus, the integrated IoT and edge cutting technology can be effectively used for data collection and processing from a wastewater treatment plant associated with algal pond technology [95, 96]. Nowadays, the open pond algal cultivation system is receiving attention for large-scale algae cultivation due to its advantages of low capital cost and easy operational processes [95]. However, the cultivation process parameters, viz. light intensity, temperature, nutrient concentration, and other physicochemical parameters affect the algal growth yield, and real-time monitoring using advanced IoT-based sensors is needed [97].

The algae-based bioprocess and biorefineries are integrated with Industry 4.0 approaches to facilitate the simultaneous production of growth-associated products and co-products with the advantages of low residual quantity and optimal downstream capital investment [98]. This involves an automated algal growth and harvesting system with integrated supervisory system via a network of IoT plug-and-play sensors with advantages of cost-effective operational costs and real-time monitoring.

The idea of Industry 4.0 takes a step forward with Industry 5.0 with an emphasis on the restorations of human hands, brains and intuitions in the manufacturing senses, with smart IoT facilities-based transformation of a production system connected via cloud servers. The industry 5.0 approach consists of both the capabilities of humans and machines, which are integrated together to enhance the process performance and manufacturing capacity. This industrial revolution can help in sound decision-making, resulting in a collective community commitment and the willingness of civic influences, thereby reducing the market risk and improving financial strength [98].

Industry 4.0 can manage the value-added products (e.g., biodiesel, biopolymers, bioethanol etc.), business strategies, and control of integrated algae-associated WWTPs. It can overcome the gaps associated with algal-based innovative manufacturing, which exploited intelligent devices for disperse manufacturing processes. However, the latest development in analytical data methods, including sensors and hyper spectral cameras, led to a paradigm shift towards application of Industry 4.0 to Industry 5.0 through machine learning-based support vector machines (SVMs) and convoluted neural networks (CNNs) [98–100]. The integrated algal pond with wastewater treatment has been reported progressively worldwide in many countries located from polar areas (North America and Europe) to the equator (Africa and South Asia) [95]. Regardless of the global presence of this technology, this cost-effective technology is facing challenges of being upgraded with advanced monitoring and control technologies to meet the standard regulations on effluent discharge. In the recent past, activated sludge-based WWTPs incorporated innovative design and controlling processes, including instrumentation, control and automation (ICA).

1.4.1 Machine Learning Approaches for Data Acquisition, Monitoring and Analysis System

The machine learning and deep learning-based artificial intelligence approach has produced tremendously powerful tools for solving complex problems in real-world applications in recent years [96]. It is noteworthy that the advanced wastewater treatment process, including microalgae-based WWTP, are complex processes and affected by diverse physical, chemical, and microbiological factors. Besides which, the stochastic perturbations and uncertainties in these processes require appropriate operational control of the system. Secondary treatment-associated microalgae cultivation

Table 1.4 Some recent applications of machine learning (ML) approaches used to understand the complex wastewater and algal cultivation systems.

S. no.	AI approach	Process studied	Findings	References
1.	ANN	Techno-economic evaluation of algae-based tertiary treatment of WWTP	- ANN-based techno-economic feasibility analysis of nutrient supplemented secondary-treated (ST) wastewater effluents integrated with pilot-scale microalgal cultivation was performed. - The study concluded with a shorter payback period of integrated wastewater- algal cultivation system than the project's lifetime.	[101]
2.	ML technique using decision tree (DT) algorithm	Exploration of significant factors of algal biomass and lipid accumulation	- The technique utilized to determine the optimum conditions of variables leading to high biomass and lipid accumulation. - Association rule mining was used to find the specific conditions leading to very high biomass and lipid levels.	[102]

(*Continued*)

Table 1.4 Some recent applications of machine learning (ML) approaches used to understand the complex wastewater and algal cultivation systems. (*Continued*)

S. no.	AI approach	Process studied	Findings	References
3.	Modeling and process optimization using artificial neural network	Biodiesel production from *Nannochloropsis salina*	- Using RSM and ANN, optimization of process parameters for biodiesel production was studied. - Maximum 86.1% of biodiesel conversion for the synthesized nanocatalyst CaO was reported under optimum process conditions.	[103]
4.	ML-based multi-objective optimization	Improved biomass and bioactive phycobiliproteins (PBPs) production by *Nostoc* sp. CCC-403	- Using hybrid ML approach, 90% and 61.76% increase in cell biomass and total PBPs production, respectively, were predicted.	[104]
5.	ML-based classification models	Classification of microalgae	- Using FlowCAM tool, two ANN models were developed for identification and classification of microalgae samples composed by *Chlorella vulgaris* and *Scenedesmus almeriensis* using microalgae cells as input data images.	[105]

(*Continued*)

Table 1.4 Some recent applications of machine learning (ML) approaches used to understand the complex wastewater and algal cultivation systems. (*Continued*)

S. no.	AI approach	Process studied	Findings	References
6.	ANN-multilayer perception	ANN model used to predict the biomass of microalgae species under different environmental conditions	- Using ANN model, it is predicted that the CO_2 and nitrogen have effects on the biomass concentration with a varying range of input parameters for different microalgae species in different environment condition.	[106]
7.	ANN	Discrimination of monoalgal and mixed algal cultures	- ANN was used to discriminate monoalgal and mixed algal cultures. - Identification of different microalgae species in the monoalgal cultures. - Estimation of approximate composition of mixed algal cultures.	[107]

(*Continued*)

Table 1.4 Some recent applications of machine learning (ML) approaches used to understand the complex wastewater and algal cultivation systems. (*Continued*)

S. no.	AI approach	Process studied	Findings	References
8.	Backpropagation neural network	Production of microalgal biomass along with the growth estimate of polyculture micro-algae in raceway pond	- Estimation of polyculture microalgae growth in a semi-continuous open raceway pond (ORP) using trained ANN model. - The structure of trained model included: eight input parameters, one hidden layer, and one output parameter with multilayer backpropagation NN algorithm.	[108]
9.	Multivariate timing-random deep belief net (MT-RDBN) modeling	MT-RDBN model for algal bloom	- Fine-tuned network parameters using back propagation NN algorithm. - The MT-RDBN model utilized time series data for improved algal bloom prediction. - A nonlinear time series model was developed for the characterization factor such as chlorophyll concentration with interaction factors (pH, water, and temperature).	[109]

(*Continued*)

Table 1.4 Some recent applications of machine learning (ML) approaches used to understand the complex wastewater and algal cultivation systems. (*Continued*)

S. no.	AI approach	Process studied	Findings	References
10.	Artificial intelligence (ANN & genetic algorithm (GA)) driven process optimization	Cleaner biomass production with co-valorization of flue gas and wastewater	- Hybrid GA-ANN used for optimization and prediction of ideal process conditions for enhanced biomass of *Scenedesmus* sp. using domestic wastewater as substrate.	[110]

is considered a tertiary treatment for nutrient recovery and is complex under natural environmental conditions. The integrated microalgae-based WWTP faces diverse environmental conditions, viz. temperature, solar radiation, nutrients availability, and culture characteristics [101]. These environmental variables are nonlinear in nature and exhibit complex relationships in this integrated system, promising nutrient uptake and bio-product formations. Thus, these systems can employ machine learning and deep learning-based AI techniques to understand the complex system. Table 1.4 shows the recent applications of artificial intelligence approaches in these processes.

Hence, these recent studies have employed artificial intelligence techniques to understand the behavior of complex algal-based systems and wastewater systems. Thus, it can be concluded that the integration of these modern intelligence approaches with integration of population-based algorithms, such as particle swarm optimization (PSO), ant colony optimization, genetic algorithm (GA), ANN and their hybrid approaches, can integrate the economic cultivation of microalgae with safe discharge of treated wastewater into the environment.

1.5 Techno-Economic Analysis of Integrated Microalgae-Based Wastewater Treatment (WWT) System

From the above discussion, it is obvious that microalgae-associated biomass is considered a promising cost-effective renewable source, since cultivation is associated with municipal wastewater treatment. Microalgae-based wastewater treatment technology requires improvement in terms of process sustainability in addition to process optimization to be considered an economic and sustainable viable option of green bioenergy. Thus, the integrated microalgae-based wastewater treatment needs to be evaluated with life cycle assessment (LCA), process input and output analysis, and material flow analysis under current perspectives. The mitigation of climate change through cleaner sustainable industrial practices with industrial energy efficiency is a global priority [111]. Although microalgae-based WWTPs have not been a major concern in relation to industrial energy use, efforts are being made to reduce energy use in integrated microalgae-WWTPs through utilizing the concept of industrial ecology [112]. The cleaner best practices and novel technologies for energy reduction in municipal WWTPs are described by Crawford and Sandino [113]. In addition, economic transparency, incentives, and accountability to stakeholders play an important role in adaptation and implementation of this novel technology. Quantitative modeling techniques, such as the material flow cost accounting (MFCA)-based economic evaluation process, which are associated with the analysis of hidden cost and material loss related to environmental impacts are extensively used [114]. In MFCA, the material and cost balance are calculated in terms of "quantity center (QC)" and the steps in the process, viz. production, recycling, and other systems are illustrated by visual models of QCs [115]. The procedure of MFCA methods has been recognized by the standardization of ISO145051 (International Organization of Standardization, 2011); however, several studies have been reported on the improvement of this method through incorporation and integration of energy flow, life cycle analysis, management control system and environmental management accounting, supply chain analysis, and "4R" circular economy principle [114].

Nowadays, sustainable environment management (SEM) is considered as a primary assessment criterion for the services provided by natural as well as man-made (industrial) processes. Life cycle assessment (LCA) has become a central instrument for SEM and has provided an international standard (IS) for modeling, assessment, and evaluation of impacts of a product/process throughout its life cycle. The aim of LCA is to evaluate the impacts on ecosystems, natural resources, and human health [116]. The LCA process accounts for the evaluation of impacts of production systems on natural ecosystem throughout the different life cycle stages (e.g., extraction of resources, incorporation into processes, and end-of-life disposal) along with the social and economic impacts. In order to minimize the amount of energy consumption and the negative impacts and cost associated with microalgae-based WWTPs; LCA can play an important role in terms of quantification and exploration of social, economic, and environmental impacts. Several studies have been reported on the LCA studies for microalgae cultivation and their various forms of energy recovery (Table 1.5). These studies not only explored the environmental impacts associated with the microalgae biomass, but also the benefits associated with microalgae cultivation (e.g., CO_2 sequestration) [117]. It is reported that incorporation of a high-rate algal pond system (in replacement of conventional activated sludge system) increased the environmental performances of WWTP [118]. Thus, the microalgae-based WWTP allows efficient recovery of pollutants (e.g., nutrients) from the effluent and can enhance economic and environmental sustainability of integrated microalgae WWTPs.

The studies mentioned in Table 1.5 show that LCA is extensively utilized as an efficient tool for feasibility analysis of microalgae-associated biofuel production with simultaneous assessment of environmental impacts in integrated WWTPs. Besides which, LCA is also able to determine the economic feasibility of microalgae cultivation integrated with different WWTPs for biofuel production. The GHG emissions from these integrated process technologies can be analyzed and modeled through LCA using suitable software tools such as SimaPro, GaBi, and OpenLCA [126].

In addition, life cycle costing (LCC) can also be performed to assess the feasibility and sensitivity of the microalgae-associated WWTPs-based biorefinery process [127]. It includes the estimations of costs associated with aggregated energy, installation, operation, downstream process, maintenance, and environmental and decommissioning over the complete lifetime of the microalgae-associated WWTP biorefinery. The details of various LCC models based on operating cost, salvage value, capital and maintenance costs are discussed by different researchers in their studies [127, 128–131].

The emergy analysis of an innovative process is also useful to evaluate its environmental sustainability in terms of availability of internal as well as external resources required for system maintenance and stability. Emergy is defined as the amount of energy consumed both directly and indirectly

Table 1.5 Recent LCA studies associated with application of microalgae and WWTPs.

S. no.	LCA approach	Process studied	Findings	References
1.	LCA - using SimaPro 9.0; Inventories - Ecoinvent v3.5	LCA for recovery of energy using briquette from microalgae biomass associated with wastewater	- Cradle-to-gate approach was used for LCA of cultivation and valorization of microalgae biomass growth in two scenarios: (i) a high-rate algal pond (HRAP), and (ii) a hybrid HRAP–biofilm reactor (BR). - LCA study focused on electric power mix and revealed about 60% improvement in total environmental impacts in both scenarios. - The environmental gains are associated with the use of wastewater for microalgae growth.	[117]
2.	LCA- life cycle inventory (LCI) for scale-up of process	Treatment, profit evaluation, and scale-up studies of microphytes growth in wastewater	- Evaluation of algal growth in wastewater for significant management of freshwater ecosystems along with wastewater treatment. - This LCA analysis elucidated the system potentiality of large-scale production of value-added product from algal associated WWTPs.	[119]

(*Continued*)

Table 1.5 Recent LCA studies associated with application of microalgae and WWTPs. (*Continued*)

S. no.	LCA approach	Process studied	Findings	References
3.	LCA - ISO14044 guidelines; Inventories - ecoinvent v3.4	Integrated side-stream microalgae process with municipal WWTP	- Improvement in environmental impacts due to integration of microalgae unit with WWTP were reported. - The proposed solution improved the overall sustainability of WWTPs through resource recovery in terms of nutrients and solar energy.	[120]
4.	Life cycle inventory (LCI) analysis	Microalgae-associated biofuels production – a concept of industrial plant	- LCI analysis compiled the real pilot-scale process data, which was used for scale-up of microalgae-associated biofuel production in an industrial plant. - Inventories for input and output were created using data of energy, nutrients, water, and materials consumption for biomass cultivation and biodiesel production for future LCA modeling. - A decision support system based on LCI inventory data was created to promote the development of sustainable pilot and large-scale algae-based industry for biofuel production.	[121]

(*Continued*)

Table 1.5 Recent LCA studies associated with application of microalgae and WWTPs. (*Continued*)

S. no.	LCA approach	Process studied	Findings	References
5.	LCA - using SimaPro 9.0.0.29; Inventories - ReCiPe 2016 Endpoint v1.02	Geospatial and LCA analyses of an integrating microalgae cultivation system	- For three different process designs, consequential LCA was used to compare four different feeds (sewage sludge, municipal biowaste, cattle and swine manure). - To identify the integration potential for microalgal cultivation system, a geospatial analysis of substrate availability was also conducted. - A significant reduction in the environmental burden of microalgae cultivation system was reported due to the uses of sewage sludge, cattle and swine manure. - The feasibility of integration of urban wastewater treatment plants to microalgae cultivation into regional economies was reported.	[122]

(*Continued*)

Table 1.5 Recent LCA studies associated with application of microalgae and WWTPs. (*Continued*)

S. no.	LCA approach	Process studied	Findings	References
6.	LCA - ISO 14044 guidelines	Bioethanol production from microalgae	- Scenario analyses based on CO_2 emission and energy balance in a microalgae-associated bioethanol production system at industrial scale were conducted. - The commercialization of microalgae-bioethanol plant along with wastewater treatments is suggested to fuel industries for CO_2 sequestration.	[123]
7.	LCA	LCA of a microalgae-based WWTP with energy balance	- Using LCA and mass and energy balances, techno-environmental performance of WWTP integrated into a high-rate algal pond were evaluated. - LCA-based performance system for microalgae-based WWTP was developed as a tool for decision-makers for biogas production under different techno-environmental aspects.	[124]

(*Continued*)

Table 1.5 Recent LCA studies associated with application of microalgae and WWTPs. (*Continued*)

S. no.	LCA approach	Process studied	Findings	References
8.	LCA - using Umberto NXT software; Inventories - Ecoinvent database v3.0	Comparative assessment of microalgae-associated biodiesel production using freshwater and wastewater as resource	- LCA-based comparative evaluation of biodiesel production in two processes, viz. algae grown in wastewater and freshwater, were performed. - Wastewater-based biodiesel production was identified as a viable sustainable solution to freshwater-based production system.	[125]

to produce a product or service [132]. The concept of emergy analysis was introduced by Odum as a method for assessing different system-based energy consumption [133]. It is widely used to evaluate the sustainability of different industrial systems related to first, second and third generation biofuel production [134], microalgae as a feedstock of bioethanol [135], oil production from microalgae [136], and supply chain related to food and agriculture production [137]. Thus, emergy analysis can be used for evaluation of sustainability of microalgae-associated WWTPs based on its energy efficiency.

1.6 Brief Case Studies of Commercially Available Microalgae-Based Wastewater Treatment (WWT) Plants

In the past decade, numerous firms have focused on algal biomass production, especially in the USA, UK, and Australia using wastewater as feedstock sources [138]. Algae Enterprises in Australia established an algae-based wastewater treatment facility which focused on the full spectrum of municipal, industrial and agricultural wastewater resources. The primary energy source of local algae type is produced in a closed PBR system through photosynthetically active radiation. The produced algal biomass is anaerobically digested to produce a methane-rich biogas which is further transformed into enriched energy (CH_4) [138, 139].

An Advanced Integrated Wastewater Pond System (AIWPS®) has been created by Oswald Green Technologies, also called Energy Ponds™, which works with a symbiotic bacterial algal consortium to be grown on organic and inorganic municipal wastewater contaminants [140, 141]. In this process, anaerobic ponds or gravity settlers are used to remove the wastewater solids in an initial pretreated stage, followed by the assimilation of microalgae into high-rate algal pools utilizing organic and inorganic material. The collected algal biomass from the Energy Ponds is processed as a fertilizer, animal feed and plastic and biofuel raw material [138]. The US company AlgaeSystems has developed a cost-efficient, floating, offshore PBR system, which is used to take nutrients from its original source under environmental and CO_2, conditions downstream [142]. It has been reported that 50,000 gal/day of raw urban wastewater was removed with an efficiency of 75% (total N), 93% (total P), or 93% (total P) (BOD). The objective of the HydroMentia Algal Turf Scrubber® (ATS), which consists of a stream pulsed in waves, is to clean wastewater [143, 144]. The removals rates of N and P were 125 mg N/m/d and 25 mg/m/d for an agricultural drainage ditch [145] with the maximum flow and continuous running of ATS. The algal biomass generated by ATS serves as compost and cattle feed to improve soil, but also can be used as a resource for the generation of biofuels [138, 144]. The approach of OneWater and Gross-Wen Technologies is based on an immobilized cell system integrated as spinning portions of the wastewater treatment system. The bacterial source and solid settling of polysaccharides are generated by the photosynthesis in this system. The bacteria may then utilize photosynthesized oxygen and create a stable ecological wastewater treatment and self-regulating system [146]. Gross-Wen Technologies' rotating algal biofilm (RAB) system is a biofilm alga connected to vertical rotating conveyor belts. The connected microalgae fix N and P of the rich liquid nutrient, while conducting photoautotrophic growth in the gaseous stage [144, 147].

1.7 Conclusion

The biorefineries of microalgae-associated WWTPs have been gaining attention in recent years due to the dual benefits of efficient removal of toxins from effluents while simultaneously getting value-added products such as bioethanol, biodiesel and biopolymers. However, there are several challenges related to these sustainable biorefineries that need to be addressed from the perspective of recent advanced technologies. This chapter focused on the recent updates on microalgae-associated biorefinery for resource recovery from WWTPs in a sustainable way. The applications of different extremophile microalgae for nutrient removal were discussed in detail. Also, different microalgae-based cultivation systems for cost-effective removal of pollutants from effluent in WWTPs were analyzed. The treatment performance of different photoreactor systems were evaluated and discussed in a concise way. In addition, the recent updates on IoT and edge computing-based monitoring and modeling of a microalgae cultivation system were evaluated. Furthermore, recent studies involving techno-economic analysis and environmental sustainability assessment in terms of material flow analysis, life cycle assessment, life cycle costing and emergy analysis were discussed in brief. Insight into commercially available integrated microalgae WWTPs technologies based on their capacity and performance was also provided at the end.

Thus, it can be concluded that the microalgae-based WWTPs can be a viable biorefinery system with multiple products recovery and will provide an economic and environmentally friendly sustainable solution to wastewater treatment system.

References

1. Nagarajan, D., Lee, D-J., Chen, C-Y., Chang, J-S. (2020) Resource recovery from wastewaters using microalgae-based approaches: A circular bioeconomy perspective. *Bioresour. Technol.*, 302, 122817.
2. Li, K., Liu, Q., Fang, F., Luo, R., Lu, Q., Zhou, W., Huo, S., Cheng, P., Liu, J., Addy, M., Chen, P., Chen, D., Ruan, R. (2019) Microalgae-based wastewater treatment for nutrients recovery: A review. *Bioresour. Technol.*, 291, 121934.
3. De Godos, I., González, C., Becares, E., García-Encina, P.A., Muńoz, R. (2009) Simultaneous nutrients and carbon removal during pretreated swine slurry degradation in a tubular biofilm photobioreactor. *Appl. Microbiol. Biotechnol.*, 82, 187-194.
4. Cai, T., Park, S.Y., Li, Y. (2013) Nutrient recovery from wastewater streams by microalgae: status and prospects. *Renew. Sustain. Energy Rev.*, 19, 360–369.
5. Lee, C.S., Lee, S.A., Ko, S.R., Oh, H.M., Ahn, C.Y. (2015) Effects of photoperiod on nutrient removal, biomass production, and algal-bacterial population dynamics in lab-scale photobioreactors treating municipal wastewater. *Water Res.* 68, 680–691.
6. Whitton, R., Ometto, F., Pidou, M., Jarvis, P., Villa, R., Jefferson, B. (2015) Microalgae for municipal wastewater nutrient remediation: mechanisms,

reactors and outlook for tertiary treatment. *Environ. Technol. Rev.*, 4 (1), 133–148.

7. Prandini, J.M., da Silva, M.L.B., Mezzari, M.P., Pirolli, M., Michelon, W., Soares, H.M. (2016) Enhancement of nutrient removal from swine wastewater digestate coupled to biogas purification by microalgae Scenedesmus spp., *Bioresour. Technol.*, 202, 67.
8. Oswald, W.J., Gotaas, H.B., Golueke, C.G., Kellen, W.R. (1957) Algae in waste treatment. *Sewage Ind. Wastes*, 29 (4), 437–455.
9. Chinnasamy, S., Bhatnagar, A., Hunt, R.W., Das, K.C. (2010) Microalgae cultivation in a wastewater dominated by carpet mill effluents for biofuel applications. *Bioresour. Technol.*, 101 (9), 3097–3105.
10. Zhou, W., Hu, B., Li, Y., Min, M., Mohr, M., Du, Z., Chen, P., Ruan, R. (2012) Mass cultivation of microalgae on animal wastewater: a sequential two-stage cultivation process for energy crop and omega-3-rich animal feed production. *Appl. Biochem. Biotechnol.*, 168 (2), 348–363.
11. Fabris M., Abriano, R.M., Pernice, M., Sutherland, D.L., Commault, A.S., Hall, C.C., Labeeuw, L., McCauley, J.I., Kuzhiuparambil, U., Ray, P., Kahlke, T., Ralph, P. J. (2020) Emerging Technologies in Algal Biotechnology: Toward the Establishment of a Sustainable, Algae-Based Bioeconomy. *Front., Plant Sci.*, 11, 279.
12. Sikder, M. T., Rahman, M. M., Jakariya, M., Hosokawa, T., Kurasaki, M., & Saito, T. (2019) Remediation of water pollution with native cyclodextrins and modified cyclodextrins: A comparative overview and perspectives. *Chem. Eng. J.*, 355, 920-941.
13. Wang, Q., Yang, Z. (2016) Industrial water pollution, water environment treatment, and health risks in China. *Environ. Poll.*, 218, 358-365.
14. Mohseni, M., Sharifi Abdar, P., & Borghei, S. M. (2016) The highest inhibition coefficient of phenol biodegradation using an acclimated mixed culture. *Water Sci. Technol.*, 73(5), 1033-1040.
15. Patel, A., Matsakas, L., Rova, U., Christakopoulos, P. (2019) A perspective on biotechnological applications of thermophilic microalgae and cyanobacteria. *Bioresour. Technol.*, 278, 424-434.
16. Wu, Y-H., Hu, H-Y., Yu, Y., Zhang, T-Y., Zhu, S.F., Zhuang, L.L., Zhang, X., Lu, Y. (2014) Microalgal species for sustainable biomass/lipid production using wastewater as resource: a review. *Renew. Sust. Energ. Rev.*, 33, 675–688.
17. Chiu, S.Y., Kao, C.Y., Chen, T.Y., Chang, Y.B., Kuo, C.M., Lin, C.S. (2015) Cultivation of microalgal Chlorella for biomass and lipid production using wastewater as nutrient resource. *Bioresour. Technol.*, 184, 179–189.
18. Mohsenpour, S. F., Hennige, S., Willoughby, N., Adeloye, A., Gutierrez, T. (2020) Integrating micro-algae into wastewater treatment: A review. *Sci. Total Environ.*, 142168.
19. Patel, A., Sartaj, K., Arora, N., Pruthi, V., & Pruthi, P. A. (2017) Biodegradation of phenol via meta cleavage pathway triggers *de novo* TAG biosynthesis pathway in oleaginous yeast. *J. Hazard. Mat.*, 340, 47-56.

20. Raddadi, N., Cherif, A., Daffonchio, D., Neifar, M., Fava, F. (2015) Biotechnological applications of extremophiles, extremozymes and extremolytes. *Appl. Microbiol. Biotechnol.*, 99, 7907.
21. Seckbach, J. (2007) Algae and Cyanobacteria in Extreme Environments. The Hebrew University of Jerusalem, Israel., pp. 813.
22. Malavasi, V., Soru, S., Cao, G. (2020) Extremophile Microalgae: the potential for biotechnological application. *J. Phycol.*, 56(3), 559-573.
23. Jagadevan, S., Banerjee, A., Banerjee, C., Guria, C., Tiwari, R., Baweja, M., Shukla, P. (2018) Recent developments in synthetic biology and metabolic engineering in microalgae towards biofuel production. *Biotechnol. Biofuels*, 11, 1–21.
24. Thao, T. Y., Linh, D. T. N., Si, V. C., Carter, T. W. & Hill, R. T. (2017) Isolation and selection of microalgal strains from natural water sources in Vietnam with potential for edible oil production. *Mar. Drugs*, 15, 1–14.
25. Duran, S. K., Kumar, P., Sandhu, S. S. (2018) A review on microalgae strains, cultivation, harvesting, biodiesel conversion and engine implementation. *Biofuels*, 7269, 1–12.
26. de-Bashan, L. E., Trejo, A., Huss, V. A., Hernandez, J. P., & Bashan, Y. (2008) *Chlorella sorokiniana* UTEX 2805, a heat and intense, sunlight-tolerant microalga with potential for removing ammonium from wastewater. *Bioresour. Technol.*, 99(11), 4980-4989.
27. Parry, M. A., Andralojc, P. J., Scales, J. C., Salvucci, M. E., Carmo-Silva, A. E., Alonso, H., Whitney, S. M. (2013) Rubisco activity and regulation as targets for crop improvement. *J. of Exp. Bot.*, 64(3), 717-730.
28. Zou, Y., Zeng, Q., Li, H., Liu, H., Lu, Q. (2021) Emerging technologies of algae-based wastewater remediation for bio-fertilizer production: a promising pathway to sustainable agriculture. *J Chem Technol Biotechnol.*, 96(3), 551-563.
29. Leddy, M.B., Plumlee, M.H., Kantor, R.S., Nelson, K. L., Miller, S.E., Kennedy, L.C., Stamps, B.W., Spear, J.R., Hasan, N.A., Colwell, R.R. (2018) High throughput DNA sequencing to profile microbial water quality of potable reuse. Newsletter, https://www.wateronline.com/doc/high-throughput-dna-sequencing-to-profile-microbial-water-quality-of-potable-reuse-0001
30. Rani, S., Gunjyal, N., Ojha, C.S.P., Singh, R.P. (2021) Review of Challenges for Algae-Based Wastewater Treatment: Strain Selection, Wastewater Characteristics, Abiotic, and Biotic Factors. *J Hazardous, Toxic, Radioact. Waste.*, 25(2), 03120004.
31. Wang, J-H, Zhang, T-Y, Dao, G-H., Xu, X-Q., Wang, X-X., Hu, H-Y (2017) Microalgae-based advanced municipal wastewater treatment for reuse in water bodies. *Appl Microbiol Biotechnol.*, 101, 2659–2675.
32. Moondra, N., Jariwala, N.D., Christian, R.A. (2020) Sustainable treatment of domestic wastewater through microalgae. *Int. J. Phytoremediation*, 22, 1480–1486.

33. Mohsenpour, S.F., Hennige, S., Willoughby, N., Adeloye, A., Gutierrez, T. (2021) Integrating micro-algae into wastewater treatment: A review. *Sci. Total Environ.*, 752, 142168.
34. Turkmenler, H., Aslan, M. (2017) An evaluation of operation and maintenance costs of wastewater treatment plants: Gebze wastewater treatment plant sample. *Desalin Water Treat.*, 76, 382–388.
35. Butcher, R.W. (1947) Studies in the Ecology of Rivers: VII. The Algae of Organically Enriched Waters. *J. Ecol.*, 35, 186.
36. Gilbert, D., Rabalais, N.N., Díaz, R.J., Zhang, J. (2010) Evidence for greater oxygen decline rates in the coastal ocean than in the open ocean. *Biogeosciences*, 7, 2283.
37. Beusen, A.H.W., Bouwman, A.F., Van Beek, L.P.H. (2016) Global riverine N and P transport to ocean increased during the 20th century despite increased retention along the aquatic continuum. *Biogeosciences*, 13, 2441–2451.
38. Van Puijenbroek, P.J.T.M., Beusen, A.H.W., Bouwman, A.F. (2019) Global nitrogen and phosphorus in urban wastewater based on the Shared Socio-economic pathways. *J. Environ. Manage.*, 231:446–456.
39. Jia, H., Yuan, Q. (2016) Removal of nitrogen from wastewater using microalgae and microalgae–bacteria consortia. *Cogent. Environ. Sci.*, 2, 1.
40. Silkina, A., Nelson, G.D., Bayliss, C.E., Pooley, C.L., Day, J.G. (2017) Bioremediation efficacy—comparison of nutrient removal from an anaerobic digest waste-based medium by an algal consortium before and after cryo-preservation. *J. Appl. Phycol.*, 29, 1331–1341.
41. Bunce, J.T., Ndam, E., Ofiteru, I.D. (2018) A review of phosphorus removal technologies and their applicability to small-scale domestic wastewater treatment systems. *Front. Environ. Sci.*, 6. https://doi.org/10.3389/FENVS.2018.00008/FULL
42. Acién Fernández, F.G., Gómez-Serrano, C., Fernández-Sevilla, J.M. (2018) Recovery of Nutrients from Wastewaters Using Microalgae. *Front. Sustain. Food Syst.*, 2, 59.
43. Molinuevo-Salces, B., Riaño, B., Hernández, D., García-González, M.C. (2019) Microalgae and wastewater treatment: Advantages and disadvantages. *Microalgae Biotechnol. Dev. Biofuel Wastewater Treat.*, 505–533. https://doi.org/10.1007/978-981-13-2264-8_20
44. Baldisserotto, C., Demaria S, Accoto O. (2020) Removal of nitrogen and phosphorus from thickening effluent of an urban wastewater treatment plant by an isolated green microalga. Plants 9, 1–23.
45. Liao, Z., Chen, Z., Xu, A., Gao, Q., Song, K., Liu, J., Hu, H-Y. (2021) Wastewater treatment and reuse situations and influential factors in major Asian countries. *J Environ. Manage.*, 282, 111976.
46. Mouget J-L, Dakhama A, Lavoie MC, Noüe J. (1995) Algal growth enhancement by bacteria: Is consumption of photosynthetic oxygen involved? FEMS *Microbiol. Ecol.*, 18, 35–43.

47. Gonçalves, A.L., Pires, J.C.M., Simões, M. (2017) A review on the use of microalgal consortia for wastewater treatment. *Algal Res.*, 24, 403–415.
48. Huete-Ortega, M., Okurowska, K., Kapoore, R.V., Johnson, M.P., Gilmour, D.J., Vaidyanathan, S. (2018) Effect of ammonium and high light intensity on the accumulation of lipids in *Nannochloropsis oceanica* (CCAP 849/10) and *Phaeodactylum tricornutum* (CCAP 1055/1). *Biotechnol. Biofuels*, 11, 60.
49. Maza-Márquez, P., González-Martínez, A., Juárez-Jiménez, B., Rodelas, B., González-López, J. (2018) Microalgae-Bacteria Consortia for the Removal of Phenolic Compounds from Industrial Wastewaters. In: *Approaches in Bioremediation. Nanotechnology in the Life Sciences*, Prasad, R., Aranda, E. (eds,), 135-184, Springer, Cham. https://doi.org/10.1007/978-3-030-02369-0_8
50. Talebi, A.F., Dastgheib, S.M.M., Tirandaz, H., Ghafari, A., Alaie, E., Tabatabaei, M (2016) Enhanced algal-based treatment of petroleum produced water and biodiesel production. *RSC Adv.*, 6, 47001–47009.
51. Sullivan Graham E.J., Dean, C.A., Yoshida, T.M., Twary, S.N., Mark, M.T., Alvarez, A., Zidenga, T., Heikoop, J.M., Perkins, G.B., Rahn, T.A., Wagner, G.L., Laur, P.M. (2017) Oil and gas produced water as a growth medium for microalgae cultivation: A review and feasibility analysis. *Algal Res.*, 24, 492–504.
52. Al-Ghouti, M.A., Al-Kaabi, M.A., Ashfaq, M.Y., Da'na, D.A. (2019) Produced water characteristics, treatment and reuse: A review. *J. Water Process. Eng.*, 28, 222-239.
53. Jaén-Gil, A., Hom-Diaz, A., Llorca, M., Vicent, T., Blánquez, P., Barceló, D., Rodríguez-Mozaz, S. (2018) An automated on-line turbulent flow liquid-chromatography technology coupled to a high-resolution mass spectrometer LTQ-Orbitrap for suspect screening of antibiotic transformation products during microalgae wastewater treatment. *J Chromatogr A.*, 1568, 57–68.
54. Cho, D.H., Choi, J.W., Kang, Z. (2017) Microalgal diversity fosters stable biomass productivity in open ponds treating wastewater. *Sci Rep.*, 7, 1979. https://doi.org/10.1038/S41598-017-02139-8
55. Lage, S., Toffolo, A., Gentili, F.G. (2021) Microalgal growth, nitrogen uptake and storage, and dissolved oxygen production in a polyculture based-open pond fed with municipal wastewater in northern Sweden. *Chemosphere*, 276, 130122. https://doi.org/10.1016/J.CHEMOSPHERE.2021.130122
56. Abo-State, M.A.M., Shanab, S. M. M., Ali, H. E. A. (2019) Effect of nutrients and gamma radiation on growth and lipid accumulation of *Chlorella vulgaris* for biodiesel production. *J. Radiat. Res. Appl. Sci.*, 12, 332–342.
57. Panahi, B., Farhadian, M., Hejazi, M.A. (2020) Systems biology approach identifies functional modules and regulatory hubs related to secondary metabolites accumulation after transition from autotrophic to heterotrophic growth condition in microalgae. *PLoS One*, 15, e0225677. https://doi.org/10.1371/journal.pone.0225677

58. Mandal, S., Shurin, J.B., Efroymson, R.A., Mathews, T.J. (2018) Functional divergence in nitrogen uptake rates explains diversity-productivity relationship in microalgal communities. *Ecosphere*, 9, e02228.
59. Yu, H., Kim, J., Lee, C. (2019) Nutrient removal and microalgal biomass production from different anaerobic digestion effluents with *Chlorella species. Sci. Rep.*, 9, 1–13.
60. Fernández, F.G.A., Reis, A., Wijffels RH, Barbosa, M., Verdelho, V., Llamas, B. (2021) The role of microalgae in the bioeconomy. *N. Biotechnol.*, 61, 99-107.
61. González-Camejo, J., Ferrer, J., Seco, A., Barat, R. (2021) Outdoor microalgae-based urban wastewater treatment: Recent advances, applications, and future perspectives. *Wiley Interdiscip. Rev. Water*, 8(3), e1518.
62. Kong, W., Shen, B., Lyu, H., Kong, J., Ma, J., Wang, Z., Feng, S. (2021) Review on carbon dioxide fixation coupled with nutrients removal from wastewater by microalgae. *J. Clean. Prod.*, 292, 125975.
63. Al-Jabri, H., Das, P., Khan, S., Thaher, M., Abdul Quadir, M. (2021) Treatment of wastewaters by microalgae and the potential applications of the produced biomass—a review. *Water*, 13, 27.
64. Nguyen, H.T., Yoon, Y., Ngo, H.H., Jang, A. (2021) The application of microalgae in removing organic micropollutants in wastewater. *Crit. Rev. Environ. Sci. Technol.*, 51(12), 1187-1220.
65. Tan, X.B., Wan, X.P., Yang, L.B., Wang, X., Meng, J., Jiang, M.J., Pi, H.J., (2021) Nutrient's recycling and biomass production from *Chlorella pyrenoidosa* culture using anaerobic food processing wastewater in a pilot-scale tubular photobioreactor. *Chemosphere*, 270, 129459.
66. Azizi, S., Bayat, B., Tayebati, H. (2021) Nitrate and phosphate removal from treated wastewater by *Chlorella vulgaris* under various light regimes within membrane flat plate photobioreactor. *Environ. Prog. Sustain. Energy.*, 40(2), e13519. https://doi.org/10.1002/ep.1351
67. Bolognesi, S., Cecconet, D., Callegari, A., Capodaglio, A.G. (2021) Combined microalgal photobioreactor/microbial fuel cell system: Performance analysis under different process conditions. *Environ. Res.*, 192, 110263. https://doi.org/10.1016/j.envres.2020.110263
68. Belohlav, V., Uggetti, E., García, J. (2021) Assessment of hydrodynamics based on Computational Fluid Dynamics to optimize the operation of hybrid tubular photobioreactors. *J. Environ. Chem. Eng.*, 9(5), 105768. https://doi.org/10.1016/j.jece.2021.105768
69. Rodríguez-Torres, M.J., Morillas-España, A., Guzmán, J.L., Acién, F.G. (2021) Modelling and pH control in raceway and thin-layer photobioreactors for wastewater treatment. *Energies*, 14, 1099. https://doi.org/10.3390/en14041099
70. Shi, J., Podola, B., Melkonian, M. (2014) Application of a prototype-scale twin-layer photobioreactor for effective N and P removal from different

process stages of municipal wastewater by immobilized microalgae. *Bioresour. Technol.*, 154, 260-266.
71. Yadavalli, R., Rao, C.S., Rao, R.S., Potumarthi, R. (2014) Dairy effluent treatment and lipids production by *Chlorella pyrenoidosa* and *Euglena gracilis*: Study on open and closed systems. *Asia-Pacific J. Chem. Eng.*, 9(3), 368-373.
72. Babatsouli, P., Fodelianakis, S., Paranychianakis, N. (2015) Single stage treatment of saline wastewater with marine bacterial-microalgae consortia in a fixed-bed photobioreactor. *J. Hazard. Mater.*, 292, 155-163.
73. Elystia, S., Muria, R.S., Anggraini, L. (2019) Removal of COD and total nitrogen from palm oil mill effluent in flat-photobioreactor using immobilised microalgae *Chlorella sp. Food Res.*, 3(2), 123 – 127.
74. Lieswito, N.A., Rinanti, A., Fachrul, M.F. (2019) Removal of heavy metal (Cu^{2+}) by immobilized microalgae biosorbent with effect of temperature and contact time. *J. Phy. Conference Series*, 12, 1402.
75. Gong, W., Xie, B., Deng, S. (2019) Enhancement of anaerobic digestion effluent treatment by microalgae immobilization: Characterized by fluorescence excitation-emission matrix coupled with parallel factor analysis in the photobioreactor. *Sci. Total Environ.*, 678, 105-113.
76. Schultze, L.K.P., Simon, M.V., Li, T. (2015) High light and carbon dioxide optimize surface productivity in a twin-layer biofilm photobioreactor. *Algal Res.*, 8, 37-44.
77. Do, T.T., Ong, B.N., Le, T.L. (2021) Growth of *Haematococcus pluvialis* on a small-scale angled porous substrate photobioreactor for green stage biomass. *Appl. Sci.*, 11(4),
78. Shakerkhatibi, M., Mosaferi, M., Zorufchi, B. K., Akbari, Z (2016) Performance evaluation of a full-scale ABS resin manufacturing wastewater treatment plant: a case study in Tabriz Petrochemical Complex. *Environ. Heal Eng. Manag.*, 3(3), 151-158.
79. Jafarinejad, S. (2019) Simulation for the performance and economic evaluation of conventional activated sludge process replacing by sequencing batch reactor technology in a petroleum refinery wastewater treatment plant. *ChemEngg.*, 3, 45.
80. Collivignarelli, M.C., Todeschini, S., Abbà, A. (2021) The performance evaluation of wastewater service: a protocol based on performance indicators applied to sewer systems and wastewater treatment plants. *Environ. Technol.*, 21, 1-18.
81. Singh, S., Kapoor, V., Siriya, K., Vashisth, M. (2021) Performance evaluation of extended aeration-based sewage treatment plants at NCT of Delhi, India. *J. Univ. Shanghai Sci. Technol.*, 23, 515–525.
82. Shen, X., Huang, D., Zhang, C., Hu, K. (2018) Performance Evaluation of Constructed Wetlands Treating Wastewater Treatment Plant Effluent in Taihu Lake, China. *Clean - Soil, Air, Water*, 46(1), 1600442.

83. Zhang, X., Huang, J., Lin, L., Wang, Y., Zhang, X. (2019) Operation Performance Evaluation of Municipal Wastewater Treatment Plant by Analytic Hierarchy Process. *IOP Conf. Ser.: Mater. Sci. Eng.*, 484, 012037.
84. Awad, S. N., Al-Obaidi, B.H.K. (2020) Performance Evaluation of Trickling Filter and Extended Aeration of Wastewater Treatment Plants. *J Eng.*, 26(3), 93-99.
85. Ayalew, A. A., Takele, T. (2020) Data set on performance evaluation of discharged wastewater quality from Guna water treatment plant for potable water purpose. *Data Br.*, 28, 104926.
86. Ashutosh, P. (2017) Performance evaluation of wastewater treatment plant based on UASB + Facultative Pond Technology- A case study of Karnal City, Haryana (India). *Int. J. Adv. Eng. Res. Dev.*, 4. https://doi.org/10.21090/IJAERD.39257
87. Ayyildiz, E., Yildiz, A., Taskin, G. A, Ozkan, C. (2021) An integrated methodology using extended Swara and Dea for the performance analysis of wastewater treatment plants: Turkey Case. *Environ. Manage.*, 67. https://doi.org/10.1007/s00267-020-01381-7
88. Minhaj, P.G.O., Adhiraga, P.M., Adityosulindro, S., Hartono, D.M. (2020) Modeling performance of industrial park wastewater treatment plant by STOAT software. *E3S Web of Conf.*, 211, 2020.
89. Mansir, I., Oertlé, E., Choukr-Allah, R. (2021) Evaluation of the performance and quality of wastewater treated by m'zar plant in Agadir, Morocco. *Water*, 13(7), 954.
90. Makuwa, S., Tlou, M., Fosso-Kankeu, E., Green, E. (2020) Evaluation of fecal coliform prevalence and physicochemical indicators in the effluent from a wastewater treatment plant in the north-west province, South Africa. *Int. J. Environ. Res. Public Health*, 17, 6381.
91. Dadar, S., Pezeschi, A., Đurin, B., Dogančić, D. (2020) Qualitative evaluation of wastewater treatment plant performance by a neural network model optimized by genetic algorithm. In: *Proceedings of 5th International Electronic Conference on Water Sciences.* MDPI, Basel, Switzerland, 8047.
92. AlAjmi A.M.A., Satyanarayana, S.V., Kasa, S.R. (2018) Performance Evaluation of MBR Technology Wastewater Treatment Plant at Sohar Port-A Case Study. *Int. J. Adv. Res.*, 6, 1277–1281.
93. Hejabi, N., Saghebian, S.M., Aalami, M.T., Nourani, V. (2021) Evaluation of the effluent quality parameters of wastewater treatment plant based on uncertainty analysis and post-processing approaches (case study). *Water Sci. Technol.*, 83(7), 1633.
94. Anaokar, G. S., Khambete, A. K., Christian, R. A. (2018) Evaluation of a performance index for municipal wastewater treatment plants using MCDM - topsis. *Int. J. Technol.*, 9.
95. Ho, L., Goethals, P.L.M. (2020) Municipal wastewater treatment with pond technology: Historical review and future outlook. *Ecol. Eng.*, 148, 105791.

96. Zhao, L., Dai, T., Qiao, Z., Sun, P., Hao, J., Yang, Y. (2020) Application of artificial intelligence to wastewater treatment: A bibliometric analysis and systematic review of technology, economy, management, and wastewater reuse. *Process Saf. Environ. Prot.*, 133, 169.
97. Zhang, W., Ma, F., Ren, F., Yang, F. (2021) Application with Internet of things technology in the municipal industrial wastewater treatment based on membrane bioreactor process. *Appl. Water Sci.*, 11, 52.
98. ElFar, O.A., Chang, C-K., Leong, H. Y., Peter, A.P., Chew, K.Y., Show, P.L. (2021) Prospects of Industry 5.0 in algae: Customization of production and new advance technology for clean bioenergy generation. *Energy Convers. Manag.*, X 10, 100048.
99. Roberts, G.W., Sturm, B.S.M., Hamdeh, U., Stanton, G.E., Rocha, A., Kinsella, T.L., Fortier, M-O. P., Sazdar, S., Detamore, M.S., Stagg-Williams, S.M. (2015) Promoting catalysis and high-value product streams by *in situ* hydroxyapatite crystallization during hydrothermal liquefaction of microalgae cultivated with reclaimed nutrients. *Green Chem.*, 17(4), 2560.
100. Terry, M. (2018) How Biotechnology and the Fourth Industrial Revolution Could Save the World. https://www.biospace.com/article/how-biotechnology-and-the-fourth-industrial-revolution-could-save-the-world/ (accessed on 18.7.2022).
101. Ansari, F.A., Nasr, M., Rawat, I., Bux. F. (2021) Artificial neural network and techno-economic estimation with algae-based tertiary wastewater treatment. *J. Water Process. Eng.*, 40, 101761.
102. Coşgun, Günay, A.M.E., Yıldırım, R. (2021) Exploring the critical factors of algal biomass and lipid production for renewable fuel production by machine learning. *Renew. Energy*, 163, 1299.
103. Raj, J.V. A., Kumar, R.P., Vijayakumar, B., Gnansounou, E., Bharathiraja, B. (2021) Modelling and process optimization for biodiesel production from Nannochloropsis salina using artificial neural network. *Bioresour. Technol.*, 329, 124872.
104. Saini, D.K., Rai, A., Devi, A., Pabbi, S., Chhabra, D., Chang, J-S., Shukla, P. (2021) A multi-objective hybrid machine learning approach-based optimization for enhanced biomass and bioactive phycobiliproteins production in Nostoc sp. CCC-403. *Bioresour. Technol.*, 329, 124908.
105. Otálora, P., Guzmán, J.L., Acién, F.G., Berenguel, M., Reul, A. (2021) Microalgae classification based on machine learning techniques. *Algal Res.*, 55, 102256.
106. Liyanaarachchi, V.C., Nishshanka, G.K.S.H., Nimarshana, P.H.V., Ariyadasa, T.U., Attalage, R.A. (2020) Development of an artificial neural network model to simulate the growth of microalga *Chlorella vulgaris* incorporating the effect of micronutrients. *J. Biotechnol.*, 312, 44.
107. Franco, B.M., Navas, L.M., Gómezb, C., Sepúlveda, C., Acién, F.G. (2019) Monoalgal and mixed algal cultures discrimination by using an artificial neural network. *Algal Res.*, 38, 101419.

108. Noguchi, R., Ahamed, T., Rani, D.S., Sakurai, K., Nasution, M.A., Wibawa, D.S., Demura, M. and Watanabe, M.M. (2019) Artificial neural networks model for estimating growth of polyculture microalgae in an open raceway pond. *Biosyst. Eng.*, 177, 122.
109. Wang, L., Zhang, T., Wang, X., Jin, X., Xu, X., Yu, J., Zhang, H., Zhao, Z. (2019) An approach of improved Multivariate Timing-Random Deep Belief Net modelling for algal bloom prediction. *Biosyst. Eng.*, 177, 130.
110. Nayak, M., Dhanarajan, G., Dineshkumar, R., Sen, R. (2018) Artificial intelligence driven process optimization for cleaner production of biomass with co-valorization of wastewater and flue gas in an algal biorefinery. *J. Clean. Prod.*, 201, 1092.
111. IPCC, 2014: Climate Change 2014: Mitigation of Climate Change. Contribution of Working Group III to the Fifth Assessment Report of the Intergovernmental Panel on Climate Change. https://doi.org/ 10.1017/cbo9781107415416
112. Basu, A.J., van Zyl, D.J.A. (2006) Industrial ecology framework for achieving cleaner production in the mining and minerals industry. *J. Clean. Prod.*, 14, 299.
113. Crawford, G., Sandino, J. (2010) Energy Efficiency in Wastewater Treatment in North America: A Compendium of Best Practices and Case Studies of Novel Approaches. IWA Publishing, London.
114. Nishimura, S., Ohtsuki, T., Goto, N., Hanaki, K. (2021) Technical-knowledge-integrated material flow cost accounting model for energy reduction in industrial wastewater treatment. *Clean. Environ. Sys.*, 3, 100043.
115. International Organization for Standardization, 2011. ISO14051, Environmental Management - Material Flow Cost Accounting - General Framework. ISO, Geneva.
116. VanderWilde, C.P., Newell, J.P. (2021) Ecosystem services and life cycle assessment: A bibliometric review. *Resour. Conserv. Recycl.*, 169, 105461.
117. Marangon, B.B., Calijuri, M.L., de Siqueira Castro, J., Assemany, P.P. (2021) A life cycle assessment of energy recovery using briquette from wastewater grown microalgae biomass. *J. Environ. Manage.*, 285, 112171.
118. Arashiro, L.T., Montero, N., Ferrer, I., Acién, F.G., Gómez, C., Garfí, M. (2018) Life cycle assessment of high-rate algal ponds for wastewater treatment and resource recovery. *Sci. Total Environ.*, 622 (623), 1118–1130.
119. El-Mekkawi, S.A., Doma, H.S., Ali, G.H., Abdo, S.M. (2021) Case study: Effective use of Microphytes in wastewater treatment, profit evaluation, and scale-up life cycle assessment. *J. Water Process. Eng.*, 41, 102069.
120. Tua, C., Ficara, E., Mezzanotte, V., Rigamonti, L. (2021) Integration of a side-stream microalgae process into a municipal wastewater treatment plant: A life cycle analysis. *J. Environ. Manage.*, 279, 111605.
121. Branco-Vieiraa, M., Costac, D., Mata, T.M., Martins, A.M., Freitas, M.A.V., Caetano, N.S. (2020) A life cycle inventory of microalgae-based biofuels production in an industrial plant concept. *Energy Rep.*, 6, 397–402.

122. Bussa, M., Zollfrank, C., Röder, H. (2020) Life-cycle assessment and geospatial analysis of integrating microalgae cultivation into a regional economy. *J. Clean. Prod.*, 243, 118630.
123. Hossain, N., Zaini, J., Mahlia, Mahlia, T.M.I. (2019) Life cycle assessment, energy balance and sensitivity analysis of bioethanol production from microalgae in a tropical country. *Renew. Sustain. Energy Rev.*, 115, 109371.
124. Lopes, A.C., Valente, A., Iribarren, D., González-Fernández, C. (2018) Energy balance and life cycle assessment of a microalgae-based wastewater treatment plant: A focus on alternative biogas uses. *Bioresour. Technol.*, 270, 138–146.
125. Raghuvanshi, S., Bhakar, V., Chava, R., Sangwan, K.S. (2018) Comparative study using life cycle approach for the biodiesel production from microalgae grown in wastewater and fresh water. *Procedia CIRP*, 69, 568 – 572.
126. Ormazabal, M., Jaca, C., Puga-Leal, R. (2014) Analysis and Comparison of Life Cycle Assessment and Carbon Footprint software. *Adv. Intel. Sys. Comput.*, 281, 1521-1530.
127. Banu, J.R., Kavitha, P.S., Gunasekaran, M., Kumar, G. (2020) Microalgae based biorefinery promoting circular bioeconomy-technoeconomic and life-cycle analysis. *Bioresour. Technol.*, 302, 122822.
128. El-Galad, M.I., El-Khatib, K.M., Zaher, F.A. (2015) Economic feasibility study of biodiesel production by direct esterification of fatty acids from the oil and soap industrial sector. *Egypt. J. Pet.*, 24, 455.
129. Strazza, C., Del Borghi, A., Costamagna, P., Gallo, M., Brignole, E., Girdinio, P. (2015) Life cycle assessment and life cycle costing of a SOFC system for distributed power generation. *Energy Convers. Manag.*, 100, 64.
130. Hanif, M., Mahlia, T.M.I., Aditiya, H.B., Chong, W.T., Nasruddin (2016) Techno-economic and environmental assessment of bioethanol production from high starch and root yield Sri kanji cassava in Malaysia. *Energy Rep.*, 2, 246-253.
131. Dasan, Y.K., Lam, M.L., Yusup, S., Lim, J.W., Lee, K.T. (2019) Life cycle evaluation of microalgae biofuels production: Effect of cultivation system on energy, carbon emission and cost balance analysis. *Sci. Total Environ.*, 688, 112-128.
132. Vassallo, P., Bastianoni, S., Beiso, I., Ridolfi, R., Fabiano, M. (2007) Emergy analysis for the environmental sustainability of an inshore fish farming system. *Ecol. Indic.*, 7, 290-298.
133. Odum, H.T. (1996). Environmental Accounting. Emergy and Environmental Decision Making. Wiley, New York.
134. Saladini, F., Patrizi, N., Pulselli, F.M., Marchettini, N., Bastianoni, S. (2016) Guidelines for emergy evaluation of first, second and third generation biofuels. *Renew. Sustain. Energy Rev.*, 66, 221-227.
135. Seghetta, M., Østergård, H., Bastianonib, S. (2014) Energy analysis of using macroalgae from eutrophic waters as a bioethanol feedstock. *Ecol. Modell.*, 288, 25-37.

136. da Cruz, R.V.A., do Nascimento, C.A.O. (2012) Emergy analysis of oil production from microalgae. *Biomass Bioenerg.*, 47, 418-425.
137. Park, Y.S., Egilmez, G., Kucukvar, M. (2016) Emergy and end-point impact assessment of agricultural and food production in the United States: A supply chain-linked Ecologically based Life Cycle Assessment. *Ecol. Indic.*, 62, 117-137.
138. Wollmann, F., Dietze, S., Ackermann, J. U., Bley, T., Walther, T., Steingroewer, J., Krujatz, F. (2019) Microalgae wastewater treatment: Biological and technological approaches. *Eng. Life Sci.*, 19(12), 860-871.
139. Montingelli, M. E., Tedesco, S. and Olabi, A. G. (2015) Biogas production from algal biomass: a review. *Renew Sust. Energ. Rev.*, 43, 961-972.
140. Green, F. B., Lundquist, T. J. and Oswald, W. J. (1995) Energetics of advanced integrated wastewater pond systems. *Water Sci. Technol.*, 31, 9–20.
141. Nurdogan, Y. and Oswald, W. J. (1995) Enhanced nutrient removal in high-rate ponds. *Water Sci. Technol.*, 31, 33–43.
142. Green, F. B., Lundquist, T. J., Quinn, N. W. T., Zarate, M. A. (2003) Selenium and nitrate removal from agricultural drainage using the AIWPS® technology. *Water Sci. Technol.*, 48, 299-305.
143. Mulbry, W., Kangas, P. and Kondrad, S. (2010) Toward scrubbing the bay: nutrient removal using small algal turf scrubbers on Chesapeake Bay tributaries. *Ecol. Eng.*, 36, 536-541.
144. Adey, W. H., Kangas, P. C. and Mulbry, W. (2011) Algal turf scrubbing: cleaning surface waters with solar energy while producing a biofuel. *Bioscience*, 61, 434–441.
145. Kangas, P. and Mulbry, W. (2014) Nutrient removal from agricultural drainage water using algal turf scrubbers and solar power. *Bioresour. Technol.*, 152, 484-489.
146. Johnson, D. B., Schideman, L. C., Canam, T. and Hudson, R. J. M. (2018) Pilot-scale demonstration of efficient ammonia removal from a high-strength municipal wastewater treatment side stream by algal bacterial biofilms affixed to rotating contactors. *Algal Res.*, 34, 143-153.
147. Gross, M., Henry, W., Michael, C. and Wen, Z. (2013) Development of a rotating algal biofilm growth system for attached microalgae growth with *in-situ* biomass harvest. *Bioresour. Technol.*, 150, 195-201.

2

The Use of Microalgae in Various Applications

Fulden Ulucan-Karnak[1,2], Mirac Sabankay[3] and M. Ozgur Seydibeyoglu[4,5]*

[1]Department of Biomedical Technologies, Graduate School of Natural and Applied Sciences, Ege University, Izmir, Turkey
[2]Department of Medical Biochemistry, Faculty of Science, Ege University, Izmir, Turkey
[3]Department of Bioengineering, Faculty of Engineering, Ege University, Izmir, Turkey
[4]Department of Materials Science and Engineering, Izmir Katip Celebi University, Izmir, Turkey
[5]University of Maine Advanced Structures and Composites Center, Orono, Maine, USA

Abstract
Microalgae are one of the impressive life forms. Since microalgae are a photosynthetic organism, they are produced in conditions which include light, water, CO_2, nutrients and with proper temperature and pH. Hence, the cultivation circumstances of microalgae are of great importance for maximum production. Lately, microalgae are in the spotlight for biofuel generations, microalgal compounds and industrial uses. Furthermore, biofuels created from microalgae are recognized as the most important renewable sources of industrial manufacturing. This chapter explicates general information about algae, and systems used to cultivate microalgae for these applications and future research.

Keywords: Microalgae, biofuel, industrial application

2.1 Introduction

Bio-based materials have been highlighted over the last few decades, due to their abundancy, environment friendliness, renewability, and biodegradability.

**Corresponding author*: seydibey@gmail.com

Charles Oluwaseun Adetunji, Julius Kola Oloke, Naveen Dwivedi, Sabeela Beevi Ummalyma, Shubha Dwivedi, Daniel Ingo Hefft and Juliana Bunmi Adetunji (eds.) Next-Generation Algae: Volume I: Applications in Agriculture, Food and Environment, (49–90) © 2023 Scrivener Publishing LLC

Among the bio-based materials, bioplastics and natural fibers have been thoroughly studied with many commercialized applications. On the other hand, marine resources present another abundant resource that has not been fully utilized for different applications. About 3/4 of the Earth is covered with oceans and seas. Thus, many different biological organisms live in the oceans, creating an important class of raw materials for industries. Among the organisms, macro and microalgae materials are of great importance. Since the number of algae on Earth is immeasurable, they have great potential for new material formulations. Therefore, this chapter focuses on microalgae systems and products.

Microalgae materials have been used for applications such as dental, nutraceutical, and cosmetics over the last few decades [1]. However, recently the energy field is another important area that utilizes microalgae to produce biodiesel, bioethanol, and hydrogen in an attempt to solve the needs of today's ever-increasing energy demand [2].

Moreover, there are still many areas waiting to be explored for the microalgae field of application. For instance, developing microalgae-based materials for use in polymer formulations and plastic materials can open new dimensions in the bioplastic material industry [3].

This chapter summarizes the previous literature on algae types, algae resources, current applications and potential future uses.

2.1.1 Algae Classification

Algae, the class of macro or micro sized photosynthetic life forms which is predicted to possess one to ten million species, are able to live in every water-based environment where solar or artificial light exists [4, 5]. Algae are grouped very diversely considering several properties, for instance, structural characteristics, photosynthetic pigment color, the cell wall structure/composition, type of energy-storing molecules, mobility mechanism and mode of reproduction. Moreover, algae classification is complex and has not yet been completed [6]. Some of them are microscopic unicellular organisms, which are single-celled. They are free floating or propelled by flagella and can grow in long chains or filaments. Others are macroscopic organisms, like *Volvox*, consisting of biflagellate cells formed in colonies. On the other hand, they can be planktonic or benthic [7].

Prokaryotic algae (microalgae) are divided into two subdivisions named *Cyanophyta* and *Prochlorophyta*, while *Eukaryotic* members of algae are divided into nine subdivisions named *Glaucophyta*, *Heterokontophyta*, *Rhodophyta*, *Haptophyta*, *Dinophyta*, *Cryptophyta*, *Chlorarachniophyta*, *Euglenophyta*, and *Chlorophyta* [7]. *Microalgae* are photosynthetic organisms including broad groups that have many diverse types. These organisms have both unicellular and multicellular forms, which are structured simply [8]. They can exist anywhere situated near sunlight and water such as oceans, fresh water, salt water, lakes, rivers, soils, ice, and hot springs. As a microalgal species, phytoplankton, which consist mainly of unicellular algae, is a major food source for various animals in aquatic habitats. Over 5000 species of microscopic algae exist in the oceans surrounding approximately 71% of the Earth's surface and are the main source of about 50% of the oxygen we breath [7]. Microalgae can convert sunlight energy to biochemical energy (ATP; adenosine triphosphate) and oxygen by absorbing carbon dioxide from air in chloroplasts. The overall process of photosynthetic reaction is represented in the equation. Thus, they can synthesize many products such as biofuel, foods or food supplements, and feed [8, 9]; and have been considered a sustainable biotechnological source for the last few decades [10].

Compared to the nearly 250,000 species of plants, there are approximately 200,000 to a few million microalgae species. Thus, more research is expected to exist owing to the biodiversity and population of microalgae. In order to be successful in algal biotechnology, the right strain should be chosen according to the properties of its products and determined culture conditions [11]. This chapter aims to give a perspective on the biotechnological applications of microalgae.

2.1.2 Cultivation of Microalgae

Large-scale algal biomass is essential and the achievement of large-scale production in industrial applications, proper quantity, and quality of supplements is needed for developing microalgae. Inorganic manures are commonly utilized as the main nutrients of fertilizer in large-scale algal cultivation and these nutrients are abundant in municipal wastewater, industrial wastewater and compost and livestock waste. Thus, harmful contaminations might be removed through industrial productions. However, the chemical and physical composition of wastes or wastewaters must be controlled to avoid inhibition of algal cultivation [12].

Microalgae could form open ponds (lakes) along with closed reactors (photobioreactors). Generally, they are cultivated in open ponds for many commercial purposes. Besides which, (i) large ponds, (ii) raceway ponds, (iii) circular ponds with turning components for blending, and (iv) huge bags could be used for cultivation according to species and desired products [13].

The open pond systems are efficient low-cost methods for cultivation of microalgae. The optimum pond depth should be enough to provide efficient light to the culture and enhance mixing [13]. However, easy contamination by other microorganisms and evaporation are the main disadvantages of these systems [14]. A photobioreactor (PBR) is used to present a system that utilizes light to cultivate by means of photosynthesis. Photobioreactors have numerous advantages over open systems, including:

- Contamination is minimum,
- Better control of environmental conditions such as pH, temperature, light, and CO_2 concentration,
- Photobioreactors have lower evaporation rates,
- Higher cell concentrations are allowed in photobioreactors,
- Complex biopharmaceuticals can be produced by photobioreactors [13].

Thus, the photobioreactor design should be considered and its effectiveness compared.

- The reactor can be provided the cultivation of various microalgal species.
- The reactor should achieve high rates of mass transfer without damage to cultured cells.
- The minimum non-illuminated part of the reactor and should work under circumstances of intense foaming because of high mass transfer [15].
- The reactor design should inhibit or reduce the fouling of the reactor, especially its light transmission effect [13, 15].

The cost of setting up and maintaining photobioreactors is the main problem. To overcome this problem, researchers and industry have started using hybrid systems. These systems are formed by a mix of open ponds and closed reactors. Initially, closed bioreactors are utilized to control or diminish contamination and support continuous high-density cell cultivation [16]. Then, dense microalgae cells are transferred to the open ponds to achieve large-scale biomass production and domination of microalgae population over the other microorganisms are the most critical criteria at the beginning of open ponds cultivation. Aquasearch currently uses this hybrid system in astaxanthin production from *Haematococcus pluvialis* [17].

2.2 End Uses of Microalgae

The first microalgal product was used by Alfred Nobel as fossil microalgal biomass of diatoms. The aim of this approach was adsorption of nitroglycerin to create dynamite. This application of microalgae product is still the most remarkable example [11].

Microalgae are typically commercially produced for a variety of other applications such as cosmetics, animal feed, pigments and human nutrition [18–20]. We have discussed this issue under three main headings, namely biofuel applications, microalgal high-value compounds and new industrial uses.

The harvesting process for algal products is the costliest and most critical point in all microalgae applications, and is designed by considering microalgae species. Water should be discarded from the algal biomass and further processes could be changed according to final production. Harvesting is divided to two steps. Firstly, biomass is isolated from bulk culture by flocculation, flotation or sedimentation; and secondly the biomass is concentrated by using centrifugation and filtration [16, 21]. After the harvesting step, various types of applications are carried out according to final products.

2.2.1 Biofuel Applications

Microalgal biomass is a strategic energy source, which is renewable, eco-friendly, and able to produce fuel for vehicles. In addition to all this, it provides socio-economic development. The quality of biofuel could be increased by various processes and has equivalent properties of the petrochemical fuels utilized in the energy industry [22].

2.2.1.1 Biodiesel

The history of biodiesel starts with the development of diesel motor by Rudolph Diesel in 1892 [23]. Today, Europe and North America are the biggest biodiesel producers and production has been increasing since 1999, with rapid acceleration [24]. The main advantages of biodiesel as a fuel are portability, accessibility, higher combustion proficiency, aromatic content, and a lower sulfur level [25].

Vegetable oils, animal fats and restaurant waste oil are mostly used to produce biodiesel in small scale as renewable oil sources. Generation of biodiesel from vegetable oils is not commercially feasible on a large scale due to the increasing competition for seed, requiring a huge amount of fresh water and production area and non-valuable byproducts [9, 25].

Apart from these oil sources, microalgae, bacteria, and fungi are used to produce biodiesel as bio-alternative oil sources [26]. Among these oil sources, microalgae have great potential for the production of biodiesel. Contrary to other oil crops, microalgae have advantages as a sustainable oil source since they are i) highly productive, ii) non-food assets, iii) able to grow non-arable land, iv) able to live in a wide variety of water sources, and v) reduce greenhouse gases and vi) produce valuable co-products along biodiesel [9, 24, 27–30].

Aquatic unicellular green algae are mostly used for biodiesel production. Dry algae biomass contains approximately 30% oil and the oil content of some species, such as such as *Botryococcus braunii,* can be as high as 70% [25, 31].

Transesterification of triglycerides with a short-chain alcohol in the presence of a base or acid-based catalyst achieves high conversion of triglycerides to fatty acid methyl esters (FAME) in a short reaction time. The oxygen existing in the biodiesel fatty chain is one of the distinctive features compared to petroleum diesel [26, 32]. Acids catalyst is not sufficient due to its corrosivity and prolonged reaction time requirement [33]. The highest biodiesel yield was achieved with *Chlorella vulgaris* by sodium methoxide used as a base catalyst at 161 °C in 51 min [34]. However, the high energy input requirement, the difficulty in removing the catalyst from the product, the alkaline water produced during the washing step, the possibility of saponification, and the difficulty in removing the glycerol are the limitations of acid-base catalyst transesterification [35]. Furthermore, this type of microalgae (*Chlorella vulgaris*) is surveyed as the fitting candidate to produce biodiesel since it has been demonstrated that it produces lipid article, maximum biomass, and lipids with fatty acid structure from a biodiesel standpoint [36].

Enzymes have been used as a green alternative in biodiesel production in the last ten years. The most valuable advantages of enzymes are (i) easy recovery of product without byproducts, (ii) not effected through free fatty acids, (iii) reusability of enzymes, (iv) conditions of moderate reaction, (v) low energy demand, and (vi) low alcohol and oil ratio [37, 38]. Extracellular and intracellular lipases whole cells (lipase producer microorganisms) were tried as biocatalyst. Immobilized lipase enzyme is more suitable than free lipase and whole cells with respect to its high stability and potential of reuse in the transesterification process. However, glycerol contamination (as a product of transesterification process) reduces the activity of immobilized enzyme and removal of contaminants from immobilized lipase becomes a more complicated process that negatively effects large-scale production financially [35, 39].

On the other hand, using a whole cell as a source of intracellular lipase source reduces the lipase cost and isolation, purification and immobilization are not required at any level. Preparation of liquid catalysis media is easier and cheaper than immobilized media. Besides, liquid phase provides the regeneration of enzyme activity [35].

Moreover, the study proposes estimating *in-situ* transesterification and enzymatic analysis of the *D. salina* in an organic solvent due to reducing the oil extraction stage and the cost. In conclusion, this method has been found to have the potential to generate biodiesel from microalgae biomass *in situ* [40].

2.2.1.2 Bioethanol

Bioethanol has been popular since the 1970s and 1980s. In 1976, the largest energy program in the world took place in Brazil. It was based on bioethanol derived from Brazilian sugar cane (Proálcool) [41]. Using bioethanol as a fuel could result in decreased levels of lead, sulfur, carbon monoxide and particulates. Additionally, it has a global benefit because of its ability to reduce CO_2 emissions [42]. The conversion of biomass materials into ethanol is called the alcoholic fermentation process. Biomass materials include sugars, starch or cellulose [16].

Ethanol has been considered a promising fuel source since the 1980s and it is produced mainly from sugar cane and starch. Brazil is regarded as the main bioethanol producer in the world [41], and currently the USA, Europe and other countries have been using ethanol instead of petrol [42, 43]. In the countries that currently produce bioethanol, blends of bioethanol and fuel are used. For instance, the blend percentage of bioethanol was 5% in Canada and around 20–25% in Brazil in 2010 [44].

Environmental sustainability is one the most vital issue within the world market of bioenergy. As a result of the use of agricultural substrates such as wheat, corn, sugar cane, sugar beet, and others to produce bioethanol, one of the most challenging questions about the sustainability of bioethanol is the competition for agricultural land for food. Consequently, identification of alternative feedstocks of bioethanol has high importance [45]. At this point, microorganisms such as bacteria, yeasts, fungi, or microalgae have become an alternative non-food feedstock. Some microalgae may have a high content of carbon compounds depending on their growth conditions. In addition, microalgae biomass as a substrate can be suitable for fermentation both without and after pretreatment. Hence, they are seen as a potential fermentable substrate for bioethanol [46].

Oscillatoria limosa, Chlamydomonas reinhardtii, Cyanothece PCC 7822, *Microcystis* PCC7806, *Oscillatoria* sp., *Microcystis aeruginosa* PCC 7806, and *Spirulina platensis* are microalgae that are able to produce ethanol photosynthetically due to tolerating extreme ethanol and salt concentrations and pH levels [47]. *Chlorella vulgaris* has high carbohydrate content; thus, it may be used as a supply of ethanol with conversion potency higher than 65.0% [48]. Ethanol can be produced and purified from the algae. Besides which, CO_2, the byproduct of the process, can be recycled. According to the existing research results, technically applicability of the ethanol production from microalgae seems reasonable [16].

2.2.1.3 Biomethane (Syngas)

Biomethane (Syngas) is generated by microalgal biomass digestion in anaerobic conditions. This biochemical process consists of hydrolysis, fermentation, acetogenesis and methanogenesis in a closed system [49]. Light energy can easily be converted into chemical energy in a short amount of time in a closed system and is formed by three main units: (i) cultivation of microalgae, (ii) bacterial growth in aerobic conditions, and (iii) digestion in anaerobic conditions [50]. Higher energy efficiency is obtained than in other energy production systems because the extraction process is not needed, and the main product in the gas phase is provided as the methane. Besides which, all parts of the microalgae structures (proteins, lipids, sugars) are utilized during the anaerobic fermentation and residual nutrients can be mineralized and reused for the next algae cultivation [50–52]. Syngas can be directly used for gas engines or gas turbines as a fuel [53, 54].

In anaerobic digestion, moisture content should be between 80.0% and 90.0%. Therefore, microalgal biomass is quite adequate. In addition, microalgae can have a high proportion of proteins. Thus, they decrease the C/N ratio that could influence anaerobic digestion efficiency [16, 46].

In regard to plant biomass, which is mostly used to produce biogas, microalgal biomass has the advantage of growing in liquid medium and including a third dimension, unlike plant biomass. Additionally, microalgae do not contain lignocelluloses, in contrast to plant biomass. Thus, pretreatment is not required before the microalgae anaerobic digestion. Besides which, the gasification of microalgal biomass includes the recovery of CO_2 [46, 55]. In addition to this, biomass residue could be utilized as the fertilizer [56].

The results of studies proved that microalgae are a unique source for biomethane production. However, the low energy ratio between microalgal oil and fossil fuels, the efficiency of microalgal biomass production systems for microalgal oil and oil extraction, and the cost of production are big obstacles in biomethane production [57–61]. Recently, integrated microalgae and wastewater/organic-wastes-based biorefinery systems promised that these barriers could be broken, and solving waste treatment and the production of biofuels will succeed in large-scale operations [56].

Photovoltaic hydrothermal pretreatment is an effective method for extending energy gain in the generation of biogas from microalgae biomass via anaerobic digestion. The laminar mixed convection features of the microalgae slurry within the solar collector definitely impact the effectiveness of the hydrothermal pretreatment due to the high viscosity and inhomogeneous heat flow. Non-uniform heat flux and laminar mixed circulation of microalgae slurry were quantitatively examined in the study. According to the study's findings, secondary flow of the microalgae slurry boosted the coefficient of heat transmission. Furthermore, it has been shown that it is advantageous for the planning and optimization of solar hydrothermal pretreatment systems [62].

2.2.1.4 Biohydrogen

Nowadays, biohydrogen production from renewable sources has attracted considerable attention among researchers who are focused on overcoming bottlenecks in the energy field. Utilization of wastes and production of clean energy sources make the biohydrogen production processes preferable. When hydrogen is burned, energy is generated in the form of heat, and the only byproduct is water; thus, it is regarded as a clean fuel. Compared to other fuels which are formed as gas, it is eco-friendly and not harmful to people [63].

Although different technologies have been improved to produce hydrogen from biomass economically, it is presently costlier than conventional energy supplies. Biohydrogen technology will play an important role in the future since it will allow for the use of sustainable energy sources [64]. There are some countries which have already attempt to establish H_2 fuel stations such as the USA, EU and Japan [65].

Biological hydrogen production processes consume less energy due to being operated under environmental temperature and pressure. In addition, these processes have low energy consumption. On the other hand, the main disadvantage of their commercial use in automobiles is the necessity of large on-board hydrogen storage tanks due to its extremely low density.

Figure 2.1 Biohydrogen production by microalgae.

Some hydrogen technologies have been developed and the most promising one is a fuel cell, which uses chemical energy of hydrogen, an electrolytic solution and a catalyst to supply energy to produce electricity with water and heat [25]. In Figure 2.1, a biohydrogen production system is presented.

Photobiological hydrogen production includes the processes such as direct and indirect biophotolysis, dark fermentation, photo-fermentations, and hybrid systems consisting of both fermentative and photosynthetic processes by using bacteria and microalgae [66, 67].

Microalgae has greater potential for biohydrogen production than bacteria because they are fast growing, require low nutrients for production, are excellent sources of valuable products such as pharmaceutical and nutraceutical compounds, can be used as a substrate for different types of fermentation, and release less toxic fermentative inhibitors during pretreatment [68]. They can turn water into H_2 (biophotolysis) photosynthetically via light energy or their biomass can be utilized as a substrate by anaerobic bacteria for dark fermentation. They also possess high theoretical light conversion efficiencies. Besides which, some species can live in both anaerobic and aerobic environments [68, 69].

To compete with fossil fuel production activity, a process should be developed and studies conducted to prove that it is possible to commercialize convenient hydrogen production by *Chlamydomonas* just with a productivity of 7% under open-air conditions [63, 70]. Although the results of microalgal hydrogen production on a laboratory scale has been promising, production on a commercial scale is still scant. Designing and scaling-up the photobioreactors is the biggest milestone for industrial-scale biohydrogen production applications [11, 71, 72]. However, industrial biorefinery applications in which wastewater is used will reduce the cost of production and make it easier to scale-up procedures [73].

2.2.1.5 Bioplastic

Bioplastics, described as plastics obtained from renewable carbon resources that are biodegradable or bio-based polymers, are biomaterials created by living organisms, like plants, animal, fungi or bacteria, ecological and sustainable [74, 75]. Besides which, nonavailability of a dense biomass as a consequence of difficulties related to cultivation is a significant problem in the production of bioplastics. In these instances, algae can be a stronger alternative than other microbial resources [76]. Because algae have great biomass, growth rate and ease of culture in natural environment.

Microalgae has a high potential as a biocomposite in plastic production and the combination products have serviceable properties. In the future, these materials will have a wide range of potential applications [77]. In the studies, various types of microalgal biomass were added to petrochemical products in different ratios (100:0 to 20:80) and the tensile strength of products was effected positively [78, 79]. However, only a few patents are related to the use of algae (US56564103 and WO00/1106 packaging in the production of the film and foam materials industry; WO 2007079719 filler in the production; Italian Patent no. RM2002A000592 in tire production) [80]. Sapalidis *et al.* [81] produced poly-(vinyl alcohol) (PVA) and Zostera-based composite films. Although this material only consists of 20% (optimal) Zostera, it gives better results in terms of thermal and mechanical properties of pure PVA. It is reported that the obtained biocomposite material can be used in the packaging industry.

Moreover, studies showed that some microalgae species can synthesize hydroxybutyrate. *Spirulina platensis* indicated potential for poly(3-hydroxybutyrate) [P(3HB)] synthesis under nitrogen-limited conditions and acetate [82].

Furthermore, one of the most recent studies assessed the possibility of microalgae generated by wastewater treatment plants for use as the foundation of a bioplastic material. A commercial *Arthrospira maxima* sample was utilized as a control. As a result, biomass including microalgae was thoroughly mixed with a plasticizer and various concentrations of glycerol, yielding a homogeneous blend that was then injection-molded. The findings demonstrated the potential of microalgae to effectively remediate wastewater through the production of bioplastic materials [83].

2.3 Microalgal High-Value Compounds

Today, many microalgal products, such as sterols, antioxidants, polysaccharides, pigments, fatty acids, biomass, and other active compounds [84], are getting considerable attention in different industrial fields (e.g., cosmetics, pharmaceuticals, functional foods, feed additives) [2, 85]. In order to produce these bioactive compounds on an industrial scale, large-scale PBR developments are required [2]. It is assumed that there will be microalgal drugs existing commercially in forthcoming years [56].

2.3.1 Polyunsaturated Fatty Acids

Two or more double bonds are found in n-3 long-chain polyunsaturated fatty acids (n-3 LC-PUFAs), with the final double bond located at the third carbon atom from the methyl end of the fatty acid chain. With regard to their beneficial effects on human health, eicosapentaenoic acid (EPA, C20:5), docosahexaenoic acid (DHA, C22:6), and arachidonic acid (AA, C20:6) are the most studied and utilized PUFAs in dietetics and therapeutics [86], in the fight against several diseases such as Alzheimer's disease, hypertension, cancer and cardiovascular diseases, or schizophrenia [87]. In addition, they have a significant role in the brain function and visual acuity in infants [88, 89]. Currently, microalgal PUFAs are very common in the food market. They are used in animal feed, maricultural products, and food additives in yogurts, cheeses, and breakfast cereals [90]. Moreover, in Europe, infant milk formulas include health-promoting purified PUFA. Additionally, a special microalga (*Schizochytrium* sp.) is added to hens' feed to generate "Omega" eggs [11]. Microalgal PUFAs trade was estimated to have expanded at the rate of 12.8% annually between 2014 and 2019, and to reach 4.3 billion USD in 2019. The general chemical structure of unsaturated fatty acid is shown in Figure 2.2.

Fish and vegetable oils are the main sources of PUFAs and seafoods are considered the most valuable source for PUFAs [91, 92]. In particular, fish are considered the most valuable source of PUFAs and contain significant PUFAs mixture for human and animal health. However, extracted fish oil

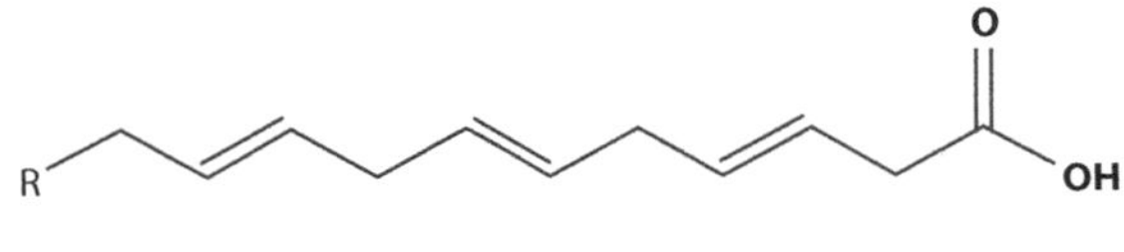

Figure 2.2 Polyunsaturated fatty acids' chemical structure.

is not sustainable enough due to its smell, undesirable taste, poor oxidative stability, and the probable toxin accumulation in fish and seasonal effects [93–95]. Furthermore, the studies proved that fish and marine organisms consumed by both humans and animals that are not able to synthesize EPA in their metabolisms. Therefore they must be taken by consuming EPA that can be transferred to humans via consumption of these organisms in the food chain [96]. In the meantime, the studies exhibited which microalgae are the real sources of fish PUFAs, and microalgae provide PUFA components in the food chain. PUFA content of microalgae was proved more abundant than other organisms [97]. They can collect up to 30–70% of lipids in their biomass [9] and have great potential as applicants in n-3 LC-PUFAs industrial production [98]. Microalgal fatty acids do not have the disadvantages of fish oils and they have better purification potential. Consequently, microalgae are an encouraging source of PUFA in both the food and feed market [11].

Some of the microalgal species, *Arthrospira* and *Chlorella*, have gained importance in the skincare trade [99]. The cosmetics companies LVMH (Paris, France) and Daniel Jouvance (Carnac, France) advanced their own microalgal production systems. Also, there are examples of sun protection and hair-care-related products that originated from microalgae in the cosmetic market (Pepha®-Ctive) [100]. Other microalgal lipid classes, such as glycolipids and phospholipids, should be noted in further research on skin care [11]. Recently, the inhibitory effects of microalgal PUFAs on allergic reactions have been demonstrated. They decrease histamine content or inhibit the release of chemical mediator [101].

In various microalgal classes (*Chrysophyceae, Eustigmatophyceae, Chlorophyceae, Prasinophyceae, Cryptophyceae, Bacillariophyceae*), especially those which contain EPA [96], apparently just a few species have economic potential. Because most microalgae have poor specific growth rates as well as cell densities while grown under standard circumstances [102], the technological necessity of developing a reliable, sustainable, and low-cost alternative source of PUFA for the public nutritional industries should be addressed [103]. The use of transgenic algae to produce EPA could be an alternative source of PUFA for the food market. Only, it does not seem possible due to the negative thoughts of consumers about transgenic food products. Consumption of seafoods which are grown by feed including transgenic algae can be an alternative way to maintain the health benefits of PUFAs in humans. That way, direct ingestion of genetically modified food would be hindered [96].

In addition to genetic ways, the efficiency of industrial-scale cultivation systems is also one of the most significant ways to achieve the production of microalgae-based EPA commercially [104]. There isn't the necessity for light for growth in heterotrophic EPA production. Thus, it is preferable to photoautotrophic conditions for cost-effective production [94, 102, 105]. Besides which, the availability of growth media and fast adaptation of microalgae are as important as the ability of microalgae for production under heterotrophic conditions [106]. Heterotrophic microalgal production process may be feasible in terms of large-scale EPA production [96].

Nannochloropsis oculata and *Phaeodactylum tricornutum* cells produce EPA while *Thalassiosira pseudonana* cells produce EPA and DHA. During the stationary phase of microalgae cells, these PUFAs are partitioned into triacylglycerols (TAGs). In comparison to *P. tricornutum* cells, *N. oculata* cells contain a higher percentage of TAGs partitioned from EPA. Besides which, a CO_2 increase and the induction of heterotrophic growth during the cultivation of *Nannochloropsis* sp. and *Crypthecodinium cohnii* result in the increment and differentiation in PUFA composition [107]. The composition of PUFA in these organisms varies both between species and growth phases [2].

Microalgal EPA and DHA production for the food market by dinoflagellate, e.g., *Crypthecodinium,* is an innovative venture [85, 108]. The European Commission (EC) Regulatory Board allowed the application of these PUFAs produced by *Ulkenia* sp. to the food market in 2003. The first company that announced their commercial production of microalgal DHA was Martek (USA). Subsequently, Nutrinova (Germany) took its place between the microalgal biotechnology-based companies that work for human health and other implementations [11].

One of the biggest problems in commercial production of PUFA is purification. There isn't any purified algal oil produced by potential EPA producer species (*Isochrysis galbana, Porphyridium purpureum, Nannochloropsis* sp., *Phaeodactylum tricornutum,* and *Nitzschia laevis*) [97, 102, 109, 110]. Thus, it cannot keep up commercially with other sources [85, 109, 111]. There is the same problem with γ-linolenic acid (GLA) and AA. Currently, the only algal oil feasible for industrial production is DHA. Furthermore, OmegaTech Corp. (USA), owned by Martek, produces the oil known as DHA Gold by *Schizochytrium* with a low-cost technology [100].

2.3.2 Carotenoids

Microalgae have many efficient protective systems because their photoautotrophic life contains high oxygen and radical stresses. These systems protect microalgae cells from cell-damaging effects of free radicals accumulation and reactive oxygen species such as superoxide ($^{\bullet}O^{-2}$), hydrogen peroxide (H_2O_2), etc. [11]. Thus, microalgae are the source of several secondary metabolites like phenolic acids, carotenoids, etc. [91, 112].

Carotenoids, which are natural pigments, have at least 600 different types. These carotenoids have importance in algae, bacteria, plants and animals' metabolism [113]. Carotenoids are the main components of photosynthetic apparatus due to their action in the reaction centers of photosystems [114]. Carotenoids have radical-scavenging property. Thus, they are in high demand particularly in the beverage market [11]. Besides which, some carotenoids have the ability to convert to vitamin A [115–117]. Due to their provitamin A activity, there are some approaches for human nutritional purposes [118]. This can be a solution for vitamin A deficiency disease, which has quickly grown in developing countries over the last few decades [96]. In addition, it has been reported that carotenoids as a dietary supplement decreases the hazards of immunological diseases like asthma and atopic dermatitis [101]. However, animals and humans cannot produce carotenoids and they have to take carotenoids and their derivative compounds via their diet [114].

Thus far, many bioactivities of carotenoids have been found such as their anti-inflammatory, anti-oxidation, anti-viral, and anti-cancer effects [119]. Natural food color additives for animal feed (fish and poultry) are the main commercial products on the food market. For instance, the nutritional supplements BioAstin, Natural Astaxanthin and Hawaiian Spirulina Pacifica are produced by Cyanotech Corporation. They distribute these microalgal products to more than 40 countries worldwide [120].

Carotenoids' quenching ability on relative oxygen species shows their anti-inflammatory property [11, 100]. Even though they have a chemopreventive anticancer effect due to these molecules [116, 121], this anticancer effect hasn't been proved *in vitro* and *in vivo* research [1]. Besides which, cosmetic applications of carotenoids are developing rapidly owing to their natural origin, microalgae [122]. When some antioxidants and bioactive compounds from microalgae are combined for cosmetic applications, demand for sun-protecting cosmetics especially increases [11]. Moreover, it has been observed that carotenoids have an inhibitory effect on antigen-induced degranulation [123].

Microalgae can synthesize diverse carotenoids. To produce high concentrations of desired carotenoids, the ideal metabolic pathway should be controlled. Fat globules increase the carotenoid level in microalgae by stimulating their production. Therefore, they are a critical factor in carotenoid production [124]. The studies showed that *Muriellopsis* sp. has the ability to accumulate high levels of lutein, a carotenoid, which has a protective effect for the treatment of degenerative diseases [2, 125].

Microalgal carotenoids compete with synthetic pigments on the market which are less expensive, notwithstanding that microalgal carotenoids have an advantage owing to their natural isomer content [115, 121]. So far, over 400 different carotenoids are known and just a few of them have become commercial products such as lutein, β-carotene, astaxanthin, zeaxanthin, bixin and lycopene, with β-carotene and astaxanthin having more importance than the others on the market [100, 126]. Research has shown that β-carotene has outstanding features compared to synthetic forms [1, 108, 127]. Despite the price of the high-value carotenoid, astaxanthin, produced by *H. pluvialis*, astaxanthin is preferred in certain instances such as in carp and chicken feed [127].

Astaxanthin maintains the protection of membranous phospholipids and other lipids against peroxidation. *H. pluvialis* is approved as a food supplement in the food market of the USA and European countries [2]. Astaxanthin is generally esterificated with fatty acids or conjugated proteins owing to its sensitivity to oxygen. Hence, even though astaxanthin can be compressed inside tablets, it can be degraded by oxidation [100]. Astaxanthin-producing companies have researched the possibility of suspending *Haematococcus* biomass in edible oils. This way, it was expected to create the barrier between the astaxanthin-rich biomass and oxygen. The trials of Cyanotech Corporation with rosemary oil were not successful because it was seen that astaxanthin is unstable in this formulation. On the other hand, Mera Pharmaceuticals chose to improve the extraction method and used edible oils as the extraction solvent during their patent oil

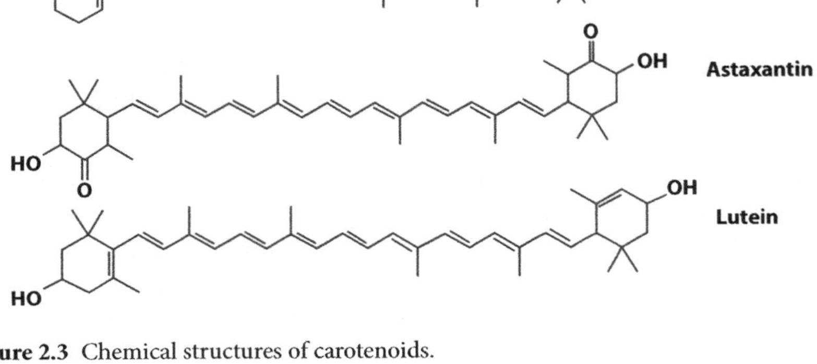

Figure 2.3 Chemical structures of carotenoids.

extraction method. Supercritical CO_2 extraction can be an alternative way to produce astaxanthin [128]. Algatech (Kibbutz Ketura, Israel) is another microalgal company which produces "crushed *Haematococcus* biomass rich in astaxanthin" and sells its products on the biopharmaceuticals market [100]. The general structure of carotenoids is shown in Figure 2.3.

2.3.3 Phycocyanin

There are many commercial products that are possible to obtain from phototrophic organisms. Phycocyanin (PC) is a blue colored, fluorescent pigment that is one of these products. When nitrogen starvation occurs in the culture medium, phycobiliprotein in the cells is selectively degraded; accordingly, it acts as an intracellular nitrogen storage compound in the case of nitrogen shortage [129]. It is just produced by microalgae and cyanobacteria [130, 131]. Figure 2.4 shows the chemical structure of phycocyanin.

Phycocyanin is commercially produced by *Spirulina platensis* that is grown in phototrophic cultures using sunlight as an energy source. It is one of the main components of the pigments contained in *Spirulina* [132, 133]. It is currently isolated from open pond cultures of *Spirulina platensis*; nevertheless, the productivity of these cultures is not enough and the open pond cultures are open to contaminating organisms [129, 131].

Phycocyanin can also be produced by *Galdieria sulphuraria*. This unicellular rhodophyte PC production by *G. sulphuraria* may offer several advantages compared to the existing manufacturing procedures for PC using *S. platensis*. The yield of heterotrophic high-cell-density fed-batch *G. sulphuraria* cultures is 1.7–13.6 times that of outdoor *S. platensis* cultures [129, 131].

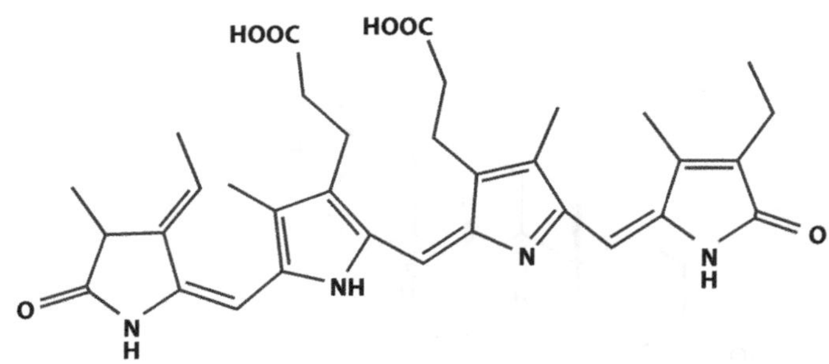

Figure 2.4 Chemical structure of phycocyanin.

Phycocyanin has many applications in the field of diagnostic histochemistry as a marker, dye in foods and cosmetics [131], and an antioxidant [129]. Recently, its anti-allergic effect was proved. Due to the antigen-specific immunoglobulin E (IgE) antibody suppression with the aid of PC, immune reaction against infectious diseases increased and allergic inflammation in mice decreased [134]. An eventual study in 2011 by Chang *et al.* has pointed out that R-phycocyanin (R-PC) has therapeutic potential against allergic airway inflammation. Last of all, phycobiliprotein, which regulates allergic inflammatory responses, seems to be another promising microalgal pigment [101].

2.3.4 Sterols

Sterol is another important component that can be obtained from microalgae. They contain some various types of sterols. For instance, clionasterol, one of the microalgal sterol types isolated from *Spirulina* sp., has expedited cardiovascular disease prevention [2]. The general chemical structure of clionasterol is shown in Figure 2.5.

Microalgae have great importance in the diet of marine-based organisms. Especially bivalves need sterols in their diet for growth due to lack of sterol bioconversion *de novo* [135, 136]. Since the microalgae mixture that contains various species is used, larval growth is accelerated and facilitated [137]. Also, microalgae sterol composition has a significant effect on bivalve larvae's phytosterol and the cholesterol composition [96]. According to the studies, in light of the suggestions of the International Council for the Exploration of the Sea (ICES), it is shown that juvenile bivalves (spat) can ingest, digest and accumulate the sterols incorporated in emulations [138]. In the research, sterol levels and both photoautotrophic and heterotrophic development of the microalga *Tetraselmis suecica* was investigated [139].

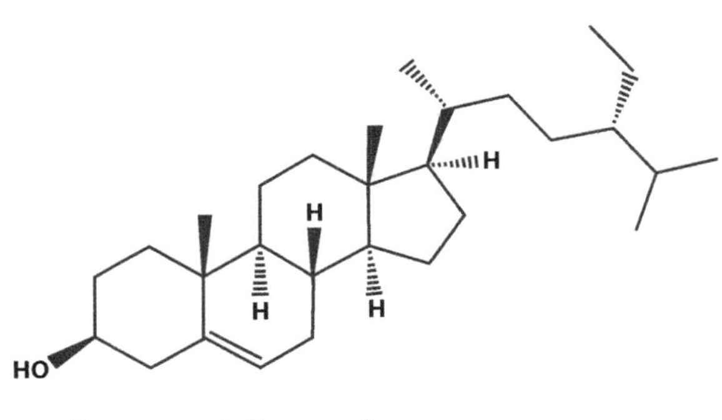

Figure 2.5 Chemical structure of clionasterol.

As a consequence, the major sterol production of photoautotrophic algae was found to be relative to heterotrophic algae.

Up to now, the data on microalgal sterols aren't very comprehensive, except for the increasing effect of microalgae as a diet supplement in aquaculture activities. Research on microalgal lipid composition should be rapidly improved due to its connection to the increase in world aquaculture production [96].

2.3.5 Polysaccharides

One of the greatest sources of polysaccharides is marine algae. Microalgal polysaccharides have anti-coagulant, anti-cancer, anti-viral, and anti-inflammatory properties. Thus, in recent years they can be found in many different markets such as food, cosmetics, and pharmacology [140].

Researchers identified that *Chlorella pyrenoidosa* and *Chlorella ellipsoidea* synthesize polysaccharides, which include glucose and any combination of galactose, mannose, N-acetyl glucosamide, rhamnose, arabinose, and N-acetylgalactosamine. The growth of *Listeria monocytogenes* and *Candida albicans* can be inhibited owing to immune stimulatory activity of microalgal polysaccharides. This approach as a chlorella extract or a food supplement may be used by humans to stimulate their immune system and enhance the growth of splenocytes and release of cytokines [141].

Chlorella sp. cells were also used to treat leprosy. A plankton soup consisting mainly of *Chlorella* sp. cells was prepared for the patients' diet. The effect of plankton soup was positive in patients, increasing their energy, weight and health [107].

Another food product including microalgae was studied by Japanese researchers. They used *C. ellipsoidea* for the production of bread and rolls, soy sauce, powdered green tea, soups, noodles, ice cream, cookies [2, 101]. One of the most recent approaches of microalgal polysaccharides that has been submitted is their use as anti-allergic agents [101].

Moreover, researchers have explored the recovery potential of carbohydrate and lipid in *Chlorella vulgaris*. The results proved that it is feasible to utilize residue of microalgae as carbohydrate stock as the substrate in biofuel and biochemical manufacturing. This represented an important wet microalgae application [142].

Furthermore, the feasibility of utilizing microalgae polysaccharides was investigated in another study. The procedure can be summarized as the application of raw extract polysaccharides from three microalgae strains to *Solanum lycopersicum* plants and a comparison of their effects on dry weight and their parts length. Consequently, this is an indication that microalgae polysaccharides, which are easy to cultivate and renewable, can be used as a plant bio-stimulant to enhance and protect agricultural crops [143].

2.3.6 Polyketides

In the pharmaceutical industry, polyketides have received great attention as a secondary metabolite, and thus their commercial value is high [144]. Polyketide products also have some applications as antibiotics and anti-coccidiosis agents in veterinary medicine. They have been included especially in the diet of poultry and cattle [96].

In marine research, microalgae were recognized as producing polyketides. It has been found that polyketides in microalgae were stored in the form of open-chain polyketides and polycyclic ether macrolides [145]. However, the toxic effects of many polyketides were determined in human therapy applications [146]. The genus *Symbiodinium* from the Dinopyhceae class could accumulate in some individual macrolides. This genus, also called "zooxanthellae," has vigorous vasoconstrictive activity [147].

On the contrary, Amphidinolide B, the class of marine macrolide, demonstrated a high level of toxicity and antitumor activity [145]. A lot of polyketides have been reported from different biological sources such as fungi and plants. However, very few polyketides have been investigated from microalgae [96].

2.4 Biomass

Microalgal biomass has two different applications in the market, which are health food products and animal feed.

2.4.1 Health Food Products

Plants have been a conventional source of nutritional supplement in the food market. Microalgae are presented as an alternative nutraceutical source superior to plants (wheat, rice and legumes) owing to their protein quality value but are not superior to animal sources (milk and meat) [2].

As a microalgal food, growing biomass is the most obvious and easiest process. Most of these products have applications in the health food market. The organisms which live under outrageous conditions, such as *Spirulina* and *Dunaliella*, or the ones which have fast growth rates, such as *Chlorella* and *Scenedesmus*, are mostly used [148]; and *Nostoc* and *Aphanizomenon are* of lesser and regional importance [11]. Among these species, *Chlorella* and *Spirulina* dominate the microalgal-based food industry [11, 149, 150].

Their component substances and biological activities have been investigated in many research studies [11]. Compared to the other *Spirulina* sp., *S. platensis* and *S. maxima* are the most popular ones for the health food market. It is reported that they have protein content of 55–70% of total dry weight [2]. The health food from *Spirulina* sp. has an enhancement effect on the immune system and supports the prevention of both viral infection and cancer. For these reasons, its industrial-scale cultivation is currently handled for human consumption. Moreover, it has potential to improve the proliferation of lactic acid bacteria in the gastrointestinal system. It has been shown that lactobacilli growth rate increased 10-fold in comparison to the control groups due to the supplementation of *Spirulina* and a reproduced aqueous-based extract added for nutrition. Due to this effect, it balances the hormone levels in adult humans when taken as a food supplement [2, 11, 151, 152]. Extracts of *Chlorella* sp. indicated enhanced hemoglobin concentrations and decreased levels of blood sugar [2, 153, 154].

In 1961, research showed that levels of *Scenedesmus* sp. extracts are acceptable as much as *Chlorella* extracts for human consumption. Generally, these extracts are accepted for inclusion in some commonly utilized foods (fruit puddings, desserts, noodles, ravioli, and soups). However, their commercial production is currently not feasible due to lack of data about on production. *Dunaliella* sp. (especially *D. salina*) have garnered attention especially due to their extraordinary talent to develop under brackish conditions [107]. Some companies located in Israel and Australia produce these microalgae dietary additions and powders, containing vitamins A and C [155]. Another company in France, INNOVALG, had the authorization in 2002 (in accordance with EC regulation number 258/97) to cultivate and produce novel food from the marine diatom *Odontella aurita* [11, 156].

Amongst the algal goods, microalgal biomass has become more appealing to consumers because of its health benefits and practicability. It is believed that functional foods are able to rapidly increase their percentage in the whole food market [11, 157]. Except for *Chlorella* fabrication systems, which are mixotrophic or heterotrophic cultivated algae, all production processes are photoautotrophic [158–160]. Half of the total world manufacturing of dry algal-based biomass, which is predicted to be around 10,000 tons per year, takes place in China and the other half takes place mostly in Japan, Taiwan, America, Australia, and India. There are also some small companies in other countries [148, 161]. Additionally, the market of functional foods from microalgae (pasta, yogurt, bread, and soft drinks) in Germany has started to be developed by food production and distribution companies. This situation has similarly occurred in France, America, Japan, China and Thailand [11, 162].

Various products with microalgal extracts have started to effectuate new trade for microalgal products, e.g., *Dunaliella* carotenoid-enriched oily extracts [70, 163, 164], *Chlorella* health drinks (growth factor of *Chlorella*), and *Spirulina* liquid CO_2 extracts (antioxidant capsules) [11, 165, 166].

Because of their biochemical structure, microalgae may be used as a food additive, potentially increasing the bioactive ingredient content of food items. Spray-dried forms of *Nannochloropsis oculata*, *Porphyridium cruentum*, and *Diacronema vlkianum* microalgae species were included in the spray formulation at various concentrations in the research. The effect that microalgae usage has on the color, sensory, flow behavior, melting, and functional properties of ice cream was investigated. The results showed that microalgae can be utilized to develop color and functional properties in ice cream making; however, the amount of use should be carefully optimized to cope with the challenges of microalgae in terms of sensory properties. These negative parameters might be minimized by encapsulation and postharvest techniques. Spray drying is a common drying method suitable for multi-purpose use due to its high efficiency and low cost. However, it might be possible to enhance microalgae colorants level and stability by applying various drying methods and/or developing encapsulation techniques [167–169].

2.4.2 Animal Feed

Feed quality has one of the most important exogenous effects on the health of animals. Up to now, many investigations have been carried out to enhance the protein ingredient in animal feed with the utilization of microalgae in small doses [170]. It is reported that the genera *Chlorella*, *Scenedesmus* and *Spirulina* had a considerable and favorable effect on animal physiology by enhancing their immunological system [171]. Thus, microalgae are the main feed additive for aquatic animals in both marine and fresh water [2, 148]. Furthermore, microalgae production is the primary and most expensive part of the process, which is the main drawback for the development of farms [148, 155]. Microalgae are also used as a live food to culture various zooplankton types. Microalgae are consumed by crustacean and finfishes as live food [155, 172]. The cost level of live microalgae biomass production correlates with the requirement of different technical equipment that is mostly inadequate.

There are some major points to consider when choosing the microalgae strain for commercial feed additive production. Firstly, the microalgal strains which produce toxins are inappropriate and many strains are toxic [173]. Secondly, biochemical constituents, including vitamins, proteins, and particularly unsaturated fatty acids, should be sufficient [2, 174]. Other criteria are the acceptable size for ingestion and digestible cell wall. Easy cultivation of selected microalgae is the last criterion for cost-effective production [2, 148].

According to the different requirements of each local seafood production, over 40 microalgae species are chosen in the aquaculture market worldwide. For instance, *Isochrysis galbana* and *Tetraselmis suecica* were added to the unfiltered seawater in which they live. As a result, they were able to grow better [11, 175]. Drum-dried *Scenedesmus* is suitable for use as Artemia food. *Spirulina* and *Chlorella* are suitable to be added into common fish feed compositions [11, 13, 176, 177]. In addition, *Chlorella* (freshwater, marine, or dried) can be used for cultivation of the rotifer *Brachionus plicatilis* [178].

Microalgae pigments (astaxanthin, phycocyanin) also have a positive effect on the cultivation of some farmed animals (fish, crustaceans and shrimps) which aren't able to synthesize pigments. *Spirulina* containing phycocyanin has been used for shrimp and *Chlorella* has been chosen as a pigment for other crustaceans [148, 171, 179]. For the pigmentation of ornamental fish, both *Spirulina* and carotenoid-containing *Dunaliella* were used [11, 180]. The addition of astaxanthin from *Haematococcus pluvialis* in the salmonid diet significantly resulted in the flesh coloration enhancement of the salmonids [181–183]. The utilization of microalgal pigments generally increases the expense of the farmer by 10 to 15% of costs [184, 185].

In addition to all these applications, the pet food market has also been encouraging. Studies carried out on minks and rabbits showed that microalgal biomass or extract has positive effects not only on the health of the pet but also the external appearance, such as shiny hair and beautiful feathers, which is one of the biggest criterion for consumers in the pet market [186, 187].

2.5 Potential Future Applications

Over the last decade, the use of bio-based materials has grown due to their abundance, eco-friendliness, renewability, and biodegradability. The latest, third generation resource is marine resources for different applications, such as biofuel and biocomposites. Microalgae are novel materials due to their high growth rate, CO_2 fixation capabilities, and lack of competition with food or feed crops. In addition, cost-effectiveness was enhanced through technological developments. In the coming decades, research on algae will increase due to the properties given above. However, sustainability is the most critical point in microalgal applications. Their product profiles could be changed in different stages. For instance, reactive oxygen species are able to damage their lipid and protein profile during cultivation by stress conditions or during harvesting/pretreatment by chemicals and oxygen in the air [188]. Besides which, costs of microalgal cultivation, harvesting processes and insufficient conversion of microalgal biomass to products are the other main obstacles in microalgal technologies [56]. Currently, promising solutions have been established at laboratory- and industrial-scale. Firstly, biorefinery systems with an integrated microalgae-based approach are the primary solution, particularly energy production such as biodiesel, biomethane and biohydrogen. These systems aim to provide impressive wastewater treatment/nutrient separation, economical microalgae production without the use of potable water or artificial nutrients, and energetic valorization of microalgal biomass through biofuel fabrication (bioH2). Their byproducts may be used as fertilizers or substrates for various fermentative approaches [73]. Besides which, biomass that is obtained from a biorefinery would be used for mixing with petrochemical-based or other bio-based plastics [77]. Secondly, genetic engineering approaches are the next solution for energy and bio-based materials such as bioplastic and nutrition supplements. The goal of genetic engineering techniques is to increase oxidative stress tolerance by adaptable laboratory developments and overexpression of antioxidative enzymes, as well as the manufacture of bioplastics. Besides which, the design of biorefinery systems starts with one genetically modified microalgal species and these systems have great potential in the next decades due to their decreasing costs and increasing algal efficiency [77, 189].

Commercialized projects on algae harvesting have been established by some companies. To begin with, Algix LLC is a bioplastics firm that has cooperated with the University of Georgia and Kimberly-Clark to commercialize the growth of aquatic biomass, such as algae, via the development of innovative thermoplastic formulations and resins. The Algix bioresins were reproduced from aquatic biomass that is produced from nitrogen and phosphorus-rich wastewater, mixed with varying concentrations of PE (polyethylene), PP (polypropylene), EVA (ethylene-vinyl acetate), PLA (polylactic acid), TPS (thermoplastic starch), PHA (polyhydroxyalkanoate), and other polymers for a variety of end uses.

Then, NASA's Offshore Membrane Enclosures for Growing Algae (OMEGA) project was carried out between 2010 and 2012, which was a creative technique to breed algae, provide clean wastewater, and capture CO_2 and to manufacture biofuel without competing with agriculture for any sources. Investigations have demonstrated that OMEGA is an important strategy to grow microalgae and process wastewater on a small scale.

Finally, the latest project is called SPLASH (Sustainable Polymers from Algae Sugars and Hydrocarbons). This EU-funded research project is based on algae derivative polymers with twenty partners, of which 55% are from SMEs and large enterprises; Avantium, Biotopic, Lankhorst Euronete, LifeGlimmer GmbH, nova-Institute GmbH, PNO Consultants B.V., Pursuit Dynamics PLC, Organic Waste Systems N.V., Paques BV, Rhodia Operations, and Value for Technology; and 45% are universities and research institutes; Ege University, Centre for Research and Technology Hellas, Fraunhofer, University of Cambridge, Bielefeld University, University of Huelva, Wageningen UR, Food & Biobased Research, Westfälische Wilhelms-Universität Münster, and Wageningen Universiteit. The aim of this 4-year-project is to generate a novel bio-based industrial cooperation using microalgae as raw material for the sustainable development and algae-based hydrocarbons and (exo)polysaccharides recovery, and also their conversion to renewable polymers. The SPLASH platform will convey knowledge, tools and technologies required for industrial biotechnology establishments based on algae and/or algal genes to produce polyesters from sugars, and polyolefins from hydrocarbons.

Though there are many applications of microalgae and other algae species in a variety of different areas, the use of microalgae for polymer applications is almost nil. The fact that microalgae is so abundant in nature provides an infinite resource for materials scientists to design new eco-friendly, degradable, and lightweight products to be used in multiple areas such as construction materials, automotive materials, packaging and the biomedical. There are plenty of fruitful areas yet to be discovered by engineers using microalgae, as they can be grown anywhere on earth or under the seas and oceans.

2.6 Conclusion

Microalgae are one of the impressive life forms and photosynthetic organisms that have great potential for biofuel generations and microalgal compounds. These products can be produced at lab scale or can be scaled-up for several industries. For industrial applications, such as microalgal compounds, biofuel production derived from microalgae needs high biomass levels in the cultivation of microalgae, contrary to lab-scale production. The scale-up process requires some special cultivation medium such as ponds or bioreactors. Development of the ponds and bioreactors involves detail information and combines biology of microalgae cells, mass transfer, and light distribution for biorefinery systems. The sustainability of biological resources is a crucial premise in conservation and development policy, and this issue is particularly concerned with efficiency, environmental hazard minimization, and socioeconomic implications. This chapter has demonstrated the potential uses of microalgal in the past, present and future when a proper amount of microalgal cultivation is obtained in ponds or photobioreactors on a large scale.

References

1. E.W. Becker, Microalgae for Human and Animal Nutrition, *Handb. Microalgal Cult. Appl. Phycol. Biotechnol. Second Ed.* (2013) 461–503. https://doi.org/10.1002/9781118567166.CH25.
2. T.M. Mata, A.A. Martins, N.S. Caetano, Microalgae for biodiesel production and other applications: A review, *Renew. Sustain. Energy Rev.* 14 (2010) 217–232. https://doi.org/10.1016/J.RSER.2009.07.020.
3. E. Bugnicourt, P. Cinelli, A. Lazzeri, V. Alvarez, Polyhydroxyalkanoate (PHA): Review of synthesis, characteristics, processing and potential applications in packaging, *Express Polym. Lett.* 8 (2014) 791–808. https://doi.org/10.3144/expresspolymlett.2014.82.
4. P. Chen, M. Min, Y. Chen, L. Wang, Y. Li, Q. Chen, C. Wang, Y. Wan, X. Wang, Y. Cheng, S. Deng, K. Hennessy, X. Lin, Y. Liu, Y. Wang, B. Martinez, R. Ruan, Review of the biological and engineering aspects of algae to fuels approach, *Int. J. Agric. Biol. Eng.* 2 (2009) 1–30. https://doi.org/10.3965/j.
5. S.A. Scott, M.P. Davey, J.S. Dennis, I. Horst, C.J. Howe, D.J. Lea-Smith, A.G. Smith, Biodiesel from algae: challenges and prospects, *Curr. Opin. Biotechnol.* 21 (2010) 277–286. https://doi.org/10.1016/J.COPBIO.2010.03.005.
6. F. Alam, A. Date, R. Rasjidin, S. Mobin, H. Moria, A. Baqui, Biofuel from Algae- Is It a Viable Alternative?, *Procedia Eng.* 49 (2012) 221–227. https://doi.org/10.1016/J.PROENG.2012.10.131.
7. R.A. Preisig, H.R., Andersen, Historical review of algal culturing techniques, in: *Algal Cult. Tech.*, Elsevier/Academic Press, 2005: pp. 1–12. http://doi.wiley.com/10.1111/j.1529-8817.2005.00114.x (accessed August 3, 2021).

8. X. Deng, Y. Li, X. Fei, Microalgae: A promising feedstock for biodiesel, African *J. Microbiol. Res.* 3 (2009) 1008–1014.
9. Y. Chisti, Biodiesel from microalgae, *Biotechnol. Adv.* 25 (2007) 294–306. https://doi.org/10.1016/J.BIOTECHADV.2007.02.001.
10. A. Demirbas, M. Fatih Demirbas, Importance of algae oil as a source of biodiesel, *Energy Convers. Manag.* 52 (2011) 163–170. https://doi.org/10.1016/j.enconman.2010.06.055.
11. O. Pulz, W. Gross, Valuable products from biotechnology of microalgae, *Appl. Microbiol. Biotechnol.* 2004 656. 65 (2004) 635–648. https://doi.org/10.1007/S00253-004-1647-X.
12. X.B. Tan, M.K. Lam, Y. Uemura, J.W. Lim, C.Y. Wong, K.T. Lee, Cultivation of microalgae for biodiesel production: A review on upstream and downstream processing, *Chinese J. Chem. Eng.* 26 (2018) 17–30. https://doi.org/10.1016/J.CJCHE.2017.08.010.
13. M.A. Borowitzka, Commercial production of microalgae: ponds, tanks, tubes and fermenters, *J. Biotechnol.* 70 (1999) 313–321. https://doi.org/10.1016/S0168-1656(99)00083-8.
14. V.O. Adesanya, E. Cadena, S.A. Scott, A.G. Smith, Life cycle assessment on microalgal biodiesel production using a hybrid cultivation system, *Bioresour. Technol.* 163 (2014) 343–355. https://doi.org/10.1016/J.BIORTECH.2014.04.051.
15. A.M. Kunjapur, R.B. Eldridge, Photobioreactor Design for Commercial Biofuel Production from Microalgae, *Ind. Eng. Chem. Res.* 49 (2010) 3516–3526. https://doi.org/10.1021/ie901459u.
16. L. Brennan, P. Owende, Biofuels from microalgae—A review of technologies for production, processing, and extractions of biofuels and co-products, *Renew. Sustain. Energy Rev.* 14 (2010) 557–577. https://doi.org/10.1016/J.RSER.2009.10.009.
17. P.M. Schenk, S.R. Thomas-Hall, E. Stephens, U.C. Marx, J.H. Mussgnug, C. Posten, O. Kruse, B. Hankamer, Second Generation Biofuels: High-Efficiency Microalgae for Biodiesel Production, *BioEnergy Res.* 2008 11. 1 (2008) 20–43. https://doi.org/10.1007/S12155-008-9008-8.
18. A.P. Batista, L. Gouveia, N.M. Bandarra, J.M. Franco, A. Raymundo, Comparison of microalgal biomass profiles as novel functional ingredient for food products, *Algal Res.* 2 (2013) 164–173. https://doi.org/10.1016/j.algal.2013.01.004.
19. M.F. de Jesus Raposo, R.M.S.C. de Morais, A.M.M.B. de Morais, Health applications of bioactive compounds from marine microalgae, *Life Sci.* 93 (2013) 479–486. https://doi.org/10.1016/j.lfs.2013.08.002.
20. B. da S. Vaz, J.B. Moreira, M.G. de Morais, J.A.V. Costa, Microalgae as a new source of bioactive compounds in food supplements, *Curr. Opin. Food Sci.* 7 (2016) 73–77. https://doi.org/10.1016/j.cofs.2015.12.006.

21. N. Pragya, K.K. Pandey, P.K. Sahoo, A review on harvesting, oil extraction and biofuels production technologies from microalgae, *Renew. Sustain. Energy Rev.* 24 (2013) 159–171. https://doi.org/10.1016/j.rser.2013.03.034.
22. E. Lois, Definition of biodiesel, *Fuel.* 86 (2007) 1212–1213. https://doi.org/10.1016/j.fuel.2006.09.028.
23. F. Ma, M.A. Hanna, Biodiesel production: a review1Journal Series #12109, Agricultural Research Division, Institute of Agriculture and Natural Resources, University of Nebraska–Lincoln.1, *Bioresour. Technol.* 70 (1999) 1–15. https://doi.org/10.1016/S0960-8524(99)00025-5.
24. A.J. Dassey, S.G. Hall, C.S. Theegala, An analysis of energy consumption for algal biodiesel production: Comparing the literature with current estimates, *Algal Res.* 4 (2014) 89–95. https://doi.org/10.1016/j.algal.2013.12.006.
25. A. Demirbas, M. Fatih Demirbas, Importance of algae oil as a source of biodiesel, *Energy Convers. Manag.* 52 (2011) 163–170. https://doi.org/10.1016/J.ENCONMAN.2010.06.055.
26. B.K. Barnwal, M.P. Sharma, Prospects of biodiesel production from vegetable oils in India, *Renew. Sustain. Energy Rev.* 9 (2005) 363–378. https://doi.org/10.1016/j.rser.2004.05.007.
27. A. Banerjee, R. Sharma, Y. Chisti, U.C. Banerjee, Botryococcus braunii : A Renewable Source of Hydrocarbons and Other Chemicals, *Crit. Rev. Biotechnol.* 22 (2002) 245–279. https://doi.org/10.1080/07388550290789513.
28. I.A. Guschina, J.L. Harwood, Lipids and lipid metabolism in eukaryotic algae, *Prog. Lipid Res.* 45 (2006) 160–186. https://doi.org/10.1016/j.plipres.2006.01.001.
29. G. Huang, F. Chen, D. Wei, X. Zhang, G. Chen, Biodiesel production by microalgal biotechnology, *Appl. Energy.* 87 (2010) 38–46. https://doi.org/10.1016/j.apenergy.2009.06.016.
30. P. Metzger, C. Largeau, Botryococcus braunii: a rich source for hydrocarbons and related ether lipids, *Appl. Microbiol. Biotechnol.* 66 (2005) 486–496. https://doi.org/10.1007/s00253-004-1779-z.
31. J. Milano, H.C. Ong, H.H. Masjuki, W.T. Chong, M.K. Lam, P.K. Loh, V. Vellayan, Microalgae biofuels as an alternative to fossil fuel for power generation, *Renew. Sustain. Energy Rev.* 58 (2016) 180–197. https://doi.org/10.1016/j.rser.2015.12.150.
32. H. Fukuda, A. Kondo, H. Noda, Biodiesel fuel production by transesterification of oils, *J. Biosci. Bioeng.* 92 (2001) 405–416. https://doi.org/10.1016/S1389-1723(01)80288-7.
33. H. Taher, S. Al-Zuhair, A.H. Al-Marzouqi, Y. Haik, M.M. Farid, A Review of Enzymatic Transesterification of Microalgal Oil-Based Biodiesel Using Supercritical Technology, *Enzyme Res.* 2011 (2011) 1–25. https://doi.org/10.4061/2011/468292.
34. A. Farooq, U.K. Amin, Y. Abdullah, Transesterification of oil extracted from different species of algae for biodiesel production, African *J. Environ. Sci. Technol.* 7 (2013) 358–364. https://doi.org/10.5897/AJEST12.167.

35. I. Rawat, R. Ranjith Kumar, T. Mutanda, F. Bux, Dual role of microalgae: Phycoremediation of domestic wastewater and biomass production for sustainable biofuels production, *Appl. Energy.* 88 (2011) 3411–3424. https://doi.org/10.1016/j.apenergy.2010.11.025.
36. R. Thirugnanasambantham, T. Elango, K. Elangovan, Chlorella vulgaris sp. microalgae as a feedstock for biofuel, *Mater. Today Proc.* 33 (2020) 3182–3185. https://doi.org/10.1016/j.matpr.2020.04.106.
37. M.K. Lam, K.T. Lee, A.R. Mohamed, Homogeneous, heterogeneous and enzymatic catalysis for transesterification of high free fatty acid oil (waste cooking oil) to biodiesel: A review, *Biotechnol. Adv.* 28 (2010) 500–518. https://doi.org/10.1016/j.biotechadv.2010.03.002.
38. A. Sivasamy, K.Y. Cheah, P. Fornasiero, F. Kemausuor, S. Zinoviev, S. Miertus, Catalytic Applications in the Production of Biodiesel from Vegetable Oils, *ChemSusChem.* 2 (2009) 278–300. https://doi.org/10.1002/cssc.200800253.
39. S. Sharma, S.S. Kanwar, Organic Solvent Tolerant Lipases and Applications, *Sci. World J.* 2014 (2014) 1–15. https://doi.org/10.1155/2014/625258.
40. S. Sugumar, R. Rajesh, P. Kannan, P. Venkatesan, Biodiesel production from the biomass of Dunaliella salina green microalgae using organic solvent, *Mater. Today Proc.* 33 (2020) 3260–3264. https://doi.org/10.1016/j.matpr.2020.04.652.
41. J. Rupprecht, From systems biology to fuel—Chlamydomonas reinhardtii as a model for a systems biology approach to improve biohydrogen production, *J. Biotechnol.* 142 (2009) 10–20. https://doi.org/10.1016/j.jbiotec.2009.02.008.
42. T. Willke, K.-D. Vorlop, Industrial bioconversion of renewable resources as an alternative to conventional chemistry, *Appl. Microbiol. Biotechnol.* 66 (2004) 131–142. https://doi.org/10.1007/s00253-004-1733-0.
43. M. FitzPatrick, P. Champagne, M.F. Cunningham, R.A. Whitney, A biorefinery processing perspective: Treatment of lignocellulosic materials for the production of value-added products, *Bioresour. Technol.* 101 (2010) 8915–8922. https://doi.org/10.1016/j.biortech.2010.06.125.
44. E. (Devlet P.T. Hatunoğlu, BİYOYAKIT Polítikalarinin Tarim Sektörüne Etkíleri, (2010) 205.
45. J. Xuan, M.K.H. Leung, D.Y.C. Leung, M. Ni, A review of biomass-derived fuel processors for fuel cell systems, *Renew. Sustain. Energy Rev.* 13 (2009) 1301–1313. https://doi.org/10.1016/j.rser.2008.09.027.
46. J.A.V. Costa, M.G. de Morais, The role of biochemical engineering in the production of biofuels from microalgae, *Bioresour. Technol.* 102 (2011) 2–9. https://doi.org/10.1016/j.biortech.2010.06.014.
47. D. Luo, Z. Hu, D.G. Choi, V.M. Thomas, M.J. Realff, R.R. Chance, Life Cycle Energy and Greenhouse Gas Emissions for an Ethanol Production Process Based on Blue-Green Algae, *Environ. Sci. Technol.* 44 (2010) 8670–8677. https://doi.org/10.1021/es1007577.
48. D. Varaprasad, N. Ragasudha, K. Paramesh, P. Chandramati Shankar, S. Nazaneen Parveen, T. Chandrasekhar, Production of Bioethanol from Green

Alga Chlorella Vulgaris: An Important Approach to Utilize Algal Feedstock or Waste, in: Bioresour. *Util. Bioprocess*, Springer Singapore, Singapore, 2020: pp. 57–65. https://doi.org/10.1007/978-981-15-1607-8_6.

49. P. Weiland, Biogas production: current state and perspectives, *Appl. Microbiol. Biotechnol.* 85 (2010) 849–860. https://doi.org/10.1007/s00253-009-2246-7.
50. C.G. Golueke, W.J. Oswald, Biological Conversion of Light Energy to the Chemical Energy of Methane, *Appl. Microbiol.* 7 (1959) 219. https://www.ncbi.nlm.nih.gov/pmc/articles/PMC1057509/ (accessed August 3, 2021).
51. L. Mendez, A. Mahdy, M. Demuez, M. Ballesteros, C. González-Fernández, Effect of high pressure thermal pretreatment on Chlorella vulgaris biomass: Organic matter solubilisation and biochemical methane potential, *Fuel.* 117 (2014) 674–679. https://doi.org/10.1016/j.fuel.2013.09.032.
52. S. Heaven, J. Milledge, Y. Zhang, Comments on 'Anaerobic digestion of microalgae as a necessary step to make microalgal biodiesel sustainable,' *Biotechnol. Adv.* 29 (2011) 164–167. https://doi.org/10.1016/j.biotechadv.2010.10.005.
53. J.H. Clark, F.E.I. Deswarte, Introduction to chemicals from biomass, (n.d.).
54. P. McKendry, Energy production from biomass (part 3): gasification technologies, *Bioresour. Technol.* 83 (2002) 55–63. https://doi.org/10.1016/S0960-8524(01)00120-1.
55. R. Harun, M. Singh, G.M. Forde, M.K. Danquah, Bioprocess engineering of microalgae to produce a variety of consumer products, *Renew. Sustain. Energy Rev.* 14 (2010) 1037–1047. https://doi.org/10.1016/j.rser.2009.11.004.
56. E. Jankowska, A.K. Sahu, P. Oleskowicz-Popiel, Biogas from microalgae: Review on microalgae's cultivation, harvesting and pretreatment for anaerobic digestion, *Renew. Sustain. Energy Rev.* 75 (2017) 692–709. https://doi.org/10.1016/j.rser.2016.11.045.
57. Y. Chisti, Constraints to commercialization of algal fuels, *J. Biotechnol.* 167 (2013) 201–214. https://doi.org/10.1016/j.jbiotec.2013.07.020.
58. Y. Chisti, Raceways-based Production of Algal Crude Oil, Green. 3 (2013). https://doi.org/10.1515/green-2013-0018.
59. Y. Chisti, Response to Reijnders: Do biofuels from microalgae beat biofuels from terrestrial plants?, *Trends Biotechnol.* 26 (2008) 351–352. https://doi.org/10.1016/j.tibtech.2008.04.002.
60. C. Posten, C. Walter, eds., Microalgal Biotechnology: Potential and Production, DE GRUYTER, 2012. https://doi.org/10.1515/9783110225020.
61. K.M. Weyer, D.R. Bush, A. Darzins, B.D. Willson, Theoretical Maximum Algal Oil Production, *BioEnergy Res.* 3 (2010) 204–213. https://doi.org/10.1007/s12155-009-9046-x.
62. C. Xiao, Q. Liao, Q. Fu, Y. Huang, A. Xia, H. Chen, X. Zhu, Numerical investigation of laminar mixed convection of microalgae slurry flowing in a solar collector, *Appl. Therm. Eng.* 175 (2020) 115366. https://doi.org/10.1016/j.applthermaleng.2020.115366.

63. M.L. Ghirardi, P.-C. Maness, C. Eckert, J. Yu, Photobiological Hydrogen Production - Prospects and Challenges, *Microbe Mag.* 4 (2009) 275–280. https://doi.org/10.1128/microbe.4.275.1.
64. C.-Y. Chen, C.-H. Liu, Y.-C. Lo, J.-S. Chang, Perspectives on cultivation strategies and photobioreactor designs for photo-fermentative hydrogen production, *Bioresour. Technol.* 102 (2011) 8484–8492. https://doi.org/10.1016/j.biortech.2011.05.082.
65. Hydrogen on the rise, *Nat. Energy.* 1 (2016) 16127. https://doi.org/10.1038/nenergy.2016.127.
66. D. Levin, Biohydrogen production: prospects and limitations to practical application, *Int. J. Hydrogen Energy.* 29 (2004) 173–185. https://doi.org/10.1016/S0360-3199(03)00094-6.
67. S. Oncel, F.V. Sukan, "Effect of light intensity and the light: dark cycles on the long term hydrogen production of Chlamydomonas reinhardtii by batch cultures," *Biomass and Bioenergy.* 35 (2011) 1066–1074. https://doi.org/10.1016/j.biombioe.2010.11.017.
68. D. Nagarajan, D.-J. Lee, A. Kondo, J.-S. Chang, Recent insights into biohydrogen production by microalgae - From biophotolysis to dark fermentation, *Bioresour. Technol.* 227 (2017) 373–387. https://doi.org/10.1016/j.biortech.2016.12.104.
69. A. Tsygankov, S. Kosourov, I. Tolstygina, M. Ghirardi, M. Seibert, Hydrogen production by sulfur-deprived Chlamydomonas reinhardtii under photoautotrophic conditions, *Int. J. Hydrogen Energy.* 31 (2006) 1574–1584. https://doi.org/10.1016/j.ijhydene.2006.06.024.
70. M.A. Borowitzka, Algal biotechnology products and processes — matching science and economics, *J. Appl. Phycol.* 4 (1992) 267–279. https://doi.org/10.1007/BF02161212.
71. S. Oncel, M. Sabankay, Microalgal biohydrogen production considering light energy and mixing time as the two key features for scale-up, *Bioresour. Technol.* 121 (2012) 228–234. https://doi.org/10.1016/j.biortech.2012.06.079.
72. X. Wang, J. Ding, W.-Q. Guo, N.-Q. Ren, Scale-up and optimization of biohydrogen production reactor from laboratory-scale to industrial-scale on the basis of computational fluid dynamics simulation, *Int. J. Hydrogen Energy.* 35 (2010) 10960–10966. https://doi.org/10.1016/j.ijhydene.2010.07.060.
73. A.P. Batista, L. Ambrosano, S. Graça, C. Sousa, P.A.S.S. Marques, B. Ribeiro, E.P. Botrel, P. Castro Neto, L. Gouveia, Combining urban wastewater treatment with biohydrogen production - An integrated microalgae-based approach, *Bioresour. Technol.* 184 (2015) 230–235. https://doi.org/10.1016/j.biortech.2014.10.064.
74. N. Rajendran, S. Puppala, S. Raj M., R. Angeeleena B., C. Rajam, Seaweeds Can Be A New Source For Bioplastics, *J. Pharm. Res.* 5 (2012) 1476–1479. http://jprsolutions.info.
75. M.M. Amar Mohanty, Murali M. Reddy, Bio-Based Materials in the New Bio Economy | AIChE, (n.d.). https://www.aiche.org/resources/publications/

cep/2012/may/bio-based-materials-new-bio-economy (accessed August 3, 2021).
76. A. Alves, S.G. Caridade, J.F. Mano, R.A. Sousa, R.L. Reis, Extraction and physico-chemical characterization of a versatile biodegradable polysaccharide obtained from green algae, *Carbohydr. Res.* 345 (2010) 2194–2200. https://doi.org/10.1016/j.carres.2010.07.039.
77. A. Rahman, C.D. Miller, Microalgae as a Source of Bioplastics, in: *Algal Green Chem.*, Elsevier, 2017: pp. 121–138. https://doi.org/10.1016/B978-0-444-63784-0.00006-0.
78. M. Bulota, T. Budtova, PLA/algae composites: Morphology and mechanical properties, *Compos. Part A Appl. Sci. Manuf.* 73 (2015) 109–115. https://doi.org/10.1016/j.compositesa.2015.03.001.
79. M.A. Zeller, R. Hunt, A. Jones, S. Sharma, Bioplastics and their thermoplastic blends from Spirulina and Chlorella microalgae, *J. Appl. Polym. Sci.* 130 (2013) 3263–3275. https://doi.org/10.1002/app.39559.
80. E. Chiellini, P. Cinelli, V.I. Ilieva, M. Martera, Biodegradable Thermoplastic Composites Based on Polyvinyl Alcohol and Algae, *Biomacromolecules.* 9 (2008) 1007–1013. https://doi.org/10.1021/bm701041e.
81. A.A. Sapalidis, F.K. Katsaros, G.E. Romanos, N.K. Kakizis, N.K. Kanellopoulos, Preparation and characterization of novel poly-(vinyl alcohol)–Zostera flakes composites for packaging applications, *Compos. Part B Eng.* 38 (2007) 398–404. https://doi.org/10.1016/j.compositesb.2006.04.005.
82. M.-H. Jau, S.-P. Yew, P.S.Y. Toh, A.S.C. Chong, W.-L. Chu, S.-M. Phang, N. Najimudin, K. Sudesh, Biosynthesis and mobilization of poly(3-hydroxybutyrate) [P(3HB)] by Spirulina platensis, *Int. J. Biol. Macromol.* 36 (2005) 144–151. https://doi.org/10.1016/j.ijbiomac.2005.05.002.
83. C.J. López Rocha, E. Álvarez-Castillo, M.R. Estrada Yáñez, C. Bengoechea, A. Guerrero, M.T. Orta Ledesma, Development of bioplastics from a microalgae consortium from wastewater, *J. Environ. Manage.* 263 (2020) 110353. https://doi.org/10.1016/j.jenvman.2020.110353.
84. S. Vijayakumar, M. Menakha, Pharmaceutical applications of cyanobacteria—A review, *J. Acute Med.* 5 (2015) 15–23. https://doi.org/10.1016/j.jacme.2015.02.004.
85. M. Olaizola, Commercial development of microalgal biotechnology: from the test tube to the marketplace, Biomol. Eng. 20 (2003) 459–466. https://doi.org/10.1016/S1389-0344(03)00076-5.
86. V.M. Ursin, Modification of Plant Lipids for Human Health: Development of Functional Land-Based Omega-3 Fatty Acids, *J. Nutr.* 133 (2003) 4271–4274. https://doi.org/10.1093/jn/133.12.4271.
87. C.H.S. Ruxton, P.C. Calder, S.C. Reed, M.J.A. Simpson, The impact of long-chain n -3 polyunsaturated fatty acids on human health, *Nutr. Res. Rev.* 18 (2005) 113–129. https://doi.org/10.1079/NRR200497.
88. W.E. Connor, Importance of n–3 fatty acids in health and disease, *Am. J. Clin. Nutr.* 71 (2000) 171S-175S. https://doi.org/10.1093/ajcn/71.1.171S.

89. I. Fedorova-Dahms, P.A. Marone, M. Bauter, A.S. Ryan, Safety evaluation of DHA-rich Algal Oil from Schizochytrium sp., *Food Chem. Toxicol.* 49 (2011) 3310–3318. https://doi.org/10.1016/j.fct.2011.08.024.
90. Z. yi Li, O.P. Ward, Production of docosahexaenoic acid by Thraustochytrium roseum, *J. Ind. Microbiol.* 13 (1994) 238–241. https://doi.org/10.1007/BF01569755.
91. V. Patil, T. Källqvist, E. Olsen, G. Vogt, H.R. Gislerød, Fatty acid composition of 12 microalgae for possible use in aquaculture feed, *Aquac. Int.* 15 (2007) 1–9. https://doi.org/10.1007/s10499-006-9060-3.
92. D.I. Sánchez-Machado, J. López-Cervantes, J. López-Hernández, P. Paseiro-Losada, Fatty acids, total lipid, protein and ash contents of processed edible seaweeds, *Food Chem.* 85 (2004) 439–444. https://doi.org/10.1016/j.foodchem.2003.08.001.
93. R. Abril, J. Garrett, S.G. Zeller, W.J. Sander, R.W. Mast, Safety assessment of DHA-rich microalgae from Schizochytrium sp. Part V: target animal safety/toxicity study in growing swine, *Regul. Toxicol. Pharmacol.* 37 (2003) 73–82. https://doi.org/10.1016/S0273-2300(02)00030-2.
94. M. Certik, S. Shimizu, Biosynthesis and regulation of microbial polyunsaturated fatty acid production, *J. Biosci. Bioeng.* 87 (1999) 1–14. https://doi.org/10.1016/S1389-1723(99)80001-2.
95. E.E.M. Luiten, I. Akkerman, A. Koulman, P. Kamermans, H. Reith, M.J. Barbosa, D. Sipkema, R.H. Wijffels, Realizing the promises of marine biotechnology, *Biomol. Eng.* 20 (2003) 429–439. https://doi.org/10.1016/S1389-0344(03)00074-1.
96. K.H.M. Cardozo, T. Guaratini, M.P. Barros, V.R. Falcão, A.P. Tonon, N.P. Lopes, S. Campos, M.A. Torres, A.O. Souza, P. Colepicolo, E. Pinto, Metabolites from algae with economical impact, *Comp. Biochem. Physiol. Part C Toxicol. Pharmacol.* 146 (2007) 60–78. https://doi.org/10.1016/j.cbpc.2006.05.007.
97. A.R. Medina, E.M. Grima, A.G. Giménez, M.J.I. González, Downstream processing of algal polyunsaturated fatty acids, *Biotechnol. Adv.* 16 (1998) 517–580. https://doi.org/10.1016/S0734-9750(97)00083-9.
98. X.-M. Sun, L.-J. Geng, L.-J. Ren, X.-J. Ji, N. Hao, K.-Q. Chen, H. Huang, Influence of oxygen on the biosynthesis of polyunsaturated fatty acids in microalgae, *Bioresour. Technol.* 250 (2018) 868–876. https://doi.org/10.1016/j.biortech.2017.11.005.
99. T. Adarme-Vega, D.K.Y. Lim, M. Timmins, F. Vernen, Y. Li, P.M. Schenk, Microalgal biofactories: a promising approach towards sustainable omega-3 fatty acid production, *Microb. Cell Fact.* 11 (2012) 96. https://doi.org/10.1186/1475-2859-11-96.
100. P. Spolaore, C. Joannis-Cassan, E. Duran, A. Isambert, Commercial applications of microalgae, *J. Biosci. Bioeng.* 101 (2006) 87–96. https://doi.org/10.1263/jbb.101.87.

101. T.-S. Vo, D.-H. Ngo, S.-K. Kim, Marine algae as a potential pharmaceutical source for anti-allergic therapeutics, *Process Biochem.* 47 (2012) 386–394. https://doi.org/10.1016/j.procbio.2011.12.014.
102. Z.-Y. Wen, F. Chen, Heterotrophic production of eicosapentaenoic acid by microalgae, *Biotechnol. Adv.* 21 (2003) 273–294. https://doi.org/10.1016/S0734-9750(03)00051-X.
103. A. Abbadi, F. Domergue, A. Meyer, K. Riedel, P. Sperling, T.K. Zank, E. Heinz, Transgenic oilseeds as sustainable source of nutritionally relevant C20 and C22 polyunsaturated fatty acids?, *Eur. J. Lipid Sci. Technol.* 103 (2001) 106–113. https://doi.org/10.1002/1438-9312(200102)103:2<106::AID-EJLT106>3.0.CO;2-V.
104. T. Lebeau, J.-M. Robert, Diatom cultivation and biotechnologically relevant products. Part II: Current and putative products, *Appl. Microbiol. Biotechnol.* 60 (2003) 624–632. https://doi.org/10.1007/s00253-002-1177-3.
105. K.H.M. Cardozo, M.A.L. Oliveira, M.F.M. Tavares, P. Colepicolo, E. Pinto, Daily Oscillation of Fatty Acids and Malondialdehyde in the Dinoflagellate Lingulodinium polyedrum, *Biol. Rhythm Res.* 33 (2002) 371–382. https://doi.org/10.1076/brhm.33.4.371.8802.
106. C.K. Tan, M.R. Johns, Screening of diatoms for heterotrophic eicosapentaenoic acid production, *J. Appl. Phycol.* 8 (1996) 59–64. https://doi.org/10.1007/BF02186223.
107. C. Barrow, F. Shahidi, eds., Marine Nutraceuticals and Functional Foods, CRC Press, 2007. https://doi.org/10.1201/9781420015812.
108. R.J. Radmer, Algal Diversity and Commercial Algal Products, *Bioscience.* 46 (1996) 263–270. https://doi.org/10.2307/1312833.
109. G. Chini Zittelli, F. Lavista, A. Bastianini, L. Rodolfi, M. Vincenzini, M.R. Tredici, Production of eicosapentaenoic acid by Nannochloropsis sp. cultures in outdoor tubular photobioreactors, *J. Biotechnol.* 70 (1999) 299–312. https://doi.org/10.1016/S0168-1656(99)00082-6.
110. E. Molina Grima, E.-H. Belarbi, F.. Acién Fernández, A. Robles Medina, Y. Chisti, Recovery of microalgal biomass and metabolites: process options and economics, *Biotechnol. Adv.* 20 (2003) 491–515. https://doi.org/10.1016/S0734-9750(02)00050-2.
111. E.H. Belarbi, E. Molina, Y. Chisti, A process for high yield and scaleable recovery of high purity eicosapentaenoic acid esters from microalgae and fish oil, *Enzyme Microb. Technol.* 26 (2000) 516–529. https://doi.org/10.1016/S0141-0229(99)00191-X.
112. G.A. Spiller, A. Dewell, Safety of an Astaxanthin-Rich Haematococcus pluvialis Algal Extract: A Randomized Clinical Trial, *J. Med. Food.* 6 (2003) 51–56. https://doi.org/10.1089/109662003765184741.
113. T. Polívka, V. Sundström, Ultrafast Dynamics of Carotenoid Excited States–From Solution to Natural and Artificial Systems, *Chem. Rev.* 104 (2004) 2021–2072. https://doi.org/10.1021/cr020674n.
114. H. Zhang, D. Huang, W.A. Cramer, Stoichiometrically Bound β-Carotene in the Cytochrome b6f complex of Oxygenic Photosynthesis Protects against

Oxygen Damage, *J. Biol. Chem.* 274 (1999) 1581–1587. https://doi.org/10.1074/jbc.274.3.1581.

115. M. García-González, J. Moreno, J.C. Manzano, F.J. Florencio, M.G. Guerrero, Production of Dunaliella salina biomass rich in 9-cis-β-carotene and lutein in a closed tubular photobioreactor, *J. Biotechnol.* 115 (2005) 81–90. https://doi.org/10.1016/j.jbiotec.2004.07.010.
116. L. Gouveia, J. Empis, Relative stabilities of microalgal carotenoids in microalgal extracts, biomass and fish feed: effect of storage conditions, *Innov. Food Sci. Emerg. Technol.* 4 (2003) 227–233. https://doi.org/10.1016/S1466-8564(03)00002-X.
117. A. Lindqvist, S. Andersson, Biochemical Properties of Purified Recombinant Human β-Carotene 15,15′-Monooxygenase, *J. Biol. Chem.* 277 (2002) 23942–23948. https://doi.org/10.1074/jbc.M202756200.
118. S.T. Mayne, Beta-carotene, carotenoids, and disease prevention in humans, *FASEB J.* 10 (1996) 690–701. https://doi.org/10.1096/fasebj.10.7.8635686.
119. H. Van Chuyen, J.-B. Eun, Marine carotenoids: Bioactivities and potential benefits to human health, *Crit. Rev. Food Sci. Nutr.* 57 (2017) 2600–2610. https://doi.org/10.1080/10408398.2015.1063477.
120. D. Leary, M. Vierros, G. Hamon, S. Arico, C. Monagle, Marine genetic resources: A review of scientific and commercial interest, *Mar. Policy.* 33 (2009) 183–194. https://doi.org/10.1016/j.marpol.2008.05.010.
121. M. Guerin, M.E. Huntley, M. Olaizola, Haematococcus astaxanthin: applications for human health and nutrition, *Trends Biotechnol.* 21 (2003) 210–216. https://doi.org/10.1016/S0167-7799(03)00078-7.
122. A.C. Guedes, H.M. Amaro, F.X. Malcata, Microalgae as Sources of Carotenoids, *Mar. Drugs.* 9 (2011) 625–644. https://doi.org/10.3390/md9040625.
123. S. Sakai, T. Sugawara, K. Matsubara, T. Hirata, Inhibitory Effect of Carotenoids on the Degranulation of Mast Cells via Suppression of Antigen-induced Aggregation of High Affinity IgE Receptors, *J. Biol. Chem.* 284 (2009) 28172–28179. https://doi.org/10.1074/jbc.M109.001099.
124. R.H. Wijffels, Potential of sponges and microalgae for marine biotechnology, *Trends Biotechnol.* 26 (2008) 26–31. https://doi.org/10.1016/j.tibtech.2007.10.002.
125. J.A. Del Campo, M. García-González, M.G. Guerrero, Outdoor cultivation of microalgae for carotenoid production: current state and perspectives, *Appl. Microbiol. Biotechnol.* 74 (2007) 1163–1174. https://doi.org/10.1007/s00253-007-0844-9.
126. C. Vílchez, I. Garbayo, M. V. Lobato, J. Vega, Microalgae-mediated chemicals production and wastes removal, *Enzyme Microb. Technol.* 20 (1997) 562–572. https://doi.org/10.1016/S0141-0229(96)00208-6.
127. R.T. Lorenz, G.R. Cysewski, Commercial potential for Haematococcus microalgae as a natural source of astaxanthin, *Trends Biotechnol.* 18 (2000) 160–167. https://doi.org/10.1016/S0167-7799(00)01433-5.

128. M.M.R. Shah, Y. Liang, J.J. Cheng, M. Daroch, Astaxanthin-Producing Green Microalga Haematococcus pluvialis: From Single Cell to High Value Commercial Products, *Front. Plant Sci.* 7 (2016). https://doi.org/10.3389/fpls.2016.00531.
129. N.T. Eriksen, The technology of microalgal culturing, *Biotechnol. Lett.* 30 (2008) 1525–1536. https://doi.org/10.1007/s10529-008-9740-3.
130. O.S. Graverholt, N.T. Eriksen, Heterotrophic high-cell-density fed-batch and continuous-flow cultures of Galdieria sulphuraria and production of phycocyanin, *Appl. Microbiol. Biotechnol.* 77 (2007) 69–75. https://doi.org/10.1007/s00253-007-1150-2.
131. J.K. Sloth, H.C. Jensen, D. Pleissner, N.T. Eriksen, Growth and phycocyanin synthesis in the heterotrophic microalga Galdieria sulphuraria on substrates made of food waste from restaurants and bakeries, *Bioresour. Technol.* 238 (2017) 296–305. https://doi.org/10.1016/j.biortech.2017.04.043.
132. J.-G. Liu, C.-W. Hou, S.-Y. Lee, Y. Chuang, C.-C. Lin, Antioxidant effects and UVB protective activity of Spirulina (Arthrospira platensis) products fermented with lactic acid bacteria, *Process Biochem.* 46 (2011) 1405–1410. https://doi.org/10.1016/j.procbio.2011.03.010.
133. C. Romay, R. Gonzalez, N. Ledon, D. Remirez, V. Rimbau, C-Phycocyanin: A Biliprotein with Antioxidant, Anti-Inflammatory and Neuroprotective Effects, *Curr. Protein Pept. Sci.* 4 (2003) 207–216. https://doi.org/10.2174/1389203033487216.
134. D. Remirez, N. Ledón, R. González, Role of histamine in the inhibitory effects of phycocyanin in experimental models of allergic inflammatory response, *Mediators Inflamm.* 11 (2002) 81–85. https://doi.org/10.1080/09629350220131926.
135. L.P. Ponomarenko, I.V. Stonik, N.A. Aizdaicher, T.Y. Orlova, G.I. Popovskaya, G.V. Pomazkina, V.A. Stonik, Sterols of marine microalgae Pyramimonas cf. cordata (Prasinophyta), Attheya ussurensis sp. nov. (Bacillariophyta) and a spring diatom bloom from Lake Baikal, *Comp. Biochem. Physiol. Part B Biochem. Mol. Biol.* 138 (2004) 65–70. https://doi.org/10.1016/j.cbpc.2004.02.007.
136. P. Soudant, J.-R. Le Coz, Y. Marty, J. Moal, R. Robert, J.-F. Samain, Incorporation of Microalgae Sterols by Scallop Pecten maximus (L.) Larvae, *Comp. Biochem. Physiol. Part A Mol. Integr. Physiol.* 119 (1998) 451–457. https://doi.org/10.1016/S1095-6433(97)00451-0.
137. N.L.C. Ragg, N. King, E. Watts, J. Morrish, Optimising the delivery of the key dietary diatom Chaetoceros calcitrans to intensively cultured Greenshell™ mussel larvae, Perna canaliculus, *Aquaculture.* 306 (2010) 270–280. https://doi.org/10.1016/j.aquaculture.2010.05.010.
138. P. Soudant, M.V. Sanles, C. Quere, J.R. Le Coz, Y. Marty, J. Moal, J.F. Samain, P. Sorgeloos, The use of lipid emulsions for sterol supplementation of spat of the Pacific oyster, Crassostrea gigas, *Aquaculture.* 184 (2000) 315–326. https://doi.org/10.1016/S0044-8486(99)00323-3.

139. Q. Jo, E. Jung Choy, D. Won Park, B. Veron, Sterol dynamics of heterotrophic Tetraselmis suecica and its nutritional implication in the bivalve aquaculture, *Aquac. Res.* 35 (2004) 371–377. https://doi.org/10.1111/j.1365-2109.2004.01025.x.
140. T.-S. Vo, S.-K. Kim, Potential Anti-HIV Agents from Marine Resources: An Overview, *Mar. Drugs.* 8 (2010) 2871–2892. https://doi.org/10.3390/md8122871.
141. J. Qi, S.M. Kim, Characterization and immunomodulatory activities of polysaccharides extracted from green alga Chlorella ellipsoidea, *Int. J. Biol. Macromol.* 95 (2017) 106–114. https://doi.org/10.1016/j.ijbiomac.2016.11.039.
142. Y. Ma, P. Wang, Y. Wang, S. Liu, Q. Wang, Y. Wang, Fermentable sugar production from wet microalgae residual after biodiesel production assisted by radio frequency heating, *Renew. Energy.* 155 (2020) 827–836. https://doi.org/10.1016/j.renene.2020.03.176.
143. F. Rachidi, R. Benhima, L. Sbabou, H. El Arroussi, Microalgae polysaccharides bio-stimulating effect on tomato plants: Growth and metabolic distribution, *Biotechnol. Reports.* 25 (2020) e00426. https://doi.org/10.1016/j.btre.2020.e00426.
144. E.M. Larsen, M.R. Wilson, R.E. Taylor, Conformation–activity relationships of polyketide natural products, *Nat. Prod. Rep.* 32 (2015) 1183–1206. https://doi.org/10.1039/C5NP00014A.
145. J. Kobayashi, M. Ishibashi, Bioactive metabolites of symbiotic marine microorganisms, *Chem. Rev.* 93 (1993) 1753–1769. https://doi.org/10.1021/cr00021a005.
146. S.-X. Yang, J.-M. Gao, Q. Zhang, H. Laatsch, Toxic polyketides produced by Fusarium sp., an endophytic fungus isolated from Melia azedarach, *Bioorg. Med. Chem. Lett.* 21 (2011) 1887–1889. https://doi.org/10.1016/j.bmcl.2010.12.043.
147. K. Onodera, H. Nakamura, Y. Oba, Y. Ohizumi, M. Ojika, Zooxanthellamide Cs: Vasoconstrictive Polyhydroxylated Macrolides with the Largest Lactone Ring Size from a Marine Dinoflagellate of Symbiodinium sp., *J. Am. Chem. Soc.* 127 (2005) 10406–10411. https://doi.org/10.1021/ja050810g.
148. J.G. Day, E.E. Benson, R.A. Fleck, *In vitro* culture and conservation of microalgae: Applications for aquaculture, biotechnology and environmental research, *Vitr. Cell. Dev. Biol. - Plant.* 35 (1999) 127–136. https://doi.org/10.1007/s11627-999-0022-0.
149. K.M. Rahman, Food and High Value Products from Microalgae: Market Opportunities and Challenges, in: *Microalgae Biotechnol. Food, Heal. High Value Prod.*, Springer Singapore, Singapore, 2020: pp. 3–27. https://doi.org/10.1007/978-981-15-0169-2_1.
150. L.M. Andrade, Chlorella and Spirulina Microalgae as Sources of Functional Foods, Nutraceuticals, and Food Supplements; an Overview, *MOJ Food Process. Technol.* 6 (2018). https://doi.org/10.15406/mojfpt.2018.06.00144.

151. Z. Khan, P. Bhadouria, P. Bisen, Nutritional and Therapeutic Potential of Spirulina, *Curr. Pharm. Biotechnol.* 6 (2005) 373–379. https://doi.org/10.2174/138920105774370607.
152. N. Seyidoglu, S. Inan, C. Aydin, A Prominent Superfood: Spirulina platensis, in: *Superfood Funct. Food - Dev. Superfoods Their Roles as Med., InTech*, 2017. https://doi.org/10.5772/66118.
153. T. Nagao, S. Meguro, T. Hase, K. Otsuka, M. Komikado, I. Tokimitsu, T. Yamamoto, K. Yamamoto, A Catechin-rich Beverage Improves Obesity and Blood Glucose Control in Patients With Type 2 Diabetes, *Obesity.* 17 (2009) 310–317. https://doi.org/10.1038/oby.2008.505.
154. C. Jong-Yuh, S. Mei-Fen, Potential hypoglycemic effects of Chlorella in streptozotocin-induced diabetic mice, *Life Sci.* 77 (2005) 980–990. https://doi.org/10.1016/j.lfs.2004.12.036.
155. M.R. Brown, S.W. Jeffrey, J.K. Volkman, G.. Dunstan, Nutritional properties of microalgae for mariculture, *Aquaculture.* 151 (1997) 315–331. https://doi.org/10.1016/S0044-8486(96)01501-3.
156. L. Gouveia, A.P. Batista, I. Sousa, A. Raymundo, N.M. Bandarra, Microalgae in novel food products, 2009.
157. P. Nova, A.P. Martins, C. Teixeira, H. Abreu, J.G. Silva, A.M. Silva, A.C. Freitas, A.M. Gomes, Foods with microalgae and seaweeds fostering consumers health: a review on scientific and market innovations, *J. Appl. Phycol.* 32 (2020) 1789–1802. https://doi.org/10.1007/s10811-020-02129-w.
158. K.N. H. Endo, H. Sanasawa, Studies on <italic>Chlorella regularis</italic>, heterotrophic fast-growing strain II. Mixotrophic growth in relation to light intensity and acetate concentration, *Plant Cell Physiol.* (1977). https://doi.org/10.1093/oxfordjournals.pcp.a075413.
159. O. Perez-Garcia, Y. Bashan, Microalgal Heterotrophic and Mixotrophic Culturing for Bio-refining: From Metabolic Routes to Techno-economics, in: Algal Biorefineries, Springer International Publishing, Cham, 2015: pp. 61–131. https://doi.org/10.1007/978-3-319-20200-6_3.
160. D. Mitra, J. (Hans) van Leeuwen, B. Lamsal, Heterotrophic/mixotrophic cultivation of oleaginous Chlorella vulgaris on industrial co-products, *Algal Res.* 1 (2012) 40–48. https://doi.org/10.1016/j.algal.2012.03.002.
161. K. Ullah, M. Ahmad, Sofia, V.K. Sharma, P. Lu, A. Harvey, M. Zafar, S. Sultana, C.N. Anyanwu, Algal biomass as a global source of transport fuels: Overview and development perspectives, *Prog. Nat. Sci. Mater. Int.* 24 (2014) 329–339. https://doi.org/10.1016/j.pnsc.2014.06.008.
162. A.P. Batista, A. Niccolai, I. Bursic, I. Sousa, A. Raymundo, L. Rodolfi, N. Biondi, M.R. Tredici, Microalgae as Functional Ingredients in Savory Food Products: *Application to Wheat Crackers, Foods.* 8 (2019) 611. https://doi.org/10.3390/foods8120611.
163. L. Novoveská, M.E. Ross, M.S. Stanley, R. Pradelles, V. Wasiolek, J.-F. Sassi, Microalgal Carotenoids: A Review of Production, Current Markets,

Regulations, and Future Direction, *Mar. Drugs.* 17 (2019) 640. https://doi.org/10.3390/md17110640.

164. I. Saadaoui, R. Rasheed, A. Aguilar, M. Cherif, H. Al Jabri, S. Sayadi, S.R. Manning, Microalgal-based feed: promising alternative feedstocks for livestock and poultry production, *J. Anim. Sci. Biotechnol.* 12 (2021) 76. https://doi.org/10.1186/s40104-021-00593-z.
165. Y. Panahi, B. Darvishi, N. Jowzi, F. Beiraghdar, A. Sahebkar, Chlorella vulgaris: A Multifunctional Dietary Supplement with Diverse Medicinal Properties, *Curr. Pharm. Des.* 22 (2015) 164–173. https://doi.org/10.2174/1381612822666151112145226.
166. A.K. Koyande, K.W. Chew, K. Rambabu, Y. Tao, D.-T. Chu, P.-L. Show, Microalgae: A potential alternative to health supplementation for humans, *Food Sci. Hum. Wellness.* 8 (2019) 16–24. https://doi.org/10.1016/j.fshw.2019.03.001.
167. Y. Durmaz, M. Kilicli, O.S. Toker, N. Konar, I. Palabiyik, F. Tamtürk, Using spray-dried microalgae in ice cream formulation as a natural colorant: Effect on physicochemical and functional properties, *Algal Res.* 47 (2020) 101811. https://doi.org/10.1016/j.algal.2020.101811.
168. F. de Farias Neves, M. Demarco, G. Tribuzi, Drying and Quality of Microalgal Powders for Human Alimentation, in: *Microalgae - From Physiol. to Appl.*, IntechOpen, 2020. https://doi.org/10.5772/intechopen.89324.
169. S. Orset, G.C. Leach, R. Morais, A.J. Young, Spray-Drying of the Microalga Dunaliella salina : Effects on β-Carotene Content and Isomer Composition, *J. Agric. Food Chem.* 47 (1999) 4782–4790. https://doi.org/10.1021/jf990571e.
170. J. Benemann, Microalgae for Biofuels and Animal Feeds, *Energies.* 6 (2013) 5869–5886. https://doi.org/10.3390/en6115869.
171. Z. Yaakob, E. Ali, A. Zainal, M. Mohamad, M. Takriff, An overview: biomolecules from microalgae for animal feed and aquaculture, *J. Biol. Res.* 21 (2014) 6. https://doi.org/10.1186/2241-5793-21-6.
172. S. Hemaiswarya, R. Raja, R. Ravi Kumar, V. Ganesan, C. Anbazhagan, Microalgae: a sustainable feed source for aquaculture, *World J. Microbiol. Biotechnol.* 27 (2011) 1737–1746. https://doi.org/10.1007/s11274-010-0632-z.
173. A.M. Contreras, I.D. Marsden, M.H.G. Munro, Effects of short-term exposure to paralytic shellfish toxins on clearance rates and toxin uptake in five species of New Zealand bivalve, *Mar. Freshw. Res.* 63 (2012) 166. https://doi.org/10.1071/MF11173.
174. C. Cahu, J.C. Guillaume, G. Stéphan, L. Chim, Influence of phospholipid and highly unsaturated fatty acids on spawning rate and egg and tissue composition in Penaeus vannamei fed semi-purified diets, *Aquaculture.* 126 (1994) 159–170. https://doi.org/10.1016/0044-8486(94)90257-7.
175. V. Andreotti, A. Chindris, G. Brundu, D. Vallainc, M. Francavilla, J. García, Bioremediation of aquaculture wastewater from Mugil cephalus (Linnaeus, 1758) with different microalgae species, *Chem. Ecol.* 33 (2017) 750–761. https://doi.org/10.1080/02757540.2017.1378351.

176. S. Radhakrishnan, C. Seenivasan, T. Muralisankar, Effect of dietary replacement of fishmeal with Chlorella vulgaris on growth performance, energy utilization and digestive enzymes in Macrobrachium rosenbergii postlarvae, *Int. J. Fish. Aquac.* 7 (2015) 62–70. https://doi.org/10.5897/IJFA15.0471.
177. U. Enyidi, Chlorella vulgaris as Protein Source in the Diets of African Catfish Clarias gariepinus, *Fishes.* 2 (2017) 17. https://doi.org/10.3390/fishes2040017.
178. F.O. Arimoro, Culture of the freshwater rotifer, Brachionus calyciflorus, and its application in fish larviculture technology, *African J. Biotechnol.* 5 (2006) 536–541. https://doi.org/10.4314/ajb.v5i7.
179. S.C. Silva, I.C.F.R. Ferreira, M.M. Dias, M.F. Barreiro, Microalgae-Derived Pigments: A 10-Year Bibliometric Review and Industry and Market Trend Analysis, *Molecules.* 25 (2020) 3406. https://doi.org/10.3390/molecules25153406.
180. M.K. Ghosh, S. Sahu, I. Gupta, T.K. Ghorai, Green synthesis of copper nanoparticles from an extract of Jatropha curcas leaves: characterization, optical properties, CT-DNA binding and photocatalytic activity, *RSC Adv.* 10 (2020) 22027–22035. https://doi.org/10.1039/D0RA03186K.
181. T.R. Sommer, W.T. Potts, N.M. Morrissy, Utilization of microalgal astaxanthin by rainbow trout (Oncorhynchus mykiss), *Aquaculture.* 94 (1991) 79–88. https://doi.org/10.1016/0044-8486(91)90130-Y.
182. S. Tolasa, S. Cakli, U. Ostermeyer, Determination of astaxanthin and canthaxanthin in salmonid, *Eur. Food Res. Technol.* 221 (2005) 787–791. https://doi.org/10.1007/s00217-005-0071-5.
183. B. Bjerkeng, K. Hamre, B. Hatlen, E. Wathne, Astaxanthin deposition in fillets of Atlantic salmon Salmo salar L. fed two dietary levels of astaxanthin in combination with three levels of α-tocopheryl acetate, *Aquac. Res.* 30 (1999) 637–646. https://doi.org/10.1046/j.1365-2109.1999.00355.x.
184. C. Grewe, S. Menge, C. Griehl, Enantioselective separation of all-E-astaxanthin and its determination in microbial sources, *J. Chromatogr. A.* 1166 (2007) 97–100. https://doi.org/10.1016/j.chroma.2007.08.002.
185. E.A. Johnson, G.-H. An, Astaxanthin from Microbial Sources, *Crit. Rev. Biotechnol.* 11 (1991) 297–326. https://doi.org/10.3109/07388559109040622.
186. S.M.A. Mobin, H. Chowdhury, F. Alam, Commercially important bioproducts from microalgae and their current applications – A review, *Energy Procedia.* 160 (2019) 752–760. https://doi.org/10.1016/j.egypro.2019.02.183.
187. S. S.M., Microalgal Biotechnology: Prospects and Applications, in: *Plant Sci., InTech*, 2012. https://doi.org/10.5772/53694.
188. L.-J. Ren, G. Sun, X.-J. Ji, X. Hu, H. Huang, Compositional shift in lipid fractions during lipid accumulation and turnover in Schizochytrium

sp., *Bioresour. Technol.* 157 (2014) 107–113. https://doi.org/10.1016/j.biortech.2014.01.078.

189. A. Barghini, V.I. Ivanova, S.H. Imam, E. Chiellini, Poly-(ε-caprolactone) (PCL) and poly(hydroxy-butyrate) (PHB) blends containing seaweed fibers: Morphology and thermal-mechanical properties, *J. Polym. Sci. Part A Polym. Chem.* 48 (2010) 5282–5288. https://doi.org/10.1002/pola.24327.

3

Arsenic Bioremoval Using Algae: A Sustainable Process

Sougata Ghosh[1,2*], Jyoti Nayak[2], Md Ashraful Islam[3] and Sirikanjana Thongmee[1]

[1]*Department of Physics, Faculty of Science, Kasetsart University, Bangkok, Thailand*
[2]*Department of Microbiology, School of Science, RK University, Rajkot, Gujarat, India*
[3]*Faculty of Veterinary Science, Bangladesh Agricultural University, Mymensingh, Bangladesh*

Abstract
Among the various hazardous elements, arsenic is considered one of the most toxic metalloids which pollute the environment either due to mining, metallurgical processes or agricultural activities. The four oxidation states of arsenic are -3, 0, 3, 5, among which the trivalent As(III) and the pentavalent As(V) forms are predominant in nature. It is important to note that As(III) can bind to the thiolate groups of cysteine residues present in enzymes with high affinity, eventually resulting in the inactivation of the enzyme. Although various physical and chemical methods are available for arsenic removal, they are often inefficient, costly and involve hazardous conditions or chemicals. Hence, biological removal of arsenic is considered to be environmentally benign, convenient, efficient and rapid. This chapter discusses the use of algae for removal of arsenic and their biotransformation. Algal biomass of *Botryococcus braunii, Chlorodesmis* sp., *Cladophora* sp., *Chlamydomonas reinhardtii, Chara braunii, Chlorella pyrenoidosa, Ostreococcus tauri, Ulva fasciata*, and others can effectively remove arsenic due to their biosorption and bioaccumulation abilities. Furthermore, the algae can be genetically modified to facilitate desired post-translational modification of the proteins responsible for removal of arsenic. In view of this background, algae mediated

**Corresponding author*: ghoshsibb@gmail.com

Charles Oluwaseun Adetunji, Julius Kola Oloke, Naveen Dwivedi, Sabeela Beevi Ummalyma, Shubha Dwivedi, Daniel Ingo Hefft and Juliana Bunmi Adetunji (eds.) Next-Generation Algae: Volume I: Applications in Agriculture, Food and Environment, (91–108) © 2023 Scrivener Publishing LLC

arsenic removal can be considered as the most powerful strategy for treatment of water contaminated with arsenic.

Keywords: Arsenic toxicity, biological removal, algae, biosorption, bioaccumulation, biotransformation

3.1 Introduction

Arsenic (As) has attracted global attention due to its predominance among the various highly toxic elements that pose a potential threat to human health and the environment. Arsenic pollutes natural water reserves from various geochemical sources. Arsenopyrite gold ores, which are the major mining material of gold mines, are considered a major source of arsenic pollution [1, 2]. Other anthropogenic sources in the environment are wood preservatives, pesticides, and herbicides that contain arsenical compounds [3]. Industrial effluents contaminated with arsenic are disposed in the environment, further contaminating the groundwater. A high risk of cancer is prevalent among an estimated 20 million Americans who are exposed to arsenic polluted water daily [4]. Approximately 57 million people consume arsenic-contaminated water from the wells in Bangladesh [5].

Trivalent arsenite [As(III)] is highly toxic compared to pentavalent arsenate [As(V)]. However, long-term exposure to arsenic may result in the occurrence of the cancer associated with organs such as lung, bladder, skin and kidney [6]. The substitution of As(V) for phosphate, affinity of As(III) for thiol groups of protein, and protein-DNA and DNA-DNA crosslinking mainly are attributed to the toxicity of arsenic [7]. This results in impairment of metabolic activities and erroneous genetic manipulations. Considering these toxicological impacts, the United States has set the current regulatory limit of arsenic at 10 ppb.

Arsenic constitutes one of the most important toxicological hazards to animals when it enters the food chain and is biomagnified in various trophic levels. In the case of cattle, mostly contaminated drinking water and arsenic deposited paddy straw are the sources of arsenic poisoning. Excretion of arsenic can also take place through the urine, milk, and dung of dairy cattle. Concentrations of arsenic in excreted materials (urine and dung) and milk increased with the increase in arsenic intake. A lesser amount of arsenic biotransformation through milk may severely affect human health. Casein is the chief site of arsenic accumulation (83%), which is attributed to the phosphoserine units present in it. A considerable amount of arsenic is also found in other livestock products such as meat, boiled egg yolk, and liver [8].

Unlike humans, animals fail to show notable arsenic toxicity-associated skin problems but the metalloid can be detected in blood, keratin tissues and feces. The ultimate result is weight loss, reduction of milk production, immune suppression, abdominal discomfort, diarrhea and other associated toxicities. In some cases, an animal may even die from acute arsenic toxicity. Symptoms mostly include lethargy, ataxia, anorexia, and diarrhea. Arsenic is mostly (about 80%) accumulated in the kidney and liver of calves and excreted in feces and urine. Long-term exposure to arsenic can lead to further accumulation, particularly in nails and hair, indicating more susceptibility of the ectodermal tissues [9].

The conventional arsenic removing technologies include approaches that are associated with polymeric ligand exchange, activated alumina sorption, and polymeric anion exchange that can remove As(V) with high efficiency compared to As(III). Hence, initially As(III) is oxidized to As(V), which is a prerequisite for a satisfactory level of arsenic removal which also adds an extra step and economic burden [10–12]. On the other hand, coagulation-like techniques are nonspecific and hence remove arsenic along with other elements. Thus, water chemistry is largely altered following such treatment processes [13].

Considering the shortcomings of the available methods for arsenic removal, there is a growing need for development of an environmentally benign green process for effective removal of arsenic in aquatic systems [14]. More recently, biologically inspired technologies are gaining more attention and preference as they provide an economical complementary and alternative strategy for arsenic removal. Biological entities like bacteria, cyanobacteria, and plants have gained tremendous acceptance and appreciation as biosorbents due to their high affinity and specificity for toxic metals [15]. In view of this background, this chapter highlights the promising role of algae for arsenic. Several arsenic tolerant microbes are isolated from arsenic-contaminated sites that can express an array of resistance proteins that are encoded in genetic circuit tightly controlled by specific metalloregulatory proteins such as ArsR [16].

3.2 Algae-Mediated Arsenic Removal

Several algae are isolated from diverse ecological niches and exploited for bioremediation of arsenic, as evident from Table 3.1. Podder *et al.* collected the green algae *Botryococcus braunii* that were cultured in BG11 culture

Table 3.1 Removal of arsenic by algae.

Algae	Arsenic removal (%)	Reference
Botryococcus braunii	85.22–88.15	[17]
Botryococcus sp.		[18]
Chara braunii		[19]
Chara vulgaris	40–45	[20]
Chlamydomonas reinhardtii		[21]
Chlorella minutissima	60–66	[22]
Chlorella pyrenoidosa		[23]
Chlorella vulgaris	73	[24]
Chlorella vulgaris		[25]
Chlorodesmis sp.	40–50	[26]
Cladophora sp.	99	[26]
Cladophora sp.	99	[27]
Dunaliella salina	95.6	[28]
Oedogonium sp.	80	[29]
Ostreococcus tauri	50	[30]
Scenedesmus acuminatus		[31]
Scenedesmus obliquus	< 98	[32]
Scenedesmus quadricauda	> 60	[33]
Scenedesmus sp. IITRIND2	70–73	[22]
Spirogyra	82.76	[34]
Ulva fasciata	80.4	[35]
Ulva reticulata	55–92.9	[36]

medium devoid of any carbon source for 7 days at 28 °C [17]. The stock solutions of arsenic were made by using $NaAsO_2$ and $Na_2HAsO_4.7H_2O$ in double-distilled water. The phycoremediation percentage of As(III) and As(V) were dependent on both the surface charge of the algal biomass and on the arsenic species. The dominant species of As(III) and As(V) were non-ionic H_3AsO_3 and $HAsO_4^{2-}$, respectively. Increase in pH enhanced the adsorbate concentration in the solution. The maximum phycoremediation of As(III) and As(V) ions were observed as 85.22% and 88.15%, respectively, at pH 9.0 after 144 h. Increase in inoculum size from 1 to 10% (v/v) resulted in the increase in the biomass concentration. Scanning electron microscope (SEM) images in Figure 3.1 showed that the surface of the native microalgal biomass was smooth, unlike those which became rough after arsenic (either As(III) or As(V)) treatment. The roughness might be attributed to the biosorption of arsenic on the surface.

In yet another study, Hubadillah *et al.* cultured the freshwater microalgae *Botryococcus* sp. at 28 to 38 °C under illumination (light intensity varying from 2.7 to 243 μ mol/m²s) for two weeks [18]. A sintering technique was used to fabricate the kaolin ceramic hollow fiber membrane (*h*-KHFM) which was hydrophobic in nature. The resulting thickness of the material was 240 μm. The river water polluted with arsenic was treated for 70 h with a *h*-KHFM module using a laboratory-scale DCMD (direct contact membrane distillation) setup. The membrane exhibited a steady flux of ~23.3 kg/m²h. A feed temperature of 60 °C, permeate temperature of 10 °C, feed flow rate (Qf) of 42 kg/h, and permeate flow rate (Qp) of 30 kg/h, were found to be optimum conditions of the DCMD process. A pH > 7 resulted in higher arsenic rejection that was attributed to the conversion of H_3AsO_3 to $H_2AsO_3^-$. However, pH did not influence the arsenic rejection efficiency of MD/*h*-KHFM that was consistent at 100%. Green color appeared on the 20th day due to the growth of *Botryococcus* sp. with DCMD retentate in water containing 112,000 ppb of arsenic.

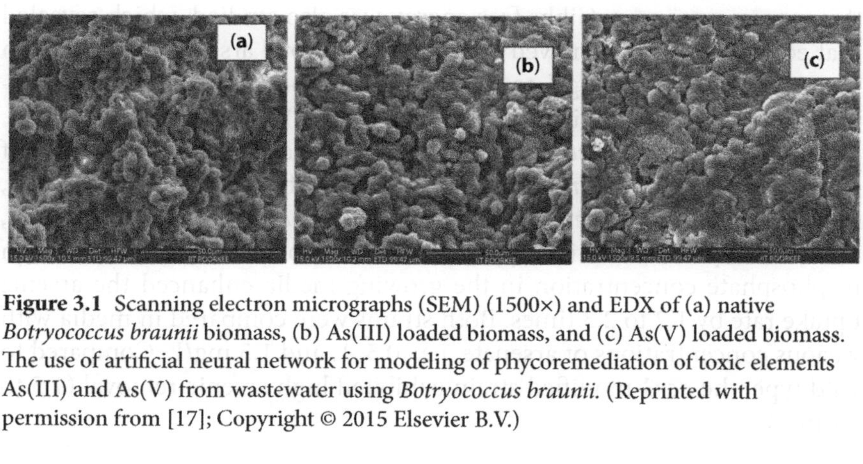

Figure 3.1 Scanning electron micrographs (SEM) (1500×) and EDX of (a) native *Botryococcus braunii* biomass, (b) As(III) loaded biomass, and (c) As(V) loaded biomass. The use of artificial neural network for modeling of phycoremediation of toxic elements As(III) and As(V) from wastewater using *Botryococcus braunii*. (Reprinted with permission from [17]; Copyright © 2015 Elsevier B.V.)

Amirnia *et al.* collected the *Chara braunii* from Japan that was investigated for deposition of manganese (Mn) and calcium (Ca) when grown in arsenic-contaminated water [19]. The variation in shoot length of *C. braunii* was evident when treated with different concentrations of As (0 and 0.5 mg L^{-1}), Ca, and Mn with addition of 1 mM $HNaCO_3$ for biomineralization. Supplementation with $CaCl_2.2H_2O$, $MnSO_4.4H_2O$, and As_2O_3 in the culture media further resulted in the expected concentrations of Ca, Mn, and As(III) (Wako Pure Chemical Industries, Ltd., Japan), respectively. The plant tissues were then digested in nitric acid, hydrogen peroxide and hydrochloric acid to evaluate the As, Ca, and Mn (mg g^{-1}) content in each plant. Accumulation of arsenic in *C. braunii* varied between ~ 61.3 and 373.8 mg kg^{-1} under variable environmental parameters. The highest uptake of arsenic by the plants *C. braunii* contained an average of ~ 373.8 mg As kg^{-1} with bicarbonate treatment.

In another such study, Taleei *et al.* selected Chara algae with two other plant species, i.e., Water hyacinth (*H. orientalis*) and Vetiver grass (*V. zizanioides*), for arsenic removal [20]. They collected the Chara algae (*Chara vulgaris*) from the Amir Kolaieh wetland in northern Iran. The collected samples were maintained in a greenhouse at a temperature of 302.13 K under pressure (87,210 Pa). The Chara algae were cultured in stock solutions of $Na_2HAsO_4.7H_2O$ with four different arsenic concentrations of 10, 40, 70 and 110 mg/L. Four pots of Chara algae were treated with the mentioned concentrations. The sampling was performed at regular time intervals. The soil containing arsenic in each pot (100 g) was measured using the atomic absorption machine. The Chara algae showed a high total reduction of arsenic with concentrations of 10 mg/Kg as compared to the other two plant species. Arsenic concentrations in Chara algae were 36.78 mg/Kg. Thermodynamic parameters such as the heat, entropy and enthalpy of the process were zero. Gibbs free energy was also studied, which revealed that equilibrium was achieved on the 60th day. Total arsenic concentration was reduced from 110 mg/L to 40 mg/L. Reduction in total arsenic concentrations increased with time (from 2 h to 60 days).

Ramírez-Rodríguez *et al.* evaluated the arsenate removal capability of *Chlamydomonas reinhardtii* against *acr3*-modified recombinant strain, *Agrobacterium tumefaciens*, using the construct pARR1 from *Pteris vittata*, a terrestrial plant which is a hyper-accumulator of arsenic [21]. Reduction in phosphate concentration in the growing media enhanced the arsenic uptake rate by 1.2 to 2.3 times. Both strains were compared in media with various concentrations of arsenate, i.e., 0.5, 1, and 1.5 mg/L. Compared to wild type, the *acr3*-modified strain exhibited high arsenic removal (1.5 to 3 times).

Arora *et al.* reported biodiesel production by the *oleaginous* green algae *Chlorella minutissima* (MCC-27), *Chlorella pyrenoidosa* (NCIM 2738) and *Scenedesmus abundans* (NCIM 2897), *Scenedesmus* sp. IITRIND2 after incubation for 6 days at 27 °C [22]. Both *C. minutissima* and *Scenedesmus* sp. IITRIND2 exhibited high tolerance against 500 mg L^{-1} for both As(III) and As(V) forms while *C. pyrenoidosa* and *S. abundans* tolerated 200 mg L^{-1} and 100 mg L^{-1} of arsenic, respectively. *Scenedesmus* sp. IITRIND2 and *C. minutissima* showed arsenic removal up to 70–73% and 60–66% respectively on the 10th day.

Zhang *et al.* cultured green algae *Chlorella pyrenoidosa* in sterile BG11 medium [23]. Bioaccumulation was observed when *C. pyrenoidosa* was incubated with varying concentrations of both As(III) and As(V). Sodium arsenite ($NaAsO_2$) and sodium arsenate ($Na_2HAsO_4^{-}.12H_2O$) were used as source of As(III) and As(V), respectively, for the bioaccumulation experiments. Extracellular polymeric substances (EPS) secreted from *C. pyrenoidosa* resulted in surface adsorption-mediated bioaccumulation in addition to enhanced intracellular uptake. EPS of the algal cells are composed of polysaccharides which can bind strongly to the arsenic through the active functional groups such as $-NH_2$, C–O–C, and C–O–H. Similarly, tyrosine-like substances at the cell-water interface are also responsible for arsenic binding and uptake. Increase in the initial concentration of the arsenic resulted in the enhancement of the uptake capacity. EPS-covered cells (EPS-C) exhibited a higher arsenic adsorption capacity compared to the EPS-free algal cells (EPS-F).

Extracellular EPS is visible in Figure 3.2a–c in the vicinity of the EPS-C. Additionally, disarranged thylakoid and disordered pyrenoid were also seen in *C. pyrenoidosa* (EPS-C and EPS-F) upon treatment with As(III) and As(V). Figure 3.2b and 3.2e indicate the appearance of lipid droplets and starch granules from chloroplast, indicating more pronounced cytotoxicity upon treatment with the more toxic form of As(III). Upon exposure to As(III) and As(V), the EPS-F showed swollen vacuoles and cytosol leakage, as evident from Figure 3.2e and 3.2f. The surface texture of the EPS-C and EPS-F upon being exposed to As(III) and As(V) exhibited disrupted plasma membrane and wrinkled surfaces, respectively, as evident from Figure 3.3. The EPS cover might be attributed to the protection of EPS-C, unlike the EPS-F, which showed more pronounced toxic responses as indicated by Figure 3.3b,e and 3.3c,f.

Pantoja *et al.* also reported arsenic removal using *C. vulgaris* [24]. The algal cells were cultured in Bold's medium without EDTA and high phosphate (1.61 mM HPO_4^{2-}). Biomass was generated by growing the cells at 25±0.5 °C in the presence of illumination at 2,646 lux (n = 33, SE = 154).

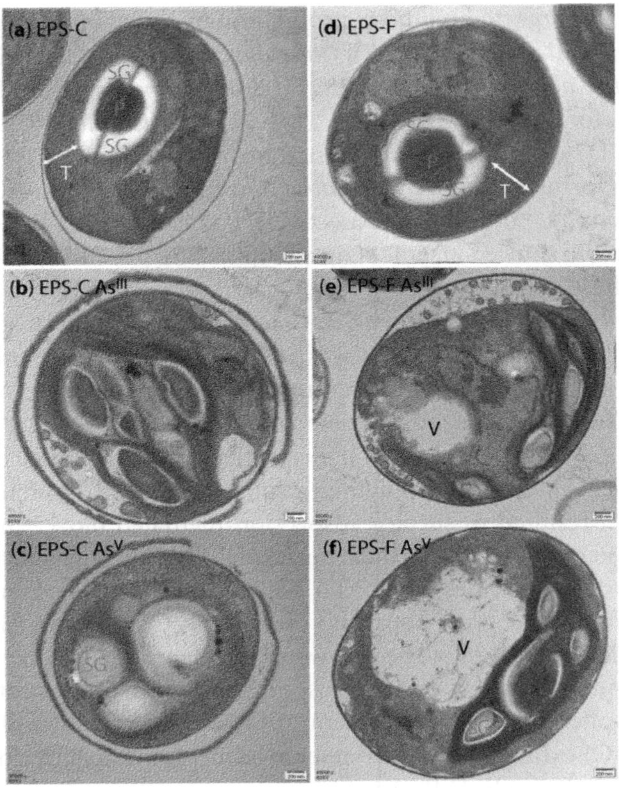

Figure 3.2 TEM images of EPS-C (a–c) and EPS-F (d–f) algal cells before and after exposure to 300 μmol L^{-1} As(III) and As(V) for 24 h: (a) EPS-C cells in an arsenic-free control, ×40,000; (b) EPS-C cells with As(III), ×40,000; (c) EPS-C cells with As(V), ×40,000; (d) EPS-F cells in an arsenic-free control, ×40,000; (e) EPS-F cells with As(III), ×40,000; (f) EPS-F cells with As(V), ×40,000. Observed structures are: thylakoid (T), pyrenoid (P), starch granule (SG) and vacuole (V). (Reprinted with permission from [23]; Copyright © 2019 Zhejiang University. Published by Elsevier B.V.)

Live *C. vulgaris* cells were grown for 3 to 5 days in Bold's medium containing 0.05% w/v dextrose in the presence of free arsenic. The As(V) inhibited the algal cultures growing in the presence of phosphate (0.003 mM) at a concentration of IC50 = 0.014 mM. The adsorption of As(V) on the living *C. vulgaris* cells marked a saturation when treated with 0.67 mM arsenic for 48 h. A 15 times higher As(III) adsorption was noticed compared to other species of arsenic. However, it is important to note that increase in arsenic concentration in the media resulted in enhancement of arsenic

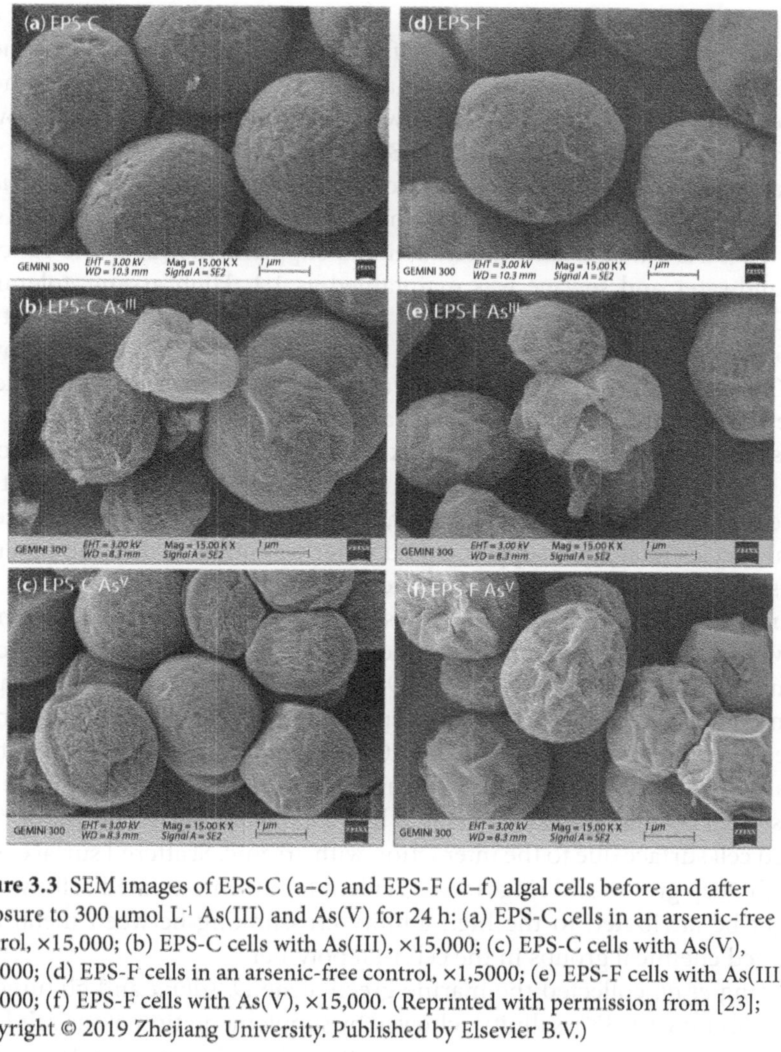

Figure 3.3 SEM images of EPS-C (a–c) and EPS-F (d–f) algal cells before and after exposure to 300 μmol L^{-1} As(III) and As(V) for 24 h: (a) EPS-C cells in an arsenic-free control, ×15,000; (b) EPS-C cells with As(III), ×15,000; (c) EPS-C cells with As(V), ×15,000; (d) EPS-F cells in an arsenic-free control, ×1,5000; (e) EPS-F cells with As(III), ×15,000; (f) EPS-F cells with As(V), ×15,000. (Reprinted with permission from [23]; Copyright © 2019 Zhejiang University. Published by Elsevier B.V.)

adsorption that was identical with all forms of arsenic. As(III) adsorption by cells was 8 times higher compared to As(V).

Awasthi *et al.* grew *C. vulgaris* in the BG11 medium at 24 ± 2 °C in the presence of illumination (60 $mmolm^{-2}$ S^{-1}) [25]. Various arsenic concentrations were maintained in the culture media to evaluate the arsenic stress tolerance of *C. vulgaris*. Reactive oxygen species like superoxide radicals ($O_2{\cdot}^-$) and hydrogen peroxide (H_2O_2) were reduced by the consortium of *P. putida* and *C. vulgaris*. This prevented membrane damage and promoted the plant growth (rice seedlings) that was attributed to the lowering of the arsenic toxicity. Upon inoculation of the consortium (*P. putida* + *C. vulgaris*) in rice seedlings, the arsenic accumulation in the root and shoot were 94 mg kg^{-1} dw and 51 mg kg^{-1} dw, respectively, which was much lower than the untreated plants (156 and 98 mg kg^{-1} dw, respectively).

Jasrotia *et al.* isolated two algal species, *Chlorodesmis* sp. and *Cladophora* sp., that were grown in local groundwater supplemented with $NaAsO_2$ and $Na_2HAsO_4.7H_2O$ and the nutritional supplements [26]. The ratio of groundwater and sewage was maintained as 1:1. High arsenic tolerance in *Cladophora* sp. and *Chlorodesmis* sp. was observed at a concentration of 6 mg/L and 4 mg/L, respectively. Nearly 50–55% COD was removed in *Chlorodesmis* sp., while 40–50% arsenic uptake in cells was evident. Interestingly, from the 11th day onwards, desorption of arsenic from the cells was observed. *Cladophora* sp. exhibited about 55–60% COD removal on the 10th day and 99% arsenic uptake was found with desorption observed from the 14th day. However, in this study the strong tolerance of algal species in arsenic-supplemented water was attributed to the superior resistance to oxidative stress-associated cellular damage.

In another study, Jasrotia *et al.* isolated an algae, *Cladophora,* which grew under the aforementioned conditions [27]. The maximum removal of arsenic achieved from water was 98.8% on the 10th day. The survival of filamentous green algae *Cladophora* species in the presence of arsenic indicated their tolerance to arsenic. A 100% arsenic removal by biosorption was evident after 10 days, after which desorption was observed. Figure 3.4 illustrates cellular uptake of arsenic by *Cladophora* algae and more ruptured cell surface due to the interaction with arsenic. Scattered surface with rough deep groves and ridges were evident for arsenic-treated cells, which might be attributed to the high affinity crosslinking between metal and charged chemical groups in the cell wall polymer.

Wang *et al.* collected the marine green algae *D. salina* and cultured in f/2 medium with 35% salinity achieved by supplementation of sea salt [28]. *D. salina* was tolerant to As(V) even after 72 h. *D. salina* could tolerate and accumulate As(V) that was largely influenced by phosphate levels. The percentage of algae-mediated arsenic removal was inversely proportional to the As(V) concentration in the medium. Around 7.7% of arsenic was removed after 9 days irrespective of PO_4^{3-} levels.

Johansson *et al.* collected *Oedogonium* and cultured in Manutec water-soluble f/2 algal growth media [29]. Stock solutions of arsenic (1 g L^{-1}) were prepared from $Na_2HAsO_4.7H_2O$ and diluted with deionized water obtained with various concentrations (0.0001 to 10 mmol L^{-1}). The yield (%) of Fe-treated *Oedogonium* biomass-generated biochar of 33.3–62.1%

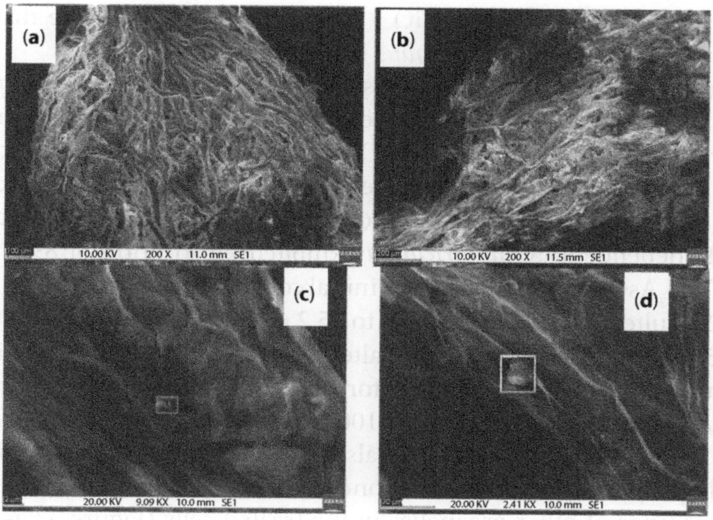

Figure 3.4 (a) Cell surface of control *Cladophora* sp. with no groves at 10 kV 200X; (b) Cell surface of test algae *Cladophora* sp. with groves at 10 kV 200X; (c, d) location of active sorptive sites for arsenic metal in *Cladophora* sp. at 20 kV. (Reprinted with permission from [27]; Copyright © 2014 Elsevier B.V.)

was produced by pyrolysis and pretreatment with Fe solutions. Almost 80% of the removal of arsenic from the AW by *Oedogonium*-modified biochar (OMB) was obtained.

In another study, Zhang *et al.* collected *Ostreococcus tauri* and cultured in Keller medium [30]. The stock solutions were prepared by $NaAsO_2$ and Na_3AsO_4. In the presence of As(III) (78 μM) and As(V) (120 μM) in the medium, 50% growth inhibition of *O. tauri* was noted, which indicated sensitivity to both As(III) and As(V). The tolerance limit of *O. tauri* for both arsenite and arsenate was 100 μM. The concentration of As(V) in the medium was directly proportional to the cellular arsenic concentration.

Lefta isolated *Scenedesmus acuminatus* and cultured in Jaworski's medium (JM) supplemented with arsenic was used to culture the algae at 27 °C for 72 h under illumination in a 14:10 h light/dark photoperiod [31]. *S. acuminatus* survived in the presence of 3.2–6.0 μg/mg of arsenic with concomitant metal uptake, while cell death was noted when the concentration of arsenic was 3.3–8.5 μg/mg.

Luo *et al.* reported uptake of arsenic in *Daphnia magna* when both nanostructured titanium dioxide (nano-TiO_2) and *Scenedesmus obliquus* were present in the media, the latter being provided as food daily [32]. *S. obliquus* as a food source increased the absorption of arsenic in MSF and

BDM. The presence of nano-TiO_2 (20 mg/L) resulted in more than 98% adsorption of arsenic by *S. obliquus* in solution.

Zhang *et al.* also reported the arsenic removing potential of *S. quadricauda* (FACHB-44) that was cultured in 500 mL of BG-11 medium [33]. Cultures were incubated at 21 ± 1 °C under a controlled light/dark cycle of 12 h/12 h. The algal cultures were reacted with various As(III) concentrations (0, 0.03, and 100 mg L^{-1}) for 96 h. The uptake and toxicity of arsenic was pH-dependent. *S. quadricauda* exhibited biosorption of 89.0 mg g^{-1} As(III) ($H_2AsO_3^-$) with pH 9.3 at an initial concentration of 100 mg L^{-1} while pH 8.2 resulted in a biosorption up to 25.2 μg g^{-1} at an initial concentration of 0.03 mg L^{-1}. The ultrastructural alterations revealed in the TEM images in Figure 3.5 exhibited cellular deformity and morphological changes in *S. quadricauda* when treated with 100 mg L^{-1} of As(III). Additionally, the partly damaged chloroplasts were also seen (Figure 3.5c–f). Toxicity was directly proportional to arsenite concentrations that were marked by the disappearance of other organelles in the treated cells (Figure 3.5g–j). It is important to note that both the chloroplasts and cell membrane of the algal cells were completely damaged at 100 mg L^{-1}.

Abioye *et al.* collected a water sample from the Kataeregi river bank where algal growth was observable [34]. The algal cells were isolated and cultured in Bold Basal Medium (BBM). Also, 5 g of pulverized poultry droppings were added into the sterile medium followed by supplementation of 0.5 g of arsenic chloride ($AsCl_3$) for two weeks. The concentration of arsenic in the Kataeregi river was 0.5196 mg/L. The *Spirogyra* exhibited 82.76% arsenic adsorption after 90 days.

Christobel and Lipton collected green, brown and red algae which were *Ulva fasciata*, *Sargassum wightii* and *Gracilaria corticata*, respectively, and investigated their arsenic removing potential [35]. Reactions parameters such as contact time, pH, biomass and initial metal ion concentration were varied to check their effect on arsenic removal. The maximum biosorption occurred around pH 6 for *U. fasciata* (84.6%) and *S. wightii* (87.1%) while pH 8 was optimum for *G. corticata* (85.4%). About 90.2% of arsenic removal was obtained with 2g/100 mL of biomass concentration. However, 80.4 % of arsenic removal was achieved with 3g/100 mL biomass of *U. fasciata*. The overall results of the experiments concluded that arsenic removal rate was high with dry biomass algae. The Freundlich arsenic adsorption capacity was in the order of *U. Fasciata* < *S. wightii* < *G. corticata*.

Senthilkumar *et al.* collected the green marine algae, *Ulva reticulata*, from Indian seashores [36]. The algal biomass was dried completely and ground to the size of 1 mm. Approximately 100 g of pulverized algal biomass were taken to produce biochar in a container sealed with aluminium

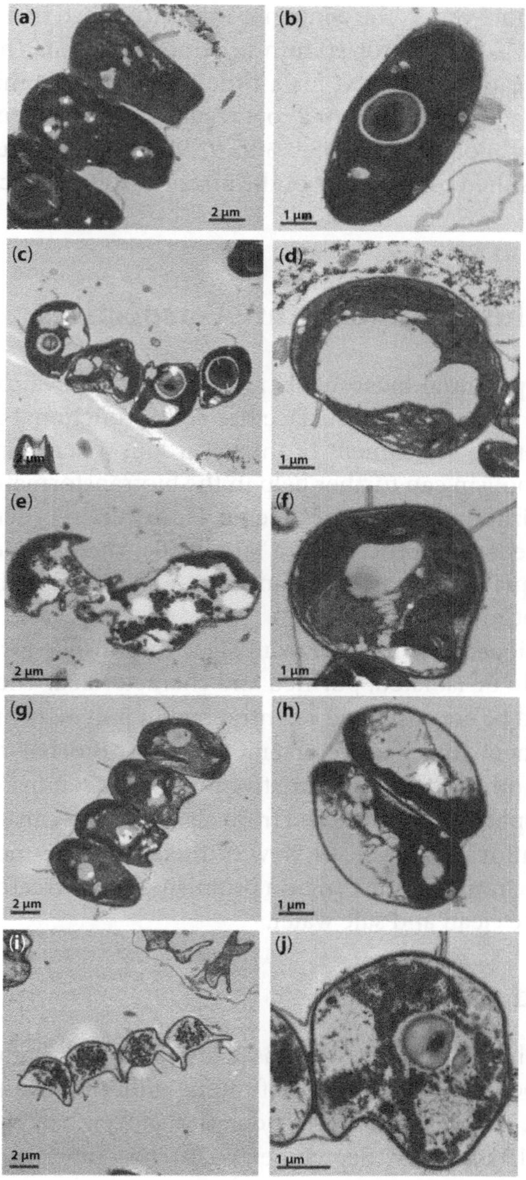

Figure 3.5 Transmission electron micrographs of *S. quadricauda* cells after the exposure time of 96 h. (a) and (b) are control; (c) and (d) are the treatments exposed to 0.03 mg L^{-1} arsenite with initial pH of 9.3; (e) and (f) are the treatments exposed to 100 mg L^{-1} arsenite with initial pH of 9.3; (g) and (h) are the treatments exposed to 0.03 mg L^{-1} arsenite with initial pH of 8.2; (i) and (j) are the treatments exposed to 100 mg L^{-1} arsenite pH of 8.2. (Reprinted with permission from [33]; Copyright © 2013 Elsevier Ltd.)

foil having minute vents. The container was introduced in a muffle furnace and nitrogen was purged for 10 min to remove oxygen. The biomass was subjected to heating at 300 °C for a time period of 120 min followed by pyrolysis. Alteration of pH from 2 to 4 resulted in enhancement of arsenic (V) removal by biochar from 55.0% to 92.9%. However, further increase of pH to 10 resulted in dropping of the arsenic (V) uptake capacity to 2.58 mg/g.

3.3 Conclusions and Future Perspectives

Algae can be potential biosorbents for effective removal of arsenic from contaminated water. The cell surface has active functional groups that are responsible for the attachment of the hazardous arsenic species [37–40]. Cellular metabolism can further help in the biotransformation, detoxification or volatilization of the toxic arsenic. However, in order to implement the phycoremediation process on a large scale, the parameters like time, temperature, pH, initial arsenic concentration and algal load should be thoroughly optimized [41–43].

Similarly, bioactive principles from algae responsible for arsenic detoxification should be isolated, purified and characterized. The purified compounds should be encapsulated or entrapped in polymers in order to reuse and recycle them for effective arsenic removal. Biochar from algae can prove to be another effective alternative for arsenic removal. Furthermore, biogenic nanoparticles synthesized from algal biomass can also be explored for arsenic removal [44, 45]. In view of this, the algae-mediated bioprocess for arsenic remediation can be a promising water treatment process in order to ensure clean and safe water.

Acknowledgment

Dr. Sougata Ghosh acknowledges Kasetsart University, Bangkok, Thailand, for the Postdoctoral Fellowship and funding under Reinventing University Program (Ref. No. 6501.0207/10870 dated 9th November, 2021 and Ref. No. 6501.0207/9219 dated 14th September, 2022).

References

1. Macy, J. M., Nunan, K., Hagen, K. D., Dixon, D. R., Harbour, P. J., Cahill, M., Sly L.I. *Chrysiogenes arsenatis* gen. nov., sp. nov., a new arsenate respiring

bacterium isolated from gold mine wastewater. *Int. J. Syst. Bacteriol.* 46(4), 1153–1157, 1996.

2. Ghosh, S., Bhagwat, T., and Webster, T.J., Arsenic removal using nanotechnology. In: Kalamdhad, A., Haq, I., (Eds.) *Emerging Treatment Technologies for Waste Management*, Springer Nature, Singapore. pp. 73-102, 2021.
3. Peryea, F. J., Kammereck, R., Phosphate-enhanced movement of arsenic out of lead arsenate-contaminated topsoil and through uncontaminated subsoil. *Water Air Soil Pollut.* 93, 243–254, 1997.
4. DeMarco, M. J., Sen Gupta, A. K., Greenleaf, J. E., Arsenic removal using a polymeric/inorganic hybrid sorbent. *Water Res.* 37(1), 164–176, 2003.
5. Nickson, R., McArthur, J., Burgess, W., Ahmed, K. M., Ravenscroft, P., Rahman, M., Arsenic poisoning of Bangladesh groundwater. *Nature* 395, 338, 1998.
6. Karagas, M. R., Tosteson, T. D., Blum, J., Morris, J. S., Baron, J. A., Klaue, B., Design of an epidemiologic study of drinking water arsenic exposure and skin and bladder cancer risk in a U.S. population. *Environ. Health Perspect.* 106 (Suppl 4),1047–1050, 1998.
7. Norman, N. C., Chemistry of arsenic, antimony, and bismuth. *J. Natl. Cancer Inst.* 40, 453–463. Springer Dordrecht, 1998.
8. Ghosh A, Majumder S, Awal MA, Rao DR, Arsenic exposure to dairy cows in Bangladesh. *Arch. Environ. Contam. Toxicol.* 64(1), 151–159, 2013.
9. Das, A., Joardar, M., Chowdhury, N.R., De, A., Mridha, D., Roychowdhury, T., Arsenic toxicity in livestock growing in arsenic endemic and control sites of West Bengal: risk for human and environment. *Environ. Geochem. Health* 43(8), 3005–3025, 2021.
10. Chwirka, J. D., Thomson, B. M., Stomp, J. M., Removing arsenic from groundwater. *J. Am. Water Works Assoc.* 92(3), 79–88, 2000.
11. Driehaus, W., Seith, R., Jekel, M., Oxidation of arsenic(III) with manganese oxides in water treatment. *Water Res.* 29(1), 297–305, 1995.
12. Wilkie, J. A., Hering, J. G., Rapid oxidation of geothermal arsenic (III) in streamwaters of the eastern Sierra Nevada. *Environ. Sci. Technol.* 32(5), 657–662, 1998.
13. Clifford, D., Subramonian, S., Sorg, T., Removing dissolved inorganic contaminants from water. *Environ. Sci. Technol.* 20 (11), 1072–1080, 1998.
14. Nickson, R. T., McArthur, J. M., Ravenscroft, P., Burgess, W. G., Ahmed, K. M., Mechanism of arsenic release to groundwater, Bangladesh, West Bengal. *Appl. Geochem.* 15(4), 403–413, 2000.
15. Say, R., Yilmaz, N., Denizli, A., Biosorption of cadmium, lead, mercury, and arsenic ions by the fungus *Penicillium purpurogenum*. *Sep. Sci. Technol.* 38(9), 2039–2053, 2003.
16. Xu, C., Shi, W., Rosen, B. P., The chromosomal *arsR* gene of *Escherichia coli* encodes a *trans*-acting metalloregulatory protein. *J. Biol. Chem.* 271(5), 2427–2432, 1996.

17. Podder, M.S., Majumdar, C.B., The use of Artificial neural network for modelling of phycoremediation of toxic elements As(III) and As(V) from wastewater using *Botryococcus braunii. Spectrochim. Acta A Mol. Biomol. Spectrosc.* 155, 130–145, 2016.
18. Hubadillah, S., Othman, M., Gani, P., Sunar, N., Tai, Z., Koo, N., Pauzan, M., Ismail, N., Integrated green membrane distillation-microalgae bioremediation for arsenic removal from Pengorak River Kuantan, Malaysia. *Chem. Eng. Process.* 153, 107996, 2020.
19. Amirnia, S., Asaeda, T., Takeuchi C., Kaneko, Y., Manganese-mediated immobilization of arsenic by calcifying macro-algae, *Chara braunii. Sci. Total Environ.* 646, 661-669, 2019.
20. Taleei, M., Ghomi, N., Jozi, S., Arsenic removal of contaminated soils by phytoremedian of Veltiver Grass, *Chara Algae and Water Hyacinth. Bull. Environ. Contam. Toxicol.* 102(1), 134-139, 2018.
21. Ramírez-Rodríguez, A., Bañuelos-Hernández, B., García-Soto, M., Govea-Alonso, D., Rosales-Mendoza, S., Torre, M., Monreal-Escalante, E., Paz-Maldonado, L., Arsenic removal using *Chlamydomonas reinhardtii* modified with the gene *acr3* and enhancement of its performance by decreasing phosphate in the growing media. *Int. J. Phytoremediation.* 21(7), 617-623, 2019.
22. Arora, N., Gulati, K., Patel, A., Pruthi, P., Poluri, K., Pruthi, V., A hybrid approach integrating arsenic detoxification with biodiesel production using oleaginous microalgae. *Algal Res.* 24, 29-39, 2017.
23. Zhang, J., Zhou, F., Liu, Y., Huang, F., Zhang, C., Effect of extracellular polymeric substances on arsenic accumulation in *Chlorella pyrenoidosa. Sci. Total Environ.* 704, 135368, 2020.
24. Pantoja Munoz, L., Purchase, D., Jones, H., Raab, A., Urgast, D., Feldmann, J., Garelick, H., The mechanisms of detoxification of As(III), dimethylarsinic acid (DMA) and As(V) in the microalgae *Chlorella vulgaris. Aquat. Toxicol.* 175, 56-72, 2016.
25. Awasthi, S., Chauhan, R., Dwivedi, S., Srivastava, S., Srivastava, S., Tripathi, R., A consortium of alga (*Chlorella vulgaris*) and bacterium (*Pseudomonas putida*) for amelioration of arsenic toxicity in rice: A promising and feasible approach. *Environ. Exp. Bot.* 150, 115-126, 2018.
26. Jasrotia, S., Kansal, A., Aradhana, M., Performance of aquatic plant species for phytoremediation of arsenic-contaminated water. *Appl. Water. Sci.* 7, 889-896, 2017.
27. Jasrotia, S., Kansal, A., Kishore, V.V.N., Arsenic phyco-remediation by *Cladophora* algae and measurement of arsenic speciation and location of active absorption site using electron microscopy. *Microchem. J.* 114, 197-202, 2014.
28. Wang, Y., Zheng, Y., Liu, C., Xu, P., Li, H., Lin, Q., Zhang, C., Ge, Y., Arsenate toxicity and metabolism in the halotolerant microalga *Dunaliella salina* under various phosphate regimes. *Environ. Sci.: Processes Impacts.* 18(6), 735-743, 2016.

29. Johansson, C., Paul, N., Nys, R., Roberts, D., Simultaneous biosorption of selenium, arsenic, and molybdenum with modified algal-based biochars. *J. Environ. Manage.* 165, 117-123, 2016.
30. Zhang, S., Sun, G., Yin, X., Rensing, C., Zhu, Y., Biomethylation and volatilization of arsenic by the marine microalgae *Ostreococcus tauri*. *Chemosphere.* 93(1), 47-53, 2013.
31. Lefta, S.N., Determine the ability of *Scenedesmus acuminatus* to uptake of arsenic by using scanning electron microscope technique. *Plant Arch.* 19 (2), 839-842, 2019.
32. Luo, Z., Li, M., Wang, Z., Li, J., Guo, J., Rosenfeldt, R. R., Seitz, F., Yan, C., Effect of titanium dioxide nanoparticles on the accumulation and distribution of arsenate in *Daphnia magna* in the presence of an algal food. *Environ. Sci. Pollut. Res.* 25, 20911–20919, 2018.
33. Zhang, J., Ding, T., Zhang, C., Biosorption and toxicity responses to arsenite (As[III]) in *Scenedesmus quadricauda*. *Chemosphere* 92(9), 1077-1084, 2013.
34. Abioye, O.P., Ezugwu, B.U., Aransiola, S.A., Ojeba, M. I., Phycoremediation of water contaminated with arsenic (As), cadmium (Cd) and lead (Pb) from a mining site in Minna, Nigeria. *Eur. J. Biol. Res.* 10(1), 35-44, 2020.
35. Christobel, J., Lipton, A.P., Evaluation of macroalgal biomass for removal of heavy metal Arsenic (As) from aqueous solution. *Int. J. Innov. Technol. Manag.* 4 (5), 94-104, 2015.
36. Senthilkumar, R., Prasad, D.M., Govindarajan, L., Saravanakumar, K., Naveen Prasad, B. S., Synthesis of green marine algal-based biochar for remediation of arsenic(V) from contaminated waters in batch and column mode of operation. *Int. J. Phytoremediation.* 22(3), 279-286, 2019.
37. Ghosh, S., Bloch, K., Webster, T.J., Bioprospecting of novel algal species with nanobiotechnology. In: Shah, M.P., Couto, S.R., Cruz S.B.V.D.L., Biswas J (Eds.) *An Integration of Phycoremediation Processes for Wastewater Treatment Plant*, Elsevier USA. pp. 41-74, 2021b.
38. Ghosh, S., Selvakumar, G., Ajilda, A.A.K., Webster, T.J., Microbial Biosorbents for Heavy Metal Removal. In: Shah, M.P., Couto, S.R., Rudra, V.K. (Eds.), *New Trends in Removal of Heavy Metals from Industrial Wastewater.* Elsevier B.V., Amsterdam, Netherlands. pp. 213-262, 2021c.
39. Ghosh, S., Sharma, I., Nath, S., Webster, T.J., Bioremediation -The natural solution. In: Shah, M.P., Couto, S.R., (Eds.) *Microbial Ecology of Wastewater Treatment Plants (WWTPs)*, Elsevier. Amsterdam, Netherlands. pp. 11-40, 2021d.
40. Ghosh, S., Joshi, K., Webster, T.J., Removal of Heavy Metals by Microbial Communities. In: Shah, M.P., Couto, S.R., (Eds.) *Wastewater Treatment Reactors*: Microbial Community Structure. Elsevier. Amsterdam, Netherlands. pp. 537-566, 2021e.
41. Ghosh, S., Sarkar, B., Kaushik, A., Mostafavi, E., Nanobiotechnological prospects of probiotic microflora: Synthesis, mechanism, and applications. *Sci. Total Environ.* 838, 156212, 2022.

42. Ghosh, S., Bhagwat, T., Kitture, R., Thongmee, S., Webster, T.J., Synthesis of graphene-hydroxyapatite nanocomposites for potential use in bone tissue engineering. *J. Vis. Exp. e63985* (In Press), 2022.
43. Nitnavare, R., Bhattacharya, J., Thongmee, S., Ghosh, S., Utilizing photosynthetic microbes in nanobiotechnology: Applications and perspectives. *Sci. Total Environ.* 841, *156457*, 2022.
44. Ghosh, S., Webster, T.J. (2021) Nanotechnology for Water Processing. In: *The Future of Effluent Treatment Plants-Biological Treatment Systems,* Shah M.P., Rodriguez-Couto, S., Mehta, K., (Eds.), Elsevier. Amsterdam, Netherlands. pp. 335-360.
45. Ghosh, S. (2020) Toxic Metal Removal Using Microbial Nanotechnology. In: *Microbial Nanotechnology,* Rai, M., Golinska, P. (Eds.), CRC Press, Boca Raton.

4

Plastics, Food and the Environment: Algal Intervention for Improvement and Minimization of Toxic Implications

Naveen Dwivedi[1]*, Pragya Sharma[2] and V.P. Sharma[3]

[1]*Department of Biotechnology, S.D. College of Engineering & Technology, Muzaffarnagar, Uttar Pradesh, India*
[2]*National Dastango (Story Teller)-Environmental Literary Aspects, Aliganj, Lucknow, UP, India*
[3]*CSIR – Indian Institute of Toxicology Research, Lucknow, Uttar Pradesh, India*

Abstract

Various changes in water, food, and the environment, including climate change, are an integral part of civilizations. The recent transitions resulting from the COVID-19 pandemic have disrupted economies, threatening the environment, circular ecosystems, and human health. As a result, state-of-the-art clean-up strategies are being developed to mitigate the potential implications of environmental toxicants/contaminants. In view of this, less plastics pollution in the environment needs to be prioritized through multidisciplinary intervention approaches. These toxic plastic pollutants are created by unsustainable use and indiscriminate disposal of polymeric products. These mismanaged waste and microplastics are a major land-based source of plastic pollution that needs to be minimized via strategic improvements and mass-level behavioral changes. Moreover, the microplastics found in food, water, and air samples have been shown to harm human health. Because the use of plastic materials is an inescapable part of everyday life, researchers have become increasingly concerned, which has led to the increasing demand for eco-friendly polymeric materials for different applications in areas such as health, food packaging, and agriculture. Plastic can be rapidly reduced to its elementary constituents with the help of biological agents. Therefore, plastic degradation using algal species has attracted the attention of researchers. This chapter focuses on

**Corresponding author*: naveen.dwivedi@gmail.com

Charles Oluwaseun Adetunji, Julius Kola Oloke, Naveen Dwivedi, Sabeela Beevi Ummalyma, Shubha Dwivedi, Daniel Ingo Hefft and Juliana Bunmi Adetunji (eds.) Next-Generation Algae: Volume I: Applications in Agriculture, Food and Environment, (109–124) © 2023 Scrivener Publishing LLC

diverse microalgae that promote biodegradation of polymers and their potential role in the removal of pollutants from the environment.

***Keywords*:** Contaminants, economic considerations, food, integrated, toxicants

4.1 Introduction

Organic polymers are used to make plastics, which are extensively employed in a variety of applications such as water bottles, food packaging, biomedical supplies, textiles, and electrical products. They have become a vital, adaptable commodity with a wide range of specifications, compositions, and uses in our daily lives. Plastics were formerly thought to be inert and innocuous, but indiscriminate disposal methods in the environment have gradually led to a slew of related concerns for society as well as a significant environmental burden. This poses a hazard to economies, ecosystems, and human health. Current clean-up initiatives are clearly attempting to mitigate the negative effects of plastic pollution, but they are unable to keep up with the growing volumes. We must prioritize it through a worldwide interdisciplinary approach, and costs may be lowered by improving the life cycle of plastics by lowering manufacturing, cutting consumption, and disposal of garbage in a scientific and sustainable manner via an integrated waste management strategy [1–15, 40–57, 66–76].

The consequences of microplastics vs. their alternatives are determined by a number of elements, including climatic or environmental circumstances, microplastic degradation kinetics, the availability of substitute materials, quality, cost, and the risks of alternative goods. The regulations must be simple to apply and narrow in scope. When alternatives demonstrate competitiveness and environmental friendliness, policymakers must strategize and assess them.

Plastic garbage has been projected to have invaded our ecosystems in the amount of 710 million metric tonnes. It is thought that coordinated global actions are needed first and foremost to reduce plastic consumption, and that effort to increase rates of reuse, recovery through waste collection, and recycling, as well as expansions of safe disposal systems, may help to speed up innovation in the plastic value chain. Plastic pollution is everywhere, and it may be found in the water system, soils, sediments, and the biomass of the atmosphere. Rapid proliferation and usage, in combination with linear economic models that neglect waste externalities, are the main factors.

The most cost-effective ways to manage plastic garbage vary by geography and social situation, and only a few corrective strategies have been offered at the local or regional level. Post-consumption management, biobased alternative goods, reuse, and the creation of novel delivery options are all viable approaches. Environmentally minded nations have implemented prohibitions on single-use plastics and microbeads in cosmetics, and the European Union has passed a rule to limit the international traffic of plastic garbage. Environmentalists, manufacturers, and nongovernmental groups are all striving to find answers, but there is still a lot that needs to be done over the next few decades. Bioplastics are gaining popularity as an environmentally acceptable alternative due to their low carbon footprint, low toxicity, and great degradability.

We might use downstream processing of microalgae for the production of bioplastics by extracting and pre-treating bioactive ingredients like lipids and cellulose. Due to economic effectiveness, little solvent usage, low energy consumption, and quick reaction rates, the intermediate processing of bioplastics via lactic acid generated from microalgae has garnered recognition for the microwave-assisted synthesis of polylactic acid. In terms of techno-economic analysis and degradation process, the dependability and efficacy of microalgae-based bioplastics have been evaluated, and enhancements may be implemented. This is made possible by genetically modifying algae strains, improving biofilm technology, lowering the cost of growth media, and combining avocado seed with microalgae bioplastics. Microalgae and cyanobacteria are well-known potential sources of polyhydroxyalkanoate (PHA), cellulose, carbohydrates, and proteins, as well as other main components of cyanobacteria for the production of bioplastics, according to new research.

The primary purpose of suitable packaging is to protect food/feed quality throughout transit, storage, and shelf-life extension. Active packaging is made up of interactions between the package and the product that limit the development of microbes and slow down the deterioration of quality.

4.2 Constituents of Chemicals in Plastics and Waste Generation

Phthalates, polyfluorinated compounds, bisphenol or equivalents, viz. A, F, and S, brominated flame retardants, and antimony trioxide are only a few of the harmful elements in plastics that can migrate or seep out and have negative impacts on the environment and human health. Because of its huge manufacturing volume and lack of management rules, plastics in electronic garbage (e-waste) have become a global environmental problem. Plastic hazardous compounds are collecting like mountains, according to several publications from industrialized and developing nations, including the EU, Japan, China, Nigeria, Africa, Bangladesh, and India. Several investigations indicate that plastic items in runoff, wastewater, and effluents have heavily contaminated aquatic bodies and seas. Approximately 80% of marine debris comes from land-based sources, with uncollected plastic packaging or polymeric items accounting for more than 60% of total waste. In current epidemic scenarios, the quantity of plastic garbage is shown to be expanding multifold since it is a matter of choice, despite knowledge and understanding that it is detrimental. It is often used for food packing, doorstep delivery, medications, blood transfusions, and personal protective equipment (PPEs). Even the remains of the dead were shrouded with polymeric sheets in specially built coffins during the pandemic. Single-use packaging is the most common type of plastic, accounting for over 40% of all plastic consumption, followed by consumer goods, building materials, automotive, and electrical applications, which account for 22%, 20%, 9%, 6%, and 3% of total plastic usage, respectively.

4.3 Packaging of Food and Minimization Through Concept of ®

Food waste reduction refers to a collection of procedures and activities aimed at reducing the quantity of waste produced. This can be accomplished through minimizing or eliminating the production of hazardous and persistent wastes, as well as encouraging a more sustainable society. It entails rethinking products and processes as well as altering society's consumption and manufacturing patterns [1–45].

4.4 Current World Production Rate of Plastics

Plastic manufacturing is expanding at an exponential rate in many regions of the world. Researchers believe that by 2050, the oceans will contain more plastic than fish in terms of weight. A total of 500 billion plastic bags are used each year, with an estimated 13 million tons ending up in the ocean, killing an estimated 100,000 marine animals. Even e-waste, municipal, and biological waste are on the rise as a result of population growth, changing usage patterns, behavioral approaches, a lack of adequate alternatives, preferences over accessible alternatives, and other societal reasons. Plastic is produced at a worldwide rate of more than 380 million tonnes per year, which may end up as toxins in our natural environment and seas. Mismanaged plastics waste content is disproportionately greater in the African and Asian continents, and by 2060, it is anticipated to have tripled (from 60–99 metric tonnes per year).

4.4.1 Plastics, Food and Packaging to Distribution in Public and Strategic National Boundaries

Packaging accounts for 40% of the entire plastics manufactured since the 1950s, with 41% being used primarily for food, drinks, and emergency supplies, as well as individuals involved in national security at border crossings. Despite worldwide policy concerns, plastics are widely utilized at various phases of the food system, including as agricultural mulch, fishing nets, and boxes for transporting products. The food system, as a whole, is expected to account for a significantly bigger fraction of the world's dependency on plastics than its share of plastic packaging alone. However, there is mounting evidence that the widespread use of plastics and a dependence on them in the global economy, particularly in single-use forms, is having negative environmental and ecological consequences on a local and global scale, with possible health consequences [38–45, 58–76].

Plastics play a crucial role in food transportation, preservation, cleanliness, and safety, extending the shelf life of goods, expanding the length of value chains, and contributing to food and nutrition security. However, the link between significant increases in plastic food packaging and rising food waste implies that, while plastic packaging may help to preserve food, it may not be enough to prevent waste. Recent calls to action on plastics have been prompted in part by observations that the extensive use of single-use or disposable plastics, along with poor recycling and waste management, is adding to visible plastic buildups in natural habitats. We need to evaluate the food system's constitutive subsectors from spade to stomach or farm to flush type phrases to examine the extent in terms of volume usage and variety of exposure outcome relationships, as well as the nature of documentation for the food system implications of plastics on food security, health, and economics at the household level and across geographical boundaries.

Mulching, growth tunnels, greenhouses, and irrigation systems are all examples of plastic in the food chain, as are fishing nets, lines, and traps, as well as product processing, storage, and distribution. Furthermore, automated devices are utilized to convey packing crates, wrapping, and food contact equipment to merchants or warehouses. Packaging is an important aspect of product sales, marketing, and consumption. Furthermore, consumer-level plastic food shopping bags, plastic dinnerware, and cooking equipment are easily accessible. They arrive at disposal collecting facilities and then waste management at the conclusion of the food system. Polymers are also employed in moving belts, suction pipes, compost storage containers, and black plastic bags dumped into landfills for next-step management by municipal integrated management innovators in large commissionaires with several queries through stakeholders.

Plastic mulch used in agriculture may enhance crop yields in the near term, but if the plastic begins to degrade in the soil, it may have an influence on plant development or the soil microbiota in future decades. Plastic shopping bags take 20 years to biodegrade in marine habitats, whereas Styrofoam cups take 50 years, plastic bottles take 450 years, and plastic fishing lines take up to 600 years. The particular human health implications include intestine damage and tissue abrasion from plastic particles, as well as endocrine dysfunction, diabetes, and reproductive difficulties caused by chemicals migrating from plastics. The chemicals that leach from plastic come from a particular human usage in the food system. Plastic food packaging has been studied in relation to foodborne illness pathways, and it is argued whether plastic packaging may help to safeguard human health by keeping food sterile and safe to consume. Ingestion, dermal exposure, or inhalation of plastic or plastic chemical additives are mechanisms by which the relationship between plastics and human health exists, and few studies demonstrate the presence (or absence) of known harmful bacteria and toxicants in foodstuffs or drink as a final outcome, and due to the complexity of determinants, plastic particles or associated chemical additives in food or drink items as an intermediate step on the pathway to human health outcomes.

Food availability, food access, food utilisation, food stability, crop or produce yield, livestock growth and welfare, soil contamination, soil temperature, soil moisture and nutrient content, nutrient leaching, pest or weed control, pesticide or fertilizer use, water use, energy use, plant growth, livestock health and management in dairy systems are all factors that must be taken into account. Food hygiene, sanitation, and preservation is also important, as well as the transportation and storage phases of the food chain, as well as quality inspections and control procedures for FSSAI/IS/ FDA/EU/ ISO WHO [70–76].

4.4.2 Future Projection on Plastic Production

Over the previous several decades, plastic output has expanded by a factor of ten. Plastics manufacturing is expected to reach 367 million metric tonnes in 2020, according to estimates. During the COVID-19 pandemic, there was a 0.3 percent reduction. Packaging accounts for around 26% of all plastics used, and it not only provides immediate economic benefits but also contributes to resource productivity by reducing food waste by increasing shelf life and reducing fuel consumption for transportation due to the low weight of packaging materials. Weathering may further breakdown collected waste plastics in the environment into minute bits such as microplastics and nanoplastics, which will cause greater harm to the environment and humans than large plastics. Plastic particles have been found in water, salts, sediments, and food items, and are consumed by creatures. More than 250 species in the marine environment have been reported to be influenced by plastic entanglement or ingestion.

Algal derivatives are expected to be a promising renewable biomass source for the production of bioplastics and biobased polymeric materials. Algae derivatives may include starch derived from microalgae, polyhydroxyalkanoate derived from cyanobacteria, or polysaccharides derived from marine and freshwater microalgae with the presence of distinct functional groups in algae, such as carboxyl, hydroxyl, and sulphate, manipulated to produce desirable bioplastics specifications suitable for food, pharmaceutical, and medical packaging. Pollution control and simultaneous bioplastics manufacturing may be achieved by standardizing strains, growth conditions, and harvesting and extracting algae in an ecologically acceptable way.

4.5 Toxic Implications of Microplastics from Food Packaging or Other Items

The fundamental driver of constant demand and production for plastics and plastic items, as well as the accumulation of plastic trash and the pollution it causes, is the growing human population. According to the 2016 World Economic Forum research titled "Rethinking the Future of Plastics," various innovations and development attempts show tremendous promise, but they have so far proven to be too fragmented and disorganized to have had a large-scale impact.

Long-term use and high-temperature exposure of polymeric materials can cause harmful chemical ingredients to migrate into food, beverages, and water. The indiscriminate dumping of plastics on land and open-air burning frequently results in harmful compounds being released into the air, posing serious public health risks. Microplastics are the most serious problem in plastic trash. Large amounts of microplastics found in cosmetics and cleaning components, such as toothpaste and microbeads in face-wash, have major environmental consequences. Microplastics are being phased out of various personal care items in numerous countries due to their negative health consequences. According to the findings, the negative effects of microplastics, particularly microbeads, microplastic fibers, and degraded polymeric goods, are many and require special attention.

Object identification or marks can occasionally be used to track macro-debris back to its source. Micro detritus contains bits of bigger plastic consumable items and containers that are difficult to identify using basic approaches. Plastic micro-debris ingestion by filter feeders at the bottom of the food chain is known to happen, but it hasn't been measured. Because polymers are known to absorb hydrophobic contaminants, ingestion of deteriorated plastic pellets and pieces raises toxicological concerns.

The possible bioavailabilities of chemicals added to plastics during manufacturing, as well as those adsorbed from the environment, are complicated challenges that require in-depth research.

The plastic products may degrade gradually, resulting in nano/microparticles with increased surface area and the ability to absorb and concentrate accessible moieties. Persistent organic pollutants (POPs) such as dichlorodiphenyltrichloroethane (DDT), polychlorinated biphenyls (PCBs), and others have the ability to transfer into animal tissues in aquatic settings. Environmental contamination and consequences on aquatic creatures may result from indiscriminate sewage system discharge, as well as health, hygiene, and sanitation concerns. Appropriate actions may be taken in concert to address a variety of socioeconomic, technical, and health concerns [1–25, 45–63].

4.5.1 Biodegradable Polymers

Due to their potential biodegradability and harmlessness, biodegradable polymers (BPs) have become the focus of study in recent years, since they would be the most effective solution to managing the problem of plastic waste accumulation. Plastic debris solutions may be achieved by a variety of measures ranging from analytical proficiency to laws against marine pollution, recycling of plastic items, control of production and consumption of biobased alternatives, and initiatives to clean up existing waste. Individual groups may not be able to complete the hurricane job on their own, and will require public support and initiative, as well as financial support and energy from many people living now and in the future, if a safer and cleaner maritime environment is to be reached for the improvement.

4.5.2 Particulate Matter from Plastics and Implications

A substantial amount of airborne microparticulate pollution is created by the abrasion of tires on road surfaces, meaning that even electric vehicles are not clean despite their elimination of tailpipe particulate matter (PM) of 2.5 and 10 emissions. This is again important from the perspective of pulmonary impacts due to the pandemic severity and its direct relation to pre- and post-COVID-19 situations and climatic drifts. The nanoplastics and microplastics are posing serious environmental issues resulting in severe consequences for ecosystems and human health. It has been established that wastewater treatment plants may act as entrance routes for microplastics to the aquifers due to degradation of larger fragments of the plastic component during the treatment process and are difficult to be removed completely during the process prior to distribution in pipelines for the stakeholders or consumers.

The emergence of nanoplastics in the environment has posed new environmental and health challenges. Focusing on the improved design, manufacture, reuse, and recycling of plastic products is important but, first of all, less consumption of plastic is necessary, particularly in the case of products which come in single-use packaging. We must realize that plastic has certain benefits and limitations in the process of packaging the food items, beverages and pharmaceuticals with few limitations and need to be efficiently managed through an integrated approach within the regulatory framework. In this effort, the role of manufacturers, regulators and stakeholders coexists for safe and innovative packaging solutions for the benefit of society.

4.6 Conclusion

From the point of production to the point of consumption, excellent food must be safely packaged and supplied in suitable packaging. To avoid associated environmental concerns connected with packaging waste, the public must be informed of the regulations and dispose of food packaging in accordance with them. This will help most contemporary food supply chains comply with ISO 22000 and HACCP food management system criteria. The food production and consumption system is gradually shifting toward sustainability by reducing food waste in order to attain a worldwide diet that is healthier and more sustainable. Land ecosystems are being degraded, marine habitats are being overexploited, and agricultural methods are being improved. A biobased value-added biopolymer with increased biodegradability, biocompatibility, and other chemo-mechanical characteristics has been discovered in a few specific formulations.

New commercial ventures are attempting to have industrial production of microbial-generated polymers using state-of-the-art extraction technologies. The algae are photoautotrophic and have very low nutritional needs for growth. Bio-based polymers, such as polyhydroxyalkanoates (PHAs) and others, offer a lot of potential as a sustainable and environmentally friendly alternative to synthetic plastics in the coming years. In light of the sustainable development goals (SDGs), they might be enhanced with new concepts and used as a preferred material. Plastics may present health concerns by indirect or direct additives migrating from food storage, transit, or usage under unfavorable environmental conditions. Plastic containers and packaging may leach toxic chemicals into stored food when heated. Bisphenol, heavy metals, and phthalates are all well-known substances that affect the endocrine system. If available over the allowed levels, they may have direct and indirect effects on the environment and human health. Recognizing the interdependence of humans and plastics, particular effort is taken to ensure that plastic manufacture, consumption, and disposal are all sustainable.

We may develop richer infrastructure by designing better policies and processes for interventions, engaging young researchers to participate in pollution mitigation initiatives, and designing better rules and procedures for increased sustainable growth. From seed to stomach or farm to forks, quality should be an important aspect of good food and packaging. Through the use of green plastic and degradable bio-based plastic goods, powerful behavioral change tactics encourage companies and people to reduce plastic contamination and increase pollution load.

References

1. Andrady A, an environmental primer. In *Plastics and the environment* (Andrady A., ed.), pp. 3–76 Hoboken, NJ: Wiley Interscience, 2003.

2. Andrady AL & Neal MA, Applications and societal benefits of plastics. *Phil. Trans. R. Soc. B* 364, 1977–1984, 2009. https://doi.org/10.1098/rstb.2008.0304.
3. APME, an Analysis of Plastics Production, Demand and Recovery in Europe-The Compelling Facts About Plastics, An analysis of plastics production, demand and recovery in Europe, 2008.
4. Barboza, L. G. A., Dick Vethaak, A., Lavorante, B. R. B. O., Lundebye, A. K., & Guilhermino, L., Marine microplastic debris: An emerging issue for food security, food safety and human health. *Marine Pollution Bulletin*, 133, 336–348, 2018. https://doi.org/10.1016/j.marpolbul.2018.05.047
5. Barnes DKA, Galgani F, Thompson RC & Barlaz M, Accumulation and fragmentation of plastic debris in global environments. *Philosophical Transactions of the Royal Society B: Biological Sciences.* 364 (1526), 1985–1998, 2009. https://doi.org/10.1098/rstb.2008.0205.
6. Batchelor, T. EU proposes total ban on plastic cutlery and straws to reduce single-use litter | The Independent. 2018. Retrieved from https://www.independent.co.uk/news/world/europe/plastic-ban-straws-eu-cutlery-cotton-buds-single-use-uk-environment-a8373351.html
7. Biello & David, Are Biodegradable Plastics Doing More Harm Than Good? Scientific American, 2011. Retrieved 1 August 2013.
8. Bokeloh, G., Gerster-Bentaya, M., & Weingärtner, L. Achieving food and nutrition security: Actions to meet the global challenge. 2005. Retrieved from www.inwent.org
9. Chae, Y., & An, Y. J. Current research trends on plastic pollution and ecological impacts on the soil ecosystem: A review. *Environmental Pollution*, 240, 387– 395, 2018. https://doi.org/10.1016/j.envpol.2018.05.008
10. Claudio, L. Our food: Packaging & public health. *Environmental Health Perspectives*, 120(6), A232– A237, 2012. https://doi.org/10.1289/ehp.120-a232
11. Dodds EC & Lawson W, Synthetic estrogenic agents without the phenanthrene nucleus. *Nature* 137, 996, 1936. https://doi.org/10.1038/137996a0.
12. Duncan, E. M., Botterell, Z. L. R., Broderick, A. C., Galloway, T. S., Lindeque, P. K., Nuno, A., & Godley, B. J. A global review of marine turtle entanglement in anthropogenic debris: A baseline for further action. *Endangered Species Research*, 34, 431–448, 2017. https://doi.org/10.3354/esr00865
13. Efferth, T., & Paul, N. W. Threats to human health by great ocean garbage patches. *The Lancet. Planetary Health*, 1(8), e301– e303, 2017. https://doi.org/10.1016/S2542-5196(17)30140-7
14. Eskenazi B, Warner M, Samuels S, Young J & Gerthoux PM, Serum dioxin concentrations and risk of uterine leiomyoma in the Seveso Women's Health Study. *American Journal of Epidemiology* 166, 79–87, 2007. https://doi.org/10.1093/aje/kwm048.
15. European Commission. Proposal for a directive of the European Parliament and of the council on the reduction of the impact of certain plastic products on the environment. 2018/0172(COD).5.

16. Ellen MacArthur Foundation. The new plastics economy catalysing action, 2017.
17. Filho, W. L. *et al.* An overview of the problems posed by plastic products and the role of extended producer responsibility in Europe. *J. Clean. Prod.* 214, 550–558, 2019.
18. FE Online. As Delhi bans plastic, look at interesting ways other states have implemented it – The financial express, 2017. Retrieved from https://www.financialexpress.com/india-news/as-delhi-bans-plastic-look-at-interesting-ways-other-states-have-implemented-it/840129/
19. Food and Agriculture Organization of the United Nations. Food systems for better nutrition. The State of Food and Agriculture, 2013. https://doi.org/10.1155/2015/638635
20. Food and Agriculture Organization of the United Nations. Submission by the Food and Agriculture Organization of the United Nations (FAO) to the United Nations Framework Convention on Climate Change (UNFCCC) on issues relating to agriculture: Adaptation measures, 2014. Retrieved from http://www.fao.org/3/a-i5188e.pdf
21. Gao, H., Yan, C., Liu, Q., Ding, W., Chen, B., & Li, Z. Effects of plastic mulching and plastic residue on agricultural production: A meta-analysis. *Science of the Total Environment*, 651, 484–492, 2019. https://doi.org/10.1016/j.scitotenv.2018.09.105
22. Geyer, R., Jambeck, J. R., & Law, K. L. Production, use and fate of all plastics ever made. *Science Advances*, 3(7), 19– 24, 2017. https://doi.org/10.1126/sciadv.1700782
23. Gray, J. M., Rasanayagam, S., Engel, C., & Rizzo, J. State of the evidence 2017: An update on the connection between breast cancer and the environment. *Environmental Health*, 16(1), 94, 2017. https://doi.org/10.1186/s12940-017-0287-4
24. Halden RU, Plastics and health risks. *Annu Rev Public Health* 31(1), 179–194, 2010. https://doi.org/10.1146/annurev.publhealth.012809.103714.
25. Haque, M. A., Jahiruddin, M., & Clarke, D. Effect of plastic mulch on crop yield and land degradation in south coastal saline soils of Bangladesh. *International Soil and Water Conservation Research*, 6(4), 317– 324, 2018. https://doi.org/10.1016/j.iswcr.2018.07.001
26. Ikezuki Y, Tsutsumi O, Takai Y, Kamei Y & Taketani Y, Determination of bisphenol a concentrations in human biological fluids reveals significant early prenatal exposure. *Hum. Report* 17, 2839–2841, 2002. https://doi.org/10.1093/humrep/17.11.2839.
27. Laurent Lebreton and Anthony Andrady, Future scenarios of global plastic waste generation and disposal, Nature Human and social science communications, *Palgrave Communications* volume 5, Article number: 6, 2019; https://www.nature.com/articles/s41599-018-0212-7(2019).
28. Wilcox C, Van Sebille E, Hardesty BD, Threat of Plastic Pollution to Seabirds is Global, Pervasive, and Increasing, PNAS Early Edition 1, 4, 2015.

29. Kamrin MA, Phthalate risks, phthalate regulation, and public health: a review. *J. Toxicol. Environ. Health B* 12,157–74, 2009. https://doi.org/10.1080/10937400902729226.
30. Kang JH, Kito K & Kondo F, Factors influencing the migration of bisphenol A from cans. *J. Food Prot.* 66, 1444–1447, 2003. https://doi.org/10.4315/0362-028X-66.8.1444.
31. Karami, A., Golieskardi, A., Keong Choo, C., Larat, V., Galloway, T. S., & Salamatinia, B. The presence of microplastics in commercial salts from different countries. *Scientific Reports*, 7, 46173, 2017. https://doi.org/10.1038/srep46173
32. Lakind JS & Naiman DQ, Bisphenol A (BPA) daily intakes in the United States: estimates from the 2003–2004, NHANES urinary BPA data. *J. Expo. Sci. Environ. Epidemiol.* 18, 608–15, 2008. https://doi.org/10.1038/jes.2008.20.
33. Law, K. L. Plastics in the marine environment. *Annual Review of Marine Science*, 9(1), 205–229, 2017. https://doi.org/10.1146/annurev-marine-010816-060409
34. Lusher, A. L., McHugh, M., & Thompson, R. C., Occurrence of microplastics in the gastrointestinal tract of pelagic and demersal fish from the English Channel. *Marine Pollution Bulletin*, 67(1–2), 94–99, 2013. https://doi.org/10.1016/J.MARPOLBUL.2012.11.028
35. Meeker JD, Sathyanarayana S & Swan SH, Phthalates and other additives in plastics: human exposure and associated health outcomes. *Phil. Trans. R. Soc. B* 364, 2097–2113, 2009. https://doi.org/10.1098/rstb.2008.0268.
36. Murray R. Gregory, Environmental Implications of Plastic Debris in Marine Settings- Entanglement, Ingestion, Smothering, Hangers- on, Hitch- Hiking and Alien Invasions, 2009, available at http://rstb.royalsocietypublishing.org/content/364/1526/2013.; Derraik, supra at 844.
37. Nielsen, T. D., Holmberg, K. & Stripple, J. Need a bag? A review of public policies on plastic carrier bags – Where, how and to what effect? *Waste Manag.* 87, 428–440, 2019.
38. OECD. Core environmental indicators. Development measurement and use, 2003. https://doi.org/10.1016/j.infsof.2008.09.005
39. Pan American Health Organization. Health, Environment and Sustainable Development: Towards the Future We Want A collection of texts based on the PAHO Seminar Series towards Rio+20 that occurred in the period between 8 February and 13 June 2012, 2013. Retrieved from https://www.paho.org/hq/dmdocuments/2013/seminario-rio-20-eng.pdf
40. Petersen, S. O., Blanchard, M., Chadwick, D., Prado, A. Del, Edouard, N., & Mosquera, J. Manure management for greenhouse gas mitigation. *Animal*, 7(s2), 266– 282, 2013. https://doi.org/10.1017/S1751731113000736
41. Plastics Europe, the compelling facts about Plastics 2007: an analysis of plastics production, demand and recovery for 2007 in Europe. Brussels, Belgium: Plastics Europe, 2008.

42. Rahman M & Brazel CS, the plasticizer market: an assessment of traditional plasticizers and research trends to meet new challenges. *Prog. Polym. Sci.* 29, 1223–48, 2004. https://doi.org/10.1016/j.progpolymsci.2004.10.001.
43. Raloff J, Food for thought: What's coming out of baby's bottle? *Sci. News Online* 156, 1–4, 1999.
44. Rancière, F., Lyons, J. G., Loh, V. H. Y., Botton, J., Galloway, T., Wang, T., Magliano, D. J. Bisphenol a and the risk of cardiometabolic disorders: A systematic review with meta-analysis of the epidemiological evidence. *Environmental Health*, 14, 2015.
45. Rastkari, N., Jeddi, M. Z., Yunesian, M., & Ahmadkhaniha, R. Effect of sunlight exposure on phthalates migration from plastic containers to packaged juices. *Journal of Environmental Health Science and Engineering*, 16(1), 27–33, 2018. https://doi.org/10.1007/s40201-018-0292-8
46. Rayner JL, Wood C & Fenton SE, Exposure parameters necessary for delayed puberty and mammary gland development in Long-Evans rats exposed in utero to atrazine. *Toxicol. Appl. Pharmacol.* 195, 23–34, 2004. https://doi.org/10.1016/j.taap.2003.11.005.
47. Relton, C., Strong, M., & Holdsworth, M. Plastic food packaging encourages obesity. *BMJ (Clinical Research Ed.)*, 344, e3824, 2012. https://doi.org/10.1136/bmj.e3824
48. Revel, M., Châtel, A., &Mouneyrac, C. Micro(nano)plastics: A threat to human health? *Current Opinion in Environmental Science & Health*, 1, 17–23, 2018. https://doi.org/10.1016/j.coesh.2017.10.003
49. Royer, S. -J., Ferrón, S., Wilson, S. T., & Karl, D. M. Production of methane and ethylene from plastic in the environment. *PLoS One*, 13(8), e0200574, 2018. https://doi.org/10.1371/journal.pone.0200574
50. Rudel RA, Dodson RE, Newton E, Zota AR & Brody JG, Correlations between urinary phthalate metabolites and phthalates, estrogenic compounds 4-butyl phenol and o-phenyl phenol, and some pesticides in home indoor air and house dust. *Epidemiology* 19, S332, 2008.
51. Sathyanarayana S, Phthalates and children's health. *Curr. Probl. Pediatr. Adolesc. Health Care*, 38, 34–49, 2008. https://doi.org/10.1016/j.cppeds.2007.11.001.
52. Schonfelder G, Wittfoht W, Hopp H, Talsness CE, Paul M &Chahoud I, Parent bisphenol A accumulation in the human maternal-fetal-placental unit. *Environ. Health Perspect.* 110, 703–707, 2002. https://doi.org/10.1289/ehp.021100703.
53. Schweitzer, J.-P., Gionfra, S., Pantzar, M., Mottershead, D., Watkins, E., Petsinaris, F., … Janssens, C. Unwrapped: How throwaway plastic is failing to solve Europe's food waste problem (and what we need to do instead). Brussels, 2018. Retrieved from www.rethinkplasticalliance.eu/
54. Smith, M., Love, D. C., Rochman, C. M., & Neff, R. A. Microplastics in seafood and the implications for human health. *Current Environmental Health Reports*, 5(3), 375– 386, 2018. https://doi.org/10.1007/s40572-018-0206-z

55. Statista. Production of plastics worldwide from 1950 to 2017 (in million metric tons), 2018. Retreived from https://www.statista.com/statistics/282732/global-production-of-plastics-since-1950
56. Statistica, NOAA, & Grant, Woods Hole Sea. Plastic can take 500 years to bio-degrade in the ocean, 2018. https://www.statista.com/chart/15905/the-estimated-number-of-years-for-selected-items-to-bio-degrade/
57. Steinmetz, Z., Wollmann, C., Schaefer, M., Buchmann, C., David, J., Tröger, J., chaumann, G. E. Plastic mulching in agriculture. Trading short-term agronomic benefits for long-term soil degradation. *Science of the Total Environment*, 550, 690– 705, 2016. https://doi.org/10.1016/j.scitotenv.2016.01.153
58. Sustainable Packaging Coalition. 101: Resin identification codes. 2017. Retrieved from https://sustainablepackaging.org/101-resin-identification-codes/
59. The Lancet Planetary Health. Microplastics and human health-an urgent problem. *The Lancet. Planetary Health*, 1(7), e254, 2017. https://doi.org/10.1016/S2542-5196(17)30121-3
60. Thompson, R. C., Moore, C. J., vom Saal, F. S., & Swan, S. H. Plastics, the environment and human health: current consensus and future trends. *Philosophical Transactions of the Royal Society of London*, 364(1526), 2153–2166, 2009. https://doi.org/10.1098/rstb.2009.0053
61. Tricco, A. C., Lillie, E., Zarin, W., O'Brien, K. K., Colquhoun, H., & Levac, D., *et al.* PRISMA Extension for scoping reviews (PRISMA-ScR): checklist and explanation. *Ann Intern Med*, 169, 467– 473, 2018. https://doi.org/10.7326/M18-0850, 2018.
62. Tyree, C., & Morrison, D. Invisibles: The plastic inside us. 2018. Retrieved from https://orbmedia.org/stories/invisibles_plastics?
63. UNEP. SINGLE-USE PLASTICS: A roadmap for sustainability, 2018. Retrieved from http://wedocs.unep.org/bitstream/handle/20.500.11822/25496/single use plastic_sustainability.pdf?sequence
64. UN resolution pledges to plastic reduction by 2030 https://www.bbc.co.uk/news/science-environment-47592111. Accessed 21st August 2019.
65. V. Guillard *et al.* The Next Generation of Sustainable Food Packaging to Preserve Our Environment in a Circular Economy Context, *Frontiers in Nutrition*, 04 December 2018. https://doi.org/10.3389/fnut.2018.00121.
66. Valderrama Ballesteros, L., Matthews, J. L., & Hoeksema, B. W. Pollution and coral damage caused by derelict fishing gear on coral reefs around Koh Tao, Gulf of Thailand. *Marine Pollution Bulletin*, 135, 1107– 1116, March 2018. https://doi.org/10.1016/j.marpolbul.2018.08.033
67. Vandenberg LN, Maffini MV, Sonnenschein C, Rubin BS & Soto AM, Bisphenol-A and the great divide: a review of controversies in the field of endocrine disruption. *Endocr. Rev.* 30, 75–95, 2009. https://doi.org/10.1210/er.2008-0021.

68. Van der Linden, S. Warm glow is associated with low- but not high-cost sustainable behavior. *Nat. Sustainability*, 1, 28–30, 2018.
69. Vogel, S. A. The politics of plastics: The making and unmaking of bisphenol a "safety". *American Journal of Public Health*, 99(Suppl 3), S559– S566, 2009.
70. Vom Saal FS & Hughes C, an extensive new literature concerning low-dose effects of bisphenol A shows the need for a new risk assessment. *Environ. Health Perspect.* 113, 926–33, 2005. https://doi.org/10.1289/ehp.7713.
71. Wagner M & Oehlmann J, Endocrine disruptors in bottled mineral water: total estrogenic burden and migration from plastic bottles. *Environ. Sci. Pollut. Res.*, 16, 278–286, 2009. https://doi.org/10.1007/s11356-009-0107-7.
72. Warner M, Eskenazi B, Mocarelli P, Gerthoux PM & Samuels S, Serum dioxin concentrations and breast cancer risk in the Seveso Women's Health Study. *Environ. Health Perspect.* 110, 625–28, 2002. https://doi.org/10.1289/ehp.02110625.
73. Welshons WV, Thayer KA, Judy BM, Taylor JA, Curran EM & Vom Saal FS, large effects from small exposures. I. Mechanisms for endocrine-disrupting chemicals with estrogenic activity. *Environ. Health Perspect.* 111, 994–1006, 2003. https://doi.org/10.1289/ehp.5494.
74. Wilson NK, Chuang JC, Morgan MK, Lordo RA & Sheldon LS, An observational study of the potential exposures of preschool children to pentachlorophenol, bisphenol-A, and nonylphenol at home and daycare. *Environ. Res.* 103, 9–20, 2007. https://doi.org/10.1016/j.envres.2006.04.006.
75. World Health Organization. WHO estimates of the global burden of food borne diseases, 2015. http://apps.who.int/iris/bitstream/handle/10665/199350/9789241565165_eng.pdf?sequence
76. World Health Organization. Food Safety, 2019. Retrieved from http://www.who.int/news-room/fact-sheets/detail/food-safety.

5

Role of Algae in Biodegradation of Plastics

Piyush Gupta[1]*, Namrata Gupta[2], Subhakanta Dash[3] and Monika Singh[2]

[1]Department of Chemistry, Faculty of Engineering and Technology, SRM Institute of Science and Technology, Ghaziabad, Uttar Pradesh, India
[2]RBS Engineering Technical Campus, Agra, Uttar Pradesh, India
[3]Department of Chemistry, Synergy Institute of Engineering and Technology, Dhenkanal, Odisha, India

Abstract
The use of plastic materials has become indispensable in our society. Therefore, plastic waste continuously accumulates in the environment, which endangers every form of life. It typically jams sewage pipes and floats across agricultural areas on rivers, streams, canals, ponds, lakes, and seas. Because a better technique to break down plastic has yet to be discovered, plastic disposal continues to be a significant concern. Chemical treatment, landfilling, and incineration are some of the techniques used in polymer degradation that have a detrimental effect on ecosystems. Algae-based biodegradation is a promising method for the environmentally benign break down of plastic waste. During the process of mineralization, algae transforms plastic trash into metabolites such as water and carbon dioxide as well as new biomass, which are important to the environment. Biodegradation is triggered by the interaction of algae enzymes with plastic polymers, which results in the weakening of chemical bonds. When it comes to deterioration, the significant factors to consider are temperature, availability of light, oxygen, and moisture. Providing insight into current practices and perspectives on bioremediation and the role played by microorganisms such as algae in the breakdown of plastics is the goal of this chapter.

***Keywords*:** Plastic, pollution, biodegradation, algae

**Corresponding author*: piyushg.1977@gmail.com

Charles Oluwaseun Adetunji, Julius Kola Oloke, Naveen Dwivedi, Sabeela Beevi Ummalyma, Shubha Dwivedi, Daniel Ingo Hefft and Juliana Bunmi Adetunji (eds.) Next-Generation Algae: Volume I: Applications in Agriculture, Food and Environment, (125–146) © 2023 Scrivener Publishing LLC

5.1 Introduction

As of today, the use of various plastics is an unavoidable aspect of human existence, and therefore they continue to accumulate in the environment and pose a global environmental danger. This accumulation may be hazardous to humans and nature if left unmonitored. Polymers like polyethylene terephthalate (PET) and polyvinyl chloride (PVC) are thermoplastic polymers that are mainly used in the manufacture of packaging goods. PET is highly recyclable material with a density of 1.41 g/cm^3, generally used in single-use plastic bottles for mineral water and soft beverages [1]. Low-density polyethylene (LDPE) and high-density polyethylene (HDPE) are the most prevalent types of plastic trash [2].

Every year, it is claimed that 500 billion polymer bags are used across the world. This extensive adoption can be due to the low cost and ease of use of these products. A majority of these bags are discarded as waste after being used only once. Plastic bags have been estimated to persist in the environment for as much as a thousand years before degrading. The accumulation of plastic bag waste results in environmental contamination, which can present itself in a variety of ways. In recent years, biological degradation has gained popularity as an innovative and eco-friendly method for plastic waste management. There are many advantages of this method over traditional decomposition, including the fact that it is usually less costly and has the potential to be considerably more efficient, as well as the fact that it does not create any dangerous chemicals. Polyethylene (PE) sheets are also susceptible to microbial deterioration and may result in the production of commercial end products from the biomass produced by the bacteria [3, 4].

Because of their non-biodegradable nature, standard means of polymer degradation, such as waste disposal, burning, and chemical treatment, are equally detrimental to nature and man's well-being. At large scale, degradative procedures, such as landfilling, burning, and recycling, are being used in conjunction with one another. The direct combustion of plastics releases harmful gases, such as heavy metals, carbon monoxide, nitrogen oxides, sulfur oxides, and dioxins, all of which contribute significantly to the loss of the ozone layer. Dioxins can cause substantial difficulties with the operation of the human endocrine system, making them a major concern for human health [5]. They can also cause extreme soil contamination, which then pollutes the environment. Moreover, as a result of the residues left behind after these treatments, considerable environmental harm has occurred. Using a landfill results in the discharge of hazardous gases into the environment, and it also has the added disadvantage of necessitating the use of enormous amounts of land space. Plastic recycling contributes to the decrease of environmental problems connected with garbage and incineration; nevertheless, the process is inefficient and results in a loss in quality of produced polymer. Moreover, the procedure is cost-effective to a lesser extent, reducing the motivation to invest in recycling plants.

Biodegradation is an effective method of degrading plastic waste in an environmentally beneficial manner. In recent years, significant attention has been paid to the biodegradability of polymer materials because of the damage done to the environment by plastic trash. However, there has yet to be developed a protocol for the feasible degradation of PE by commercial biodegradation, which would be beneficial. PE polymers are composed of C and H atoms and are extremely sturdy to microbial degradation. Thermal degradation is highly reliant on temperature, light exposure, oxygen, and the presence or absence of moisture in the environment. The two options for reducing "plastic waste" are: (a) the creation of biodegradable commodities produced from fossil fuels and/or renewable resources, as well as the re-engineering of commodity polymers with a complete carbon backbone (oxo-biodegradable); and (b) the identification of potential microalgae and bacteria, as well as their toxins, in order to establish the protocol for effective biodegradation of a variety of plastics.

Plastic may be rapidly degraded to its constituent parts and recycled with the help of biological agents [6]. Various mechanisms, including chemical, thermal, optical, and biological degradation, are involved in this process [7, 8]. When microbes cling to plastic surfaces, extracellular enzymes produced by the organisms degrade the latter, causing physical and chemical changes [9]. Developing microorganisms, such as indigenous microalgae, that use the polymer of a plastic as a carbon source will lead plastic to completely degrade in the presence of oxygen, resulting in the creation of CO_2 and biomass, which will be utilized in the manufacturing of an end product [10–13].

In this chapter, the use of different microalgae is presented to convert plastics into product as CO_2, H_2O, and cell biomass, which is known as mineralization. The methods can be classified into two groups: (i) metabolism dependent, also known as bioaccumulation; and (ii) metabolism independent, also known as passive biosorption [14].

5.2 What are Microalgae?

Microalgae are microorganisms that can photosynthesize and yield proteins, carbohydrates, and lipids. They can be prokaryotic or eukaryotic in nature. Because of their unicellular or basic multicellular structure, they may develop quickly and survive in severe environments. Eukaryotic microalgae include diatoms (Bacillariophyta) and green algae (Chlorophyta), whereas prokaryotic include Cyanobacteria (Cyanophyceae). Microalgae may be found in all existing earth ecosystems, not just in terrestrial but also in aquatic environments, and they constitute a diverse spectrum of species that can survive in an extensive range of circumstances. There are more than 50,000 species on the globe, but only roughly 30,000 of them have been studied, tested and evaluated [15]. The most important needs for developing algae are sunlight, water, nutrients, and fertile land. Algae are more active at using sunlight than terrestrial plants, and they absorb hazardous pollutants. They have minimal resource requirements, and they also do not compete with food/agriculture for limited precious resources [16].

5.3 Some Biodegradable Pollutants

Fuels, polycyclic aromatic hydrocarbons (PAHs), polychlorinated biphenyls (PCBs), insecticides, and colors are all examples of chemicals that fall under this category. Hydrocarbons are the organic molecules having structures made up of hydrogen and carbon, according to their chemical composition. Alkanes are the most common form of the aliphatic group, whereas alkenes (C_6H_6) are the most common form of the aromatic group [17]. PAHs as organic pollutants, are known as hydrophobic organic contaminants (HOCs). Industries are responsible for a large portion of PAH contamination [18]. PAHs have the potential to attach to organically rich soils and sediments, accumulate in fish and other aquatic organisms, and can be transmitted to people through seafood consumption.

PCBs have been used in different commercial and industrial applications. They are potentially harmful, endocrine disrupting chemicals that may lead to cancer if they are consumed in large quantities [19].

Chemical compounds or mixtures of chemicals used to prevent, kill, repel, or mitigate the presence of pests are referred to as "pesticides." Pesticides that dissolve quickly are referred to as nonpersistent, while pesticides which are resistant to degradation are referred to as persistent pesticides. A frequent form of degradation is carried out by microorganisms in soil, particularly fungi and bacteria, which consume pesticide like sources of nutrition.

A number of industries, including textile, rubber product manufacturing sectors, paper and printing industries, pharmaceutical and cosmetic businesses, utilize dyes [20]. Azo dyes are the aromatic compounds with one or more azo (–N=N–) groups. Because of their structural characteristics, these azo dyes are poorly biodegradable, and treatment of wastewater containing dyes often includes physical and/or chemical techniques such as coagulation-flocculation, adsorption, filtration, oxidation, electrochemical methods, etc. [21]. When it comes to color removal from effluent, the effectiveness of a biological process relies on use of microorganisms which are capable of decolorizing synthetic dyes with a variety of chemical structures.

Heavy metals, in contrast to organic contaminants, cannot be eliminated; instead, they must be converted into a stable form or removed from the environment in order to be eradicated. Biotransformation is the process by which metals are cleaned up by bacteria. Heavy metals are acted upon by microorganisms through various mechanisms, including biosorption (binding of metal to cell surface by the mechanism of physicochemical reactions), bioleaching (heavy metals mobilization by the excretion of non-minerals acids or by the process of methylation), biomineralization (heavy metals immobilization throughout the formation of insoluble sulfides) [22, 23].

5.4 Overview of Plastics

Chemically synthesized organic polymers, often known as plastics, are long molecules (monomers) connected via covalent bonding and have properties such as hydrophobicity, inertness, and high molecular weight [24]. Single-use products, such as household goods, packaging and shopping bags, may be manufactured using their flexible and moldable qualities. They are also robust and durable while being lightweight and affordable. Using heavy crude oil to make traditional plastics might contribute to concerns such as resource exhaustion, changes in the climate and greenhouse gas emissions [25]. Although plastics' biodegradability is determined by their chemical structure, it is not determined by the sources from which the monomers are extracted. The commonly used plastics, such as PET, PVC, HDPE, LDPE, PS, and PP, as well as miscellaneous plastics, are non-biodegradable. These polymers have a high molecular weight and are hence durable due to the repetition of small monomer units [7].

Currently available on the market are four different kinds of degradable plastics, including compostable plastics, biobased plastics, photodegradable plastics, and biodegradable plastics. The breakdown of biodegradable plastics is typically triggered by interactions with water, enzymes, ultraviolet light, and pH changes. Polyhydroxyalkanoate (PHA) is an example of a biodegradable material that is entirely biodegradable and possesses characteristics that are comparable to those of conventional polymers.

5.5 Bioremediation of Plastics

The process of biodegradation uses microbial activity to transform organic substrate (polymers) into tiny molecular weight fragments which may be degraded to H_2O and CO_2 [26–29]. The physicochemical characteristics of a polymer are essential in the process of biodegradation. The biodegradation capability obtained by the microbes is closely linked to the molecular weight and crystalline behavior of polymers. Exo-enzymes, which are enzymes that work outside the cell and have a broad range of activity, from conversion oxidative to hydrolytic capabilities, are involved in polymer breakdown. Their effect on the polymer may be characterized as depolymerization in broad terms. Exo-enzymes typically breakdown complicated polymer structures into smaller, simpler components that may be taken up by a microbial cell to finish the degradation process (Figure 5.1).

Identifying prospective algae and their toxins that are capable of degrading polymeric materials biologically is one method that may be used to reduce "white pollution." Septic water contains algae that have been shown to grow on artificial substrata such as polythene surfaces (Table 5.1).

These colonizing algae were less harmful and nontoxic [13]. In order for plastic to degrade, algae must adhere to the surface of the water and produce ligninolytic and exopolysaccharide enzymes. The synthesis of these enzymes is critical for the decomposition of plastic [32]. Biodegradation occurs when algal enzymes in the liquid medium interact with macromolecules on the plastic surface [36]. Because algae that grow on PE surfaces have higher cellular contents (carbohydrates and proteins) and a higher specific growth rate, it's assumed that the polymer is being used as a carbon source by the algae [31, 34]. The biodegradation of leaching components, as well as the coloring of pigments through diffusion into polymers have all been reported in previous studies. These include fouling,

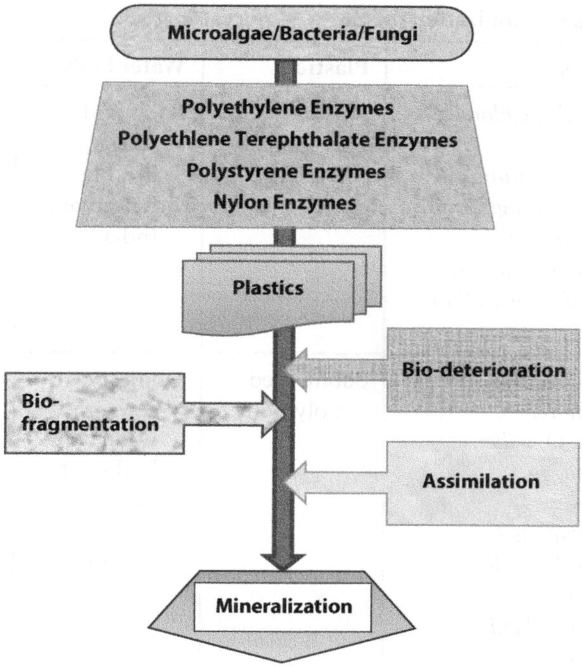

Figure 5.1 Algal colonization and biodegradation process.

Table 5.1 Algae colonization on plastic surface.

Algae species	Plastic	Water body	Reference
Coleochaete scutata, Coleochaete soluta, Chaetophora, Aphanochaete, Gloeotaenium, Oedogonium, Oocystis, Oscillatoria, Phormidium, and *Chroococcus*	Polythene	Oligotrophic water bodies of Lucknow, Uttar Pradesh	[30]
Anabaena spiroides (blue-green alga), *Navicula pupula,* and *Scenedesmus dimorphus* (green alga)	Dumped waste polyethylene bags	Sewage in Chennai City, Tamil Nadu	[31]

(*Continued*)

Table 5.1 Algae colonization on plastic surface. (*Continued*)

Algae species	Plastic	Water body	Reference
Amphora ovalis, Chlorella vulgaris, Closterium constatum, Microcystis aeruginosa, Monoraphidium contortum, Oscillatoria tenuis, and *Phormidium tenue*	Waste polythene Materials	Pools and lakes in Kota Rajasthan, India	[13]
Oscillatoria princeps, O. acuminate, O. subbrevis, O. willei, O. amoena, O. splendida, O. vizagapatensis, O. limnetica, O. earlei, O. peronata, O. formosa, O. okeni, O. geitleriana, O. limosa, O. chalybea, O. salina, O. rubescens, O. curviceps, O. tenuis, and *O. laete-virens*	Submerged polythene	Domestic sewage water bodies of Silchar town, Assam	[32–35]
Phormidium lucidum, Oscillatoria subbrevis, Lyngbya diguetii, Nostoc carneum, and *Cylindrospermum muscicola*	Dumped waste polythene bags	Domestic sewage water drains of Silchar town, Assam	[32–35]

corrosion, hydrolysis, and penetration. The blue-green alga *Anabaena spiroides* (Cyanobacterium) showed highest percentage of LDPE breakdown (8.18%) [31]. Sarmah and Rout [35] discovered that readily available, easily isolable, and fast-growing freshwater nontoxic cyanobacteria (*Oscillatoria subbrevis* and *Phormidium lucidum*) are capable of colonizing the surface of PE and biodegrading LDPE efficiently without any pretreatment or use of pro-oxidant additives.

Bisphenol A (BPA), a polymer that is extensively used in the plastics sector, was biodegraded employing *Aeromonas hydrophila* bacteria and *Chlorella vulgaris* microalgae. BPA was rapidly degraded using algae, and after 168 hours the levels were below detection limits, showing that it did not have estrogenic effects, as previously reported [37]. Hirooka *et al.* [38] discovered that the green alga *Chlorella fusca* var. *vacuolata* degraded BPA into non-estrogenic compounds. In addition to the various applications, microalgae can be genetically modified to function as a microbial cell factory capable of generating and secreting enzymes that break down plastic waste. The green microalgae *Chlamydomonas reinhardtii* was transformed for PETase, and its cell lysate was incubated with PET, resulting in dents and holes on the surface of the PET film, as well as the formation of TPA, which is the PET's completely decomposed form [39]. The researchers utilized *P. tricornutum* to develop PETase effectively, which demonstrated catalytic activity against copolymer polyethylene terephthalate glycol and PET plastics. These research studies have offered potential ecologically acceptable methods for biologically degrading PET via microalgae and synthetic biology, which is now being explored.

5.6 Microalgae's Effect on Microplastics

In aquatic environments, interactions between microalgae and plastic waste have the potential to cause considerable changes in the characteristics of these polymers, which could have major implications for their long-term fate [40]. The various processes that have been documented in the literature can be divided into two categories: plastic polymer alteration and/or biodegradation, and changes in polymer density and sinking behavior (Figure 5.2a).

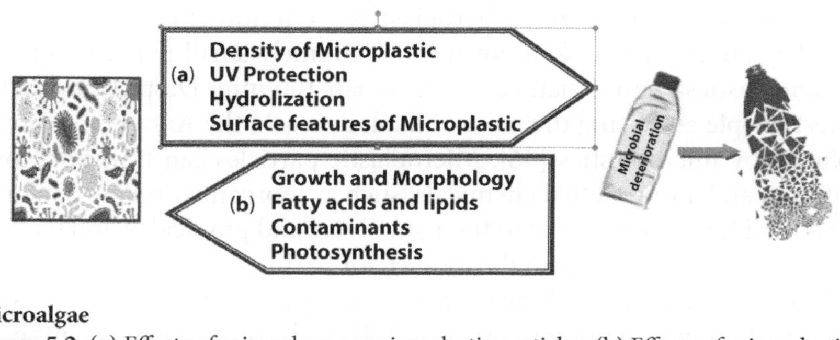

Microalgae

Figure 5.2 (a) Effects of microalgae on microplastic particles; (b) Effects of microplastics on microalgae.

Several studies have shown that the biofouling processes of microplastics can significantly alter their properties, with the adsorption capacity of microplastics in particular often increasing [41]. The ions adsorbed from water onto biofilms were found to be less strongly bonded than those adsorbed to anion-exchange polymers, resulting in the ions being more easily desorbed (leached) from the biofilm, according to Kurniawan *et al.* [42]. As a result, microalgae adherence to microplastic surfaces is crucial for pollutant adsorption and desorption from these surfaces. As a result, given that Cyanobacteria are photosynthetic organisms, the benefit of enhanced sunlight exposure when floating in water on floating plastic particles could be the primary cause of their enrichment on floating plastic waste [43]. Despite the fact that microalgae do not directly contribute to the breakdown of plastic, their presence may be crucial in determining the existence of other hydrocarbon-degrading bacteria, and diatoms have been suggested as a possible habitat for these microorganisms [44, 45]. Biofouling of large surfaces may cause opposite effects, shielding the plastics from ultraviolet light and so delaying the photodegradation process that occurs on the surface, as previously stated [44]. As far as plastic materials are concerned, algae are significant because they have the capacity to modify the density of the polymers that they have colonized, which in turn influences the vertical fluxes of the material. It is crucial to note that microalgae play a part in this since they have the ability to alter the density of the polymers that they have colonized.

5.7 Microplastics' Effect on Microalgae

Small-sized microplastics (those with a diameter of less than 5 mm) are either the result of abiotic breakdown of larger-sized macroplastics or are manufactured intentionally for this purpose (for example, pharmaceuticals or personal care items). Microplastics are found almost everywhere, and they are further broken down into extremely small particles, known as nanoplastics, with a diameter of less than 100 mm. Despite having the fewest people and being the most difficult to reach, the Antarctic region is polluted by microplastics [46]. Microplastic particles can transport toxic additives and absorb lethal chemicals, such as inorganic/organic contaminants, and heavy metals due to their small size and physical characteristics and they could even get into the food chain [47].

Mixotrophic algae, such as *Cryptomonas* sp., feed on microbiota colonized polypropylene microplastic (PE-MP) and store carbon to synthesize necessary polyunsaturated fatty acids, according to a study by Taipale and colleagues [48]. Direct contact with PE-MP or its releasing chemicals proved damaging to mixotrophic algae (*Cryptomonas* sp.). Khoironi and Anggoro [49] discovered that higher concentrations of microplastics resulted in lower rates of microalgae growth in their study on the effects of microplastics on *Spirulina* sp. growth. Microplastics have the ability to generate shading effects in culture, lowering light intensity and so severely impacting microalgal photosynthesis [50]. According to Zhang *et al.* [51], the negative impact of microplastic on microalgae was not due to shadowing, but rather to interactions between microalgae and microplastic, such as accumulation and adsorption.

The effect of microplastics on microalgae depends on the size of the microplastic's particles. Larger microplastic particles had major effects because they interfered with light transport and disrupted photosynthesis, while smaller particles harmed the surface of microalgal cell wall by absorbing into the surface of the cell wall [52]. *Raphidocelis subcapitata*, on the other hand, grew faster in exposure media including plastic microbeads of 63–75 mm size, than in control medium [53]. Chae *et al.* [54] discovered that when microplastics were employed to evaluate the influence of microplastics that were larger than the algal cells, cell growth and photosynthetic capability of the marine microalga *Dunaliella salina* were boosted without changing cell shape (about 200 mm diameter). Several chemicals, including stabilizers, phthalates, and endocrine disruptors, were observed responsible for the increased growth. However, it should be investigated whether algae may utilize microplastic as a carbon source for their growth. In general, more study is required to understand how microplastics impact algae, which are significant primary producers in ecosystems and thus need more in-depth investigation.

5.8 Techniques Used for Analysis of Plastic Biodegradation

In order to evaluate polymer biodegradability, it is necessary to measure the rate of CO_2 evolution, the rate of O_2 absorption, the change in polymer properties (chemical and physical), and the rate of growth of organisms when the polymer is in the presence of the organism [55]. Thus, it is essential to perform a number of tests in lieu of assessing plastic deterioration for the reasons listed below:

(i) Weight loss of microplastics may be induced by the leaching of compounds, such as plasticizers, which could explain their weight reduction.
(ii) The generation of carbon dioxide may be due to the breakdown of the polymer's low molecular weight component, rather than the deterioration of the polymer's longer chains.

The degree and kind of degradation may be determined using different methods, which are described below in Table 5.2 and Figure 5.3.

Table 5.2 Analytical techniques for properties of polymers.

SN	Changes in properties of polymer	Type of technique	Reference
1	Mechanical: Tensile strength - Elongation at fail and modulus of the polymer	DMR (dynamic mechanical analysis)	[56]
2	Physical: Morphology-Microcracks	SEM	[57]
	Density, Contact angle, Viscocity, Molecular Weight Distribution	HT-GPC (high-temperature gel permeation chromatography)	[58]
	Melting and Glass Transition temperature	Thermogravimetric analysis, DSC	[59, 60]
	Crystalline and amorphous region	X-diffraction, Small- and Wide-angle X-ray Scattering	[61]
3	Chemical properties	Fourier transform infrared spectroscopy (FTIR)	[57, 62]
4	Molecular Weight	Thin layer chromatography (TLC), Gas chromatography (GC),	

(*Continued*)

Table 5.2 Analytical techniques for properties of polymers.

SN	Changes in properties of polymer	Type of technique	Reference
		Nuclear magnetic resonance (NMR), Matrix-assisted laser desorption ionization–time of flight mass spectrometry (MALDI-TOF MS), Chemiluminescence (CL), Gas chromatography-mass spectrometry (GC-MS)	[61, 63, 64]
5	CO_2 evolution test	GC	[61, 65]
6	Metabolic activity of the cell	Adenosine triphosphate (ATP), Fluorescein diacetate (FDA), Protein analysis	[66, 67]

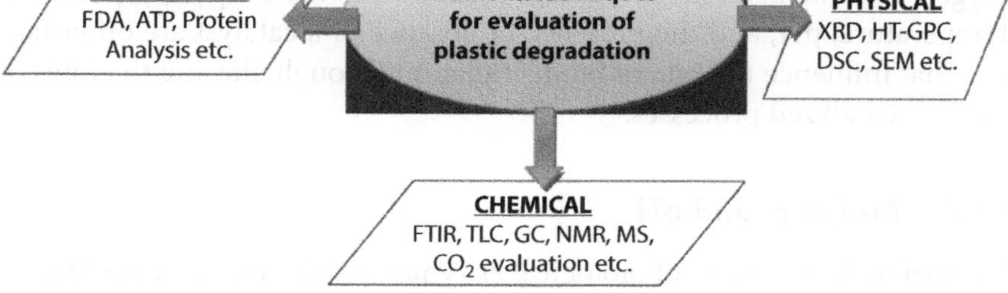

Figure 5.3 Evaluation of degree of plastics deterioration.

5.9 Factors Influencing the Deterioration of Plastics Using Microorganisms

Microorganisms can degrade a wide spectrum of organic pollutants due to their metabolic machinery and capacity to adapt to harsh environments. Microbes are essential players because of their role in site cleansing. The chemical composition and amount of pollutants, their availability to microorganisms, and physicochemical characteristics of the surrounding environment are all factors that impact the efficiency of these approaches [68]. As a result, variables that determine the pace at which pollutants are degraded by microorganisms are either related to the microorganisms and their nutritional requirements (biological factors) or to environmental factors.

5.9.1 Biological Factors

The metabolic ability of microorganisms is one example of a biotic factor. Microorganisms that degrade organic compounds are affected by biotic factors that include the direct suppression of enzymatic activities and the growth processes of the bacteria that degrade the organic compounds [69]. Microbes may be hindered in a number of ways, including competition for scarce carbon sources, antagonistic interactions amongst microbes, and predation of microbes by bacteriophages and protozoa. The rate of contamination degradation is usually determined by the amount of "catalyst" available and the concentration of contaminant present. The quantity of "catalyst" in this context refers to both the number of organisms capable of metabolizing the contaminant and the amount of enzymes (or enzyme combinations) generated by each individual cell. The degree to which pollutants are metabolized is influenced by the individual enzymes involved, as well as their "affinity" for the contamination and the availability of the contaminant. Furthermore, in order for unrestricted microbial growth to occur in a controlled environment, adequate amounts of nutrients and oxygen must be accessible in usable form and in the proper proportions. Temperature, pH, and the presence or absence of a catalyst are other factors that influence the rate of biodegradation through altering the rates of enzyme-catalyzed processes.

5.9.2 Moisture and pH

The metabolic turnover of biological enzymes involved in the degradation process is dependent on the temperature at which they are active, and this turnover will not be the same at all temperatures. According to this, every 10 °C reduction in ambient temperature reduces the rate of biodegradation by nearly half. Despite the fact that biodegradation can occur at any pH, most aquatic and terrestrial systems, including the environment, have a pH of 6.5 to 8.5, which is regarded to be ideal for biodegradation. Moisture has an impact on the rate of contaminant metabolism because it determines the types and amounts of soluble elements that are available, as well as the osmotic pressure and pH of terrestrial and aquatic systems.

5.9.3 Environmental Factors

The kind of soil and the amount of organic matter in the soil mixture impact the adsorption of inorganic chemicals on the surface of a solid. A contaminant enters the bulk mass of a soil matrix and is permanently integrated into the soil matrix through an absorption process similar to leaching. Both adsorption and absorption are beneficial in the treatment of pollution because they lower the availability of the contaminant to most bacteria as well as the pace at which the chemical is digested. Differences in porosity between the unsaturated and saturated zones of an aquifer matrix may have an influence on the flow of fluids and the migration of pollutants across the aquifer matrix in some instances. Because fine-grained sediments make it more difficult for the matrix to transmit gases like oxygen, methane, and carbon dioxide as soils become more saturated, gas transfer becomes even more difficult. As a result, the pace and kind of biodegradation that takes place may be affected. The oxidation-reduction potential of the soil must be assessed in order to calculate the electron density of a system. It is possible to get biological energy by oxidizing certain compounds. Electrons are carried to more oxidized substances known as electron acceptors and then transferred back in this process. If your electron density is low, you're in an anaerobic environment; if your electron density is high, you're in an aerobic environment [70–74].

5.10 Future Prospects

Plastic waste biodegradation is aided by microalgae because of their ability to produce enzymes which break chemical bonds present in polymeric plastic waste. Great interest has been piqued in the effort to utilize microalgae to alter these polymers into metabolites like water, carbon dioxide, and biomass. Because this subject is still in its infancy, more research is needed. It is necessary to do additional biodegradation experiments using algae in order to determine their effectiveness, especially given the wide range of plastics under investigation. It is also yet unknown how microalgae break down microplastics, which is an area of investigation. More in-depth research into the impacts of microplastics on microalgae would greatly aid in the knowledge of the effect of particles size on biodegradation study. As a result of the lack of convincing results from current research sttudies, there is a significant study gap about whether the microplastics would kill microalgae or, conversely, if microalgae will decompose the microplastics. The presence of hydrocarbons, lipids, proteins, and other high-value-added substances in microalgae is another significant concern since they are mostly utilized in food items and should be free from contaminants, including microplastics. Microplastics have important consequences on the health of aquatic creatures and humans, and it is critical to study these implications. Because of concerns about harmful health effects on humans, purifications of microalgal products are required to ensure the safety of microalgae intended for food. Research should be performed in order to have a better understanding of the health and environmental repercussions of plastic wastes of all sizes.

5.11 Conclusion

Microbial activities are critical for environmental regeneration and the global carbon cycle's maintenance, and they should not be disregarded. All of these processes are referred to as "biodegradation." Microorganisms may destroy or convert a wide range of synthetic compounds and other substances having ecotoxicological consequences, such as hydrocarbons and heavy metals. However, in the vast majority of cases, this statement relates to potential degradabilities assessed in the lab using specific cultures and under ideal growth conditions, rather than actual deterioration. Due to a multitude of factors, including competition with microbes, insufficient supply of essential substrates, unfavorable environmental conditions (aeration, moisture, pH, temperature), and poor bioavailability of the contaminant, biodegradation happens at a slower rate in natural settings. The hydrophilicity of plastic plays an important role in the adhesion of microorganism cells to the plastic surface, which has an influence on plastic biodegradation. Environmental biotechnology aims to address and alleviate these difficulties, allowing microorganisms to be used in bioremediation procedures. As a result, supporting the activities of indigenous microorganisms in polluted biotopes and increasing their degradative capabilities by bioaugmentation or biostimulation is critical.

References

1. Mortula, M. M. Health risk assessment of PET bottles in GCC. *Int. J. Environ. Ecol. Eng.* 7(3), 141-146, 2013.
2. Andrady, A. L., Neal, M. A. Applications and societal benefits of plastics. *Philos. Trans. R. Soc. Lond., B, Biol. Sci.* 364(1526), 1977-1984, 2009.
3. Iranzo, M., Sainz-Pardo, I., Boluda, R., Sanchez, J., Mormeneo, S. The use of microorgansims in environmental remediation. *Ann. Microbiol.* 51, 135-143, 2001.
4. Michaud, L., Di Marco, G., Bruni, V., Giudice, A. L. Biodegradative potential and characterization of psychrotolerant polychlorinated biphenyl-degrading marine bacteria isolated from a coastal station in the Terra Nova Bay (Ross Sea, Antarctica). *Mar. Pollut. Bull.* 54, 1754-1761, 2007.
5. Pilz, H., Brandt, B., Fehringer, R. The impact of plastics on life cycle energy consumption and greenhouse gas emissions in Europe. Summary Report, Denkstatt GmbH, Vienna, Austria, 2010.
6. Ghosh, S. K., Pal, S., Ray, S. Study of microbes having potentiality for biodegradation of plastics. *Environ. Sci. Pollut. Res.* 20, 4339-4355, 2013.
7. Alshehrei, F. Biodegradation of synthetic and natural plastic by microorganisms. *J. Appl. Environ. Microbiol.* 5(1), 8-19, 2017.
8. Gewert, B., Plassmann, M. M., MacLeod, M. Pathways for degradation of plastic polymers floating in the marine environment. *Environ. Sci.: Process. Impacts*, 17(9), 2015.
9. Lucas, N., Bienaime, C., Belloy, C., Queneudec, M., Silvestre, F., Edmundo, J., Saucedo, N. Polymer biodegradation: Mechanisms and estimation techniques. *Chemosphere*, 73, 429-442, 2008.
10. Arkatkar, A., Arutchelvi, J., Sudhakar, M., Bhaduri, S., Uppara, P.V., Doble, M., Approaches to enhance the biodegradation of polyolefins. *The Open Environ. Eng. J.* 2, 68-80, 2009.
11. Arutchelvi, J., Sudhakar, M., Arkatkar, A., Doble, M., Bhaduri, S., Uppara, V.P. Biodegradation of polyethylene and polypropylene. *Indian J. Biotechnol.* 7, 9-22, 2008.
12. Muthukumar, A., Veerappapillai, S. Biodegradation of plastics-A brief review. *Int. J. Pharm. Sci. Rev. Res.* 31(2), Article No. 36, 204-209, 2015.
13. Sharma, M., Dubey, A., Pareek, A. Algal flora on degrading polythene waste. *CIB Tech. J. Microbiol.*, 3, 43-47, 2014.
14. Wenten, L. G., Khoiruddin, K., Harimawan, A., Ting, Y. P., Boopathy, R. Membrane biosorption: Recent advances and challenges. *Curr. Pollut. Rep.* 6, 152-172, 2020.
15. Richmond, A. Handbook of Microalgal Culture: Biotechnology and Applied Phycology (Blackwell). Blackwell Publishing Company, Iowa, 2004.
16. Brennan, L., Owened, P. Biofuels from microalgae - A review of technologies for production, processing and extractions of biofuels and co-products. *Renew. Sustain. Energy Rev.* 14(2), 557-577, 2010.

17. McMurry, J. In Aromatic hydrocarbons. Organic Chemistry. 5th ed brooks/cole: Thomas learning. New York, pp. 120-180, 2000.
18. Mrozik, A., Piotrowska-Seget, Z., Labuzek, S., Bacterial degradation and bioremediationof polycyclic aromatic hydrocarbons. *Pol. J. Environ. Stud.* 12(1), 15-25, 2003.
19. Seeger, M., Hernández, M., Méndez, V., Ponce, B., Córdoval, M., González, M., Bacterial degradation and bioremediation of chlorinated herbicides and biphenyls. *J. Soil Sci. Plant Nutr.* 10(3), 320-332, 2010.
20. Raffi, F., Hall, J. D., Cernigila, C. E., Mutagenicity of azo dyes used in foods, drugs and cosmetics before and after reduction by *Clostridium species* from the human intestinal tract. *Food Chem. Toxicol.* 35, 897-901, 1997.
21. Verma, P., Madamwar, D. Decolorization of synthetic dyes by a newly isolated strain of Serratia marcescers. *World J. Microbiol. Biotechnol.* 19, 615-618, 2003.
22. Lesley, A., Ogilvie, Penny, R. Hirsch, Microbial ecological theory: Current perspectives. Publisher: Caister Academic Press University of Brighton, UK and Rothamsted Research, Harpenden, UK, 2012.
23. Lloyd, J. R., Lovley, D. R. Microbial detoxification of metals and radionuclides. *Curr. Opin. Biotechnol.* 12, 248-253, 2001.
24. Brigham, C. Biopolymers: biodegradable alternatives to traditional plastics. In: B. Torok, T. Dransfield (Eds.), Green Chem, Elsevier, pp. 753-770, 2018.
25. Latif, A., Ong, M. Y., Nomanbhay, S., Salman, B., Show, P. L. Estimation of carbon dioxide (CO_2) reduction by utilization of algal biomass bioplastic in Malaysia using carbon emission pinch analysis (CEPA), Bioengineered, 11(1), 154-164, 2020.
26. Ahmed, T., Shahid, M., Azeem, F., Rasul, I., Shah, A. A., Noman, M. Biodegradation of plastics: Current scenario and future prospects for environmental safety. *Environ. Sci. Pollut. Res.* 8, 7287-7298, 2018.
27. Banerjee, A., Chatterjee, K., Madras, G. Enzymatic degradation of polymers: A brief review. *Mater. Sci. Technol.* 30(5), 567-573, 2014.
28. Pathak, V. M. Review on the current status of polymer degradation: A microbial approach. *Bioresour. Bioprocess.* 4(1),15, 2017.
29. Zheng, Y., Yanful, E. K., Bassi, A. S. A review of plastic waste biodegradation. *Critical Reviews in Biotechnology* 25(4), 243-250, 2005.
30. Suseela, M., Toppo, K. Algal biofilms on polythene and its possible degradation. *Curr. Sci.* 92, 285-287, 2007.
31. Kumar, R. V., Kanna, G., Elumalai S. Biodegradation of polyethylene by green photosynthetic microalgae. *J. Biorem. Biodegrad.* 8, 2, 2017.
32. Sarmah, P., Rout J. Algal colonization on polythene carry bags in a domestic solid waste dumping site of Silchar town in Assam. Phykos. 48, 67-77, 2018.
33. Sarmah, P., Rout, J. Biochemical profile of five species of cyanobacteria isolated from polythene surface in domestic sewage water of Silchar town, Assam (India). *Curr. Trends Biotechnol. Pharm.* 12(1), 7-15, 2018.

34. Sarmah, P., Rout, J. Efficient biodegradation of low-density polyethylene by *Cyanobacteria* isolated from submerged polyethylene surface in domestic sewage water. *Environ. Sci. Pollut. Res.* 25(33), 33508-33520, 2018.
35. Sarmah, P., Rout, J. Colonization of *Oscillatoria* on submerged polythenes in domestic sewage water of Silchar town, Assam (India). *J. Algal Biomass Utln.* 8(4), 135-144, 2017.
36. Chinaglia, S., Tosin, M., Degli-Innocenti, F. Biodegradation rate of biodegradable plastics at molecular level. *Polym. Degrad. Stabil.* 147, 237-244, 2018.
37. Gulnaz, O., Dincer, S. Biodegradation of bisphenol a by *Chlorella vulgaris* and *Aeromonas hydrophilia*. *J. Appl. Biol. Sci.* 3, 79-84, 2009.
38. Hirooka, T., Nagase, H., Uchida, K., Hiroshige, Y., Ehara, Y., Nishikawa, J., Nishihara, T., Miyamoto, K., Hirata, Z. Biodegradation of bisphenol a and disappearance of its estrogenic activity by the green alga *Chlorella fusca var. Vacuolata*. *Environ. Toxicol. Chem.* 24, 1896-1901, 2005.
39. Kim, J. W., Park, S. B., Tran, Q. G., Cho, D. H., Choi, D. Y., Lee, Y. J., Kim, H. S. Functional expression of polyethylene terephthalate-degrading enzyme (PETase) in green microalgae. *Microb. Cell Factories*, 19, 97, 2020.
40. Yokota, K., Waterfield, H., Hastings, C., Davidson, E., Kwietniewski, E., Wells, B. Finding the missing piece of the aquatic plastic pollution puzzle: Interaction between primary producers and microplastics. *Limnol. Oceanogr. Lett.* 2, 91-104, 2017.
41. Kalcíkova, G., Skalar, T., Marolt, G., Jemec Kokalj, A. An environmental concentration of aged microplastics with adsorbed silver significantly affects aquatic organisms. *Water Res.* 175, 115644, 2020.
42. Kurniawan, A., Yamamoto, T., Tsuchiya, Y., Morisaki, H. Analysis of the ion adsorption-desorption characteristics of biofilm matrices. *Microbes Environ.* 27, 399-406, 2012.
43. Roager, L., Sonnenschein, E.C., Bacterial candidates for colonization and degradation of marine plastic debris. *Environ. Sci. Technol.* 53, 11636-11643, 2019.
44. Andrady, A. L. Microplastics in the marine environment. *Mar. Pollut. Bull.* 62, 1596-1605, 2011.
45. Dudek, K.L., Cruz, B.N., Polidoro, B., Neuer, S. Microbial colonization of microplastics in the Caribbean Sea. *Limnol. Oceanogr. Lett.* 5, 5–17, 2020.
46. Waller, C. L., Griffiths, H. J., Waluda, C. M., Thorpe, S. E., Loaiza, I., Moreno, B., Pacherres, C. O., Hughes, K. A. Microplastics in the Antarctic marine system: an emerging area of research. *Sci. Total Environ.* 598, 220-227, 2017.
47. Law, K. L. Plastics in the marine environment. *Ann. Rev. Mar. Sci.* 9, 205-229, 2017.
48. Taipale, S. J., Peltomaa, E., Kukkonen, J. V. K., Kainz, M. J., Kautonen, P., Tiirola, M., Tracing the fate of microplastic carbon in the aquatic food web by compound-specific isotope analysis. *Sci. Rep.* 9, 19894, 2019.

49. Khoironi, A., Anggoro, S. Evaluation of the interaction among microalgae *Spirulina sp.* plastics polyethylene terephthalate and polypropylene in freshwater environment. *J. Ecol. Eng.* 20, 2019.
50. Yurtsever, M., Sevindik, T. O., Tunca, H. The impact of PS microplastics on green algae *Chlorella vulgaris* growth. In: *15th International Conference on Environmental Science and Technology, Rhodes*, Greece, 2017.
51. Zhang, C., Chen, X., Wang, J., Tan, L. Toxic effects of microplastic on marine microalgae *Skeletonema costatum*: Interactions between microplastic and algae. *Environ. Pollut.* 220, 1282-1288, 2017.
52. Liu, G., Jiang, R., You, J., Muir, D. C. G., Zeng, E. Y., Microplastic impacts on microalgae growth: Effects of size and humic acid. *Environ. Sci. Technol.* 54, 1782-1789, 2020.
53. Canniff, P. M., Hoang, T. C. Microplastic ingestion by *Daphnia magna* and its enhancement on algal growth. *Sci. Total Environ.* 633, 500-507, 2018.
54. Chae, Y., Kim, D., An, Y. J. Effects of micro-sized polyethylene spheres on the marine microalga *Dunaliella salina:* Focusing on the algal cell to plastic particle size ratio. *Aquat. Toxicol.* 216, 105296, 2019.
55. Mohan, K. S., Srivastava, T. Microbial deterioration and degradation of polymeric materials. *J. Biochem. Tech.* 2(4), 210-215, 2010.
56. Huang, C. Y., Roan, M. L., Cuo, M. C., Lu, W. L. Effect of compatibilizer on the biodegradation and mechanical properties on high content starch/low density polyethylene. *Polym. Degrade. Stab.* 90, 95-105, 2005
57. Da Luz, J. M. R., Paes, S. A., Nunes, M. D., da Silva, Md C. S., Kasuya, M. C. M. Degradation of oxo-biodegradable plastic by *Pleurotus ostreatus*. PLoS ONE 8(8), 69386, 2013.
58. Kathiresan, K. Polythene and Plastics-degrading microbes from the mangrove soil. *Rev. Biol. Trop.* 51(3), 629-634, 2003.
59. Ramis, X., Cadenato, A., Salla, J. M., Morancho, J. M., Valles, A. Thermal degradation of polypropylene/starch based materials with enhanced biodegradability. *Polym. Degrad. Stab.* 86, 483-491, 2004.
60. Zuchoswka, D., Hlavata, D., Steller, R., Adamiak, W., Meissner, W. Physical Structure of polyolefin starch blends after ageing. *Polym. Degrad. Stab.* 64(3), 39-347, 1999.
61. Albertson, A. C., Barenstedt, C., Karlsson, S., Lindberg, T. Degradation product pattern and morphology changes as means to differentiate abiotically and biotically aged degradable polyethylene. *Polymer*, 36(16), 3075-3083, 1995.
62. Doble, S. M., Sriyutha, M. P., Venkatesan, R. Marine microbe mediated biodegradation of low and high density polyethylenes. *Int. Biodeterior. Biodegrad.* 61(3), 203–213, 2008.
63. Deguchi, T., Kakezawa, M., Nishida, T. Nylon biodegradation by lignin-degrading fungi. *Appl. Environ. Microb.* 63(1), 329–331, 1997.
64. Cacciari, P., Quatrini, G., Zirletta, E., Mincione, V., Vinciguerra, P. Isotactic polypropylene biodegradation by a microbial community: Physicochemical

characterization of metabolites produced. *Appl. Environ. Microbiol.* 59(11), 3695-3700, 1993.
65. Seneviratne, G., Tennkoon, N. S., Weerasekara, M. L. M. A. W., Nandasena, K. A. Polythene biodegradation by a developed *Penicillium-Bacillus* biofilm. *Curr. Sci.* 90, 20-21, 2006.
66. Kounty, M., Lemaire, Delort, A. M. Biodegradation of polyethylene films with pro-oxidant additives. *Chemosphere*, 64, 1243-1252, 2006.
67. Gilan, I., Hadar, Y., Sivan, A. Colonization and biofilm formation and biodegradation of polythene by a strain of *Rhodococcus rubber. Appl. Microbiol. Biotechnol.* 65, 97-104, 2004.
68. Fantroussi, S., Agathos, S. N. Is bioaugmentation a feasible strategy for pollutant removal and site remediation? *Curr. Opin. Microbiol.* 8, 268-275, 2005.
69. Riser-Roberts, E. Remediation of petroleum contaminated soils. Biological, physical and chemical process. Lewis Publishers Inc. (ed). CRC Press LLC. USA, pp. 1-542, 1998.
70. Cases, I., de Lorenzo, V. Genetically modified organisms for the environment: stories of success and failure and what we have learned from them. *Int. Microbiol.* 8, 213-222, 2005.
71. Moog, D. J., Schmitt, J., Senger, J., Zarzycki, K. H., Rexer, U., Linne, T., Erb, U., Maier, G. Using a marine microalga as a chassis for polyethylene terephthalate (PET) degradation. *Microb. Cell Factories*, 18, 171, 2019.
72. Sayler, G. S., Ripp, S. Field applications of genetically engineered microorganisms for bioremediation processes. *Curr. Opin. Microbiol.* 11, 286-289, 2000.
73. Thielen, M. Bioplastics: Plants and Crops Raw Materials Products, Fachagentur Nachwachsende Rohstoffe eV (FNR) Agency for Renewable Resources, Gülzow, Germany, 2014.
74. Van der Heul, R. M. Environmental Degradation of petroleum hydrocarbons (Master's thesis), 2009.

6

Application of Algae and Bacteria in Aquaculture

Anne Bhambri[1,2], Santosh Kumar Karn[1]* and Arun Kumar[2]

[1]*Department of Biochemistry and Biotechnology, Sardar Bhagwan Singh University, Dehradun, Uttarakhand, India*
[2]*Department of Biotechnology, Shri Guru Ram Rai University, Dehradun, Uttarakhand, India*

Abstract

In the aquaculture industry, nitrogen pollution from ammonia and nitrite is the major environmental issue. For the aquaculture industry, the high concentration of ammonia, nitrite and nitrate are one of the biggest challenges to produce sustainable fisheries and consequently needs to be treated in an economically feasible and environmentally friendly way. Microalgae play an important role in the aquatic food chain and are also used in the rearing of aquatic animals such as fishes, shrimps and mollusks at different stages of growth. Microalgae are one of the promising alternatives for the replacement of fish oil and meal and confirm sustainability standards in the aquaculture. The most frequently used microalgae genera in aquaculture include *Scenedesmus, Thalassiosira, Chlorella, Pavlova, Tetraselmis, Nannochloropsis, Chaetoceros, Phaeodactylum,* etc. In aquaculture, microalgae have potential as they are the sources of lipids, minerals, pigments, proteins, vitamins, etc. Microalgae have a possibility to decrease the dependence on conventional raw materials in aquafeed. There are significant beneficial effects for the use of microalgae that might potentially replace or reduce the common feed stuff because of their positive effect and nutritional quality on the aquatic species growth rate because of elevated triglycerides and deposition of protein in muscles, content of omega-3-fatty acid, reduced output of nitrogen into the environment and also improved resistance to disease. There are various biotechnological tools used to enhance the removal of ammonia, nitrite and nitrate by potential microalgae strains.

**Corresponding author*: santoshkarn@gmail.com

Charles Oluwaseun Adetunji, Julius Kola Oloke, Naveen Dwivedi, Sabeela Beevi Ummalyma, Shubha Dwivedi, Daniel Ingo Hefft and Juliana Bunmi Adetunji (eds.) Next-Generation Algae: Volume I: Applications in Agriculture, Food and Environment, (147–162)

Keywords: Aquaculture, microalgae, fish oil, aquaculture industry, fish meal

6.1 Introduction

Marine plants have been used by coastal populations from very ancient times as fuel, soil fertilizers, fodder for animals and also as food complement. In the ashes of prehistoric fireplaces, traces of algae have been found, which allow us to think that algae were used as food stuff in very early times. Algae are photosynthetic organisms which contain chlorophyll and acquire energy from the sun and their carbon from carbon dioxide. All the organisms which use carbon dioxide for their requirement of carbon are known as autotrophs. The size of algae ranges from one micron to many meters. In aquaculture, algae are generally beneficial by supplying a natural food base and oxygen for cultured animals like the dinoflagellates that cause red tides. In the chain of food nutrition and manufacture, algae are the first link and are also the culture diet for rotifers and give direct nutrition for larvae. The production of algal for feed is separated into the intensive monoculture for bivalves' larval stages, certain fish species and shrimp as well as general culture for grow out of bivalves, shrimp and carp. For larval feeds, the favored genera of microalgae are *Nannochloropsis*, *Chaetoceros*, *Isochrysis*, *Thalassiosira* and *Tetraselmis*.

These cultured larval organisms get nourished directly or indirectly. Indirectly means algae are providing via rotifers, daphnia and artemia that sequentially nourish the target larval organisms. In the rearing of all stages of marine gastropods, such as conch and abalone; marine bivalve mollusks like oysters, scallops and clams; larvae of various marine fish species and penaeid shrimp and zooplankton, microalgae are an extremely important food source. Marine microalgae are the basic source of nutrients which are extremely important for larval development for almost all marine invertebrates that allow transformation as well as growth via adult and juvenile stages to proceed. For the feed of aquaculture commodities in Samoa namely sea urchin *Tripneustes gratilla,* diatoms *Chaetoceros gracilis* and *Navicula ramoissima* are currently cultured.

The largest but most poorly understood kingdom of microorganisms are the microalgae, which are phytoplankton or single-celled algae on the earth.

Microalgae represent the primary source in the aquatic food chain and the natural nutritional base of all the phytonutrients [1]. Algae support the production of renewable resources by some 100×10^6 t per year from fishing. Consequently, the microalgae that compose the phytoplankton play an important role in nurturing aquatic animals such as shrimp, fish and mollusks and are also of interest for aquaculture. Besides that, there are various applications for molecules from these phototrophic microorganisms in health, human and animal food and cosmetology, whereas some of their properties also concern the environment, the production of renewable energy and also supporting life in space [2]. Microalgae play a vital role in the watery environment of aquaculture. They are essential to maintain the pond ecosystem's natural function and also stabilize the environment of the pond. The chemical as well as physical factors of water, community structure and algae population density are closely associated. Simultaneously, the physical as well as the chemical factors of water changes due to the direct influence of the species and quality of microalgae [3, 4].

Diatoms are a beneficial microalga which promote the transformation and decomposition of nutrient salt in water and also eliminate as well as reduce concentration of organic pollutants, nitrite nitrogen, ammonia nitrogen as well as additional toxic substances in the presence of their growth. In the aquaculture system, they also elevate the dissolved oxygen by photosynthesis and promote the decomposition as well as oxidation of organic matter. These diatoms are also beneficial due to their biochemical composition as they have a huge amount of calcium, iron, sterols, magnesium, inorganic salts as well as polyunsaturated fatty acids, in addition to a variety of vitamins that can be digested well by aquatic animals, but they do not contain cellulose. They also have good effects on the ecological balance of the water body and on water purification.

Consequently, the best water environments in the freshwater aquaculture are the water bodies with diatom-dominated phytoplankton communities whose growth is predisposed by chemical as well as environmental factors; whereas the silicon concentration is an essential factor because diatom requires silicon in the formation of their cell wall [5].

In bodies of freshwater, the commonly known diatom species is *Synedra* sp. [6], known as the dominant species in aquaculture ponds. As the primary producers in the aquatic food chain, microalgae are the source of numerous phytonutrients, such as arachidonic acid (ARA) and docosahexaenoic acid (DHA), which are precursors to the high nutritional component that extensively increase Omega 3 fatty acids. Microalgae that are a natural pigment source for the culture of salmonid fish, ornamental fish as well as prawns are *Haematococcus pluvialis*, *Spirulina* as well as *Dunaliella salina*. Over the last four decades, hundreds of microalgae have been tested as food but only less than twenty gained extensive use in aquaculture.

Algaculture is the form of aquaculture which involves the farming of algal species. In open ocean, microalgae play an essential nutritional role for aquatic animals and also in marine aquaculture. According to the Food and Agriculture Organization of the United Nations, the growth of the aquaculture industry is three times faster than land-based animal agriculture and will become even more extensive as the natural fisheries become exhausted [7]. There are various companies which harvest feed from aquaculture by using *Spirulina* as well as *Chlorella* or their mixture. For aquaculture, the use of microalgae such as microalgae species *Cryptonemia crenulate* and *Hypnea cervicornis* are rich in protein found in the diets of shrimp [8].

6.2 The Major Problem of Nitrite and Ammonia in Aquaculture

It is unfortunate that there are various aquaculture systems that generate a large amount of wastewater with a huge amount of suspended solids, total phosphorus, organic matter and total nitrogen that are discharged into the lakes, rivers and oceans [9]. To check the quality of water, there are non-conservative parameters, such as biological oxygen demand, total ammonia nitrogen, chemical oxygen demand, nitrate, nitrite and phosphate, which are very essential but at increased levels are harmful to aquatic life [10]. Furthermore, when the environment receives harmful aquaculture effluent with a high nitrogenous load with an inadequate capacity to incorporate as well as process nutrients into the biogeochemical cycles and food chains, significant amounts of nutrients tend to gather, causing an ecological imbalance through hyper-nitrification and eutrophication that require remediation [11].

Microbial-based farming systems are one of the strategies with the greatest possibility of attaining sustainable aquaculture as the favorable microbes may improve the water quality and also decrease the discharge of farming waste as well as the occurrence of disease [12–14]. In farming water management, nitrite and ammonia nitrogen are the two very essential indexes as their toxicity may directly impact on the productivity of the aquaculture [15, 16] and also on the survival of aquatic animals [17, 18].

So, for successful culture a key target is to sustain low concentration of ammonia nitrogen and nitrite in the farming water. In the aquaculture, ammonia nitrogen has two main sources, i.e., the decomposition of feed by microbes and also the metabolic waste of breeding animals [19]. Nitrite is mainly generated by the oxidation metabolism of ammonia and accumulates due to the imbalances in the nitrifying microbial activity [20].

Aquaculture plays a very important role in improving the living standards of humans. Aquaculture is the fastest growing agricultural sector and provides around 47% of human fish consumption worldwide. In 2011, the per capita supply of fish increased to around 19 kg and has regularly grown, whereas aquaculture reported about 43% of the production of fish globally and the production of shrimp culture reached up to 21% [7]. Over the next 20 to 30 years, the output of aquaculture is set to further rise from 60 to 100% based on population growth as well as elevated per capita consumption of fish [9]. The elevation of aquaculture production needs the use of more inputs, which leads to increased waste generation.

6.3 Techniques for Nitrite, Nitrate and Ammonia Removal

Various physical, chemical as well as biological processes are used for remediation of ammonia, nitrite and nitrate such as reverse osmosis [21], electrocoagulation [22], chemical precipitation [23], ion-exchange [24], chemical reduction [25], membrane process [26], adsorption method [27], etc. Among all these methods, the biological method is the most significant and commonly used process for nitrite, nitrate and ammonia removal from the aquaculture previously discharged into the surface or groundwater [28, 29], as it is an easy, reliable and economically attractive method used for organic pollutants biodegradation which is accepted by the public [30, 31].

6.4 Beneficial Application of Algae in Aquaculture

Microalgae are the accepted food source for various small aquaculture animals such as fish fry and larval shrimp; as well as the many other various microalgal species commercially used as live prey as well as dried whole algal feed [32]. The most frequently used species are those of *Thalassiosira, Pavlova, Nannochloropsis, Skeletonema, Chlorella, Phaeodactylum, Tetraselmis, Isochrysis* and *Chaetoceros,* with a combination of different species often employed to deliver the right balance of essential micronutrients, proteins and lipids [33]. In older animals, supplements of grain-based or fishmeal-based feed with algal-derived biomass supply the necessary pigments like red astaxanthin, which colors the shellfish flesh as well as that of salmonids, and also contributes additional nutritional value [33]. Algal supplements are reported to contain several compounds which serve as nonspecific immunostimulants in addition to having inherent nutritional benefits which improve the distinctive defense mechanisms in animals and thus provides higher resistance to pathogens [33]. Antimicrobial compounds have been produced by the microalgal species which could be effective against the viral, protists pathogens as well as bacterial [34]. Consequently, microalgae signify a valuable component of shellfish as well as finfish feed. These phototrophic organisms are cultured in low-cost photobioreactors, either on-site or off-site at a more appropriate location for the cultivation of algal, including the production of concentrates of frozen algal and algal pastes suitable for shipping as well as storage, and dried algae. Therefore, both the whole cells as well as the crude algal extracts are used as a supplement or as a replacement in fish or else shrimp fish feed to help address the microbial infections issues [35, 36]. *Tetraselmis suecica* microalgal homogenate shows good antimicrobial activity *in vitro* compared to various *Vibrio* species like notable shrimp pathogens, including *V. parahaemolyticus, V. alginolyticus, V. vulnificus* and *V. anguillarum* [37]. There are other microalgal species like *Phaeodactylum tricornutum, Stichochrysis immobilis, Euglena viridis, Chaetoceros lauderi* and *Dunaliella tertiolecta* which are also generally used as aquaculture feed which have similar antimicrobial activities *in vitro* as compared to the fish and shrimp pathogens [34, 38, 39]. Microalgal feeding also is seen as a way to increase the tolerance of the host animal to disease. It is noted that the fish oil replacement with algal meal consists of large amounts of LC-PUFAs arachidonic acid (AA) as well as docosahexaenoic acid (DHA), which notably improves the immune parameters like superoxide dismutase activity, total hemocyte count, bactericidal activity at post-larval stage, phenoloxidase activity and *Litopenaeus vannamei*, and increases the survival rate from *V. harveyi* infection [40].

Microalgae used as a probable constituent to aquafeeds might suggest ecologically attractive alternatives for innovative ingredients like yeasts, insects and seaweeds, along with traditional ingredients of plants and marine origin [41]. The insertion of microalgae is becoming widespread in aquafeeds as their use increases in the aquaculture sector along with their basic nutritional values [42, 43]. The main genera of cultured microalgal are *Nannochloropsis, Tisochrysis, Tetraselmis, Chlorella, Arthrospira* and *Scenedesmus* [44, 45].

Microalgae have already been used to improve the water quality as well as the removal of inorganic molecules [46]. Additionally, non-axenic microalgae culture wastewater from fresh markets and fish farms have also been used as a medium [47–49]. In a eutrophic water body, microalgae that might assimilate nutrients efficiently are proven to be an excellent way of remediating wastewater [50, 51]. It is observed that for the assimilation of nutrients, microalgae show great performance in agricultural waste stream remediation, municipal wastewater, food industry effluent, etc. [52–54]. In aquaculture or in the production of food supplements, the most commonly used microalgae that do not produce toxins are *Skeletonema costatum*, *Nannochloropis* sp., *Dunaliella tertiolecta*, *P. tricornutum*, *Isochrysis*, *T. suecica*, *P. lutheri*, *Chaetoceros gracilis* [55]. Nevertheless, even within the same species it is considered that large differences exist between the toxic as well as nontoxic microalgae strains. Cyanobacteria, dinoflagellates and diatoms produce toxins that may affect humans. Consequently, at the strain level, it is essential to know the algae applications for the feed or food as it plays a vital role in knowing their safety levels. Even if there are some microalgae groups, like some dinoflagellates that are potentially harmful, some of their strains are indeed used as feed or food supplements.

The high-cost of aquafeed production continues to place constraints on the aquaculture industry. Large-scale production can reduce the sale price, which is economically feasible by using microalgae [41, 56]. However, on the basis of their processing as well as growth conditions, microalgal biomass consists of high concentrations of toxins as well as trace elements which can constrain their incorporation into aquafeeds.

6.5 Algae and Bacteria for Nitrite, Nitrate and Ammonia Transformation

In the environment, nitrite is the ephemeral forms of nitrogen. In both primary (physical settling of solids) and secondary (various forms of oxidation like activated sludge or trickling filters) wastewater treatment systems as well as in surface waters, it occurs as the least prevalent form of inorganic nitrogen. By using the nitrifying bacteria as well as algae, removal of nitrogen elevates the nitrogen conversion capacity of the system. Algae substantially elevates the overall removal capacity of ammonium [57] by providing oxygen to nitrifiers [58], while nitrifiers decrease the level of oxygen underneath the inhibition thresholds for algae [59]. *Nannochloropsis* sp. of algae converts nitrate into ammonium ion via the use of nitrate reductase enzyme [60, 61]. Certain cyanobacteria like *Microcystis* produce neurotoxins or cyanotoxins, which may persist in the column of water long after the bloom of algae has faded [62]. Nearly all the algae can grow on low to high concentrations of nitrate; nevertheless, various strains have acute sensitivity to high ammonia or ammonium concentration. In the blood, nitrate reacts with hemoglobin and forms methemoglobin that reduces the capacity of red blood cells to release the oxygen to the tissues, and this lack of oxygen results in methemoglobinemia (blue baby syndrome) [63]. If the maximum level of nitrate polluted groundwater reaches above the 10 mg/l NO_3^--N mentioned by the World Health Organization (WHO) as well as United States Environmental Protection Agency (EPA), and 12 mg/l NO_3^- according to Europe, it must be treated before being used as drinking water [64].

The concentrations of nitrate are usually very high compared to the ammonia concentrations in various waterways like estuarine and marine systems. Even though high levels of nitrate can be tolerated both by algae and plants, there are issues of toxicity related to ammonium. There are various strains of algae which show high ammonium tolerance whereas others have discrete sensitivity even to the low concentration of ammonium [65]. This sensitivity or toxicity to ammonia happens because of the shift in pH, which happens when the carbon dioxide becomes limited and bicarbonate ions are taken up by algae. Furthermore, these ions get broken down into hydroxyl ion as well as carbon dioxide. Next, carbon dioxide has been used for photosynthesis, whereas hydroxyl ions are excreted back into the water. Due to this, the pH becomes elevated, resulting in the conversion of ammonium ions into ammonia [66]. Furthermore, the treatment strategy of ammonia and nitrite are mentioned in Figure 6.1.

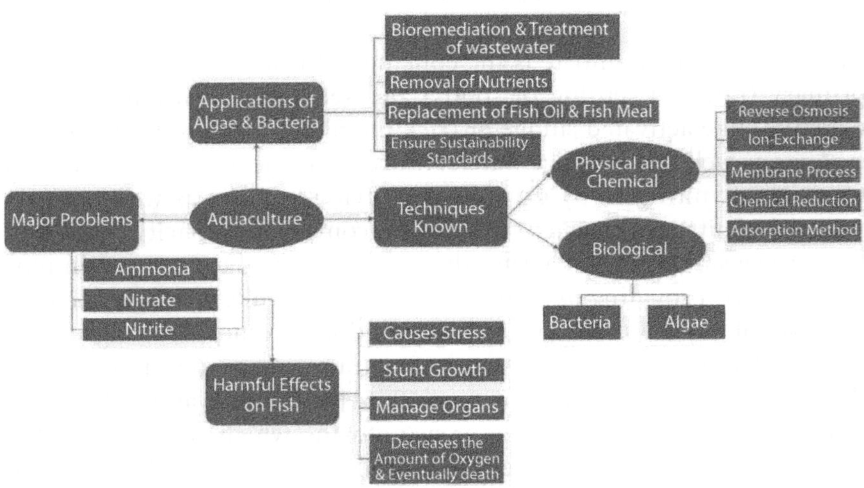

Figure 6.1 Outline of problem and remediation in aquaculture based on algae and bacteria.

In a study of six microalgae classes, Collos and Harrison [67] showed that *Dinophyceae* (dinoflagellates) had the least tolerance to ammonium whereas *Cyanophyceae* (blue-green algae) had the highest tolerance.

Three common species of algae *Scenedesmus rubescens, Cyanobacterium phormidium* sp., and *Chlamydomonas reinhardtii* were compared by Su *et al.* [68] using the effluent from a secondary clarifier with total Kjeldahl nitrogen of 95.5 mg/l comprised of ammonia (95.5%) and the balance by nitrite as well as nitrate. They cultured the cells in photobioreactors with a light/dark cycle of 12:12 h, with 7000 lux. *C. reinhardtii* removed the nitrate within 4 days whereas *C. Phormidium* sp. and *S. rubescens* removed nitrate within 7and 6 days, respectively.

The uptake of nitrate by 20 different strains of microalgae was compared that were cultured under like conditions by Sydney *et al.* [69]. They showed that *Chlorella vulgaris* and *Botryococcus braunii* had the maximum removal efficiency of nitrate with an uptake rate of 20.28 and 22.2 mg/l/d, respectively.

In primary-treated wastewater, the complete removal of nitrate was observed by Kang *et al.* with the growth of *Haematococcus pluvialis* [70]. They observed that the initial concentration of nitrate was 42.4 mg/l with an uptake rate of 8.48 mg/l/d.

Wang *et al.* [71] suggested that *Chlorella vulgaris* removed 62.5% of nitrate with a removal rate of 2.65 mg/l/d and also at a removal rate of 0.001 mg/l/d from the effluent without the ammonium ions. Li *et al.* [72] observed that *Neochloris oleoabundans* was able to completely remove the nitrate having an initial concentration of 452 mg/l with an uptake rate of 150 mg/l/d, whereas Hulatt *et al.* [73] observed that *Chlorella vulgaris* and *Dunaliella tertiolecta* have a high uptake rate of nitrate, i.e., 103.3 and 155 mg/l/d, respectively. Hence, *Neochloris oleoabundans, Chlorella vulgaris* and *Dunaliella tertiolecta* are excellent candidates for the bioremediation of nitrate.

6.6 Conclusion

Microalgae and bacteria have been used for the removal of nitrite, nitrate and ammonia from the aquaculture and have potential as an alternative technique to conventional removal methods of nutrients. For microalgae, aquaculture may be considered as an available, cost-effective medium. Algae growth delivers an effective means of retrieving the nutrients in aquaculture by the assimilation and transformation of nitrate, nitrite and ammonia and converts them into biomass. There are various physiochemical techniques used for the remediation of nitrite, nitrate and ammonia from the aquaculture. In this chapter we included the major problems associated with nitrite, nitrate and ammonia in aquaculture and also mentioned the beneficial applications of algae and bacteria as well as the transformation of nitrite, nitrate and ammonia in the aquaculture by using microalgae and bacteria.

Acknowledgments

The authors are thankful to Dr. Gaurav Deep Singh, Chancellor, Sardar Bhagwan Singh University, Dehradun, India, for providing space to facilitate the completion of this chapter.

References

1. Priyadarshani, I., Sahu, D., Rath, B., Algae in Aquaculture. *International Journal of Health Sciences and Research.*, 2(1), 108-114, 2012.
2. Muller-Feuga, A., Microalgues Marine, Les Enjeux de la Recherche. *Ifremer*, Plouzane, France, 35, 1977.
3. Johnston, P. and Santillo, D., Chemical usage in aquaculture: Implications for residues in market products. Technical Note 06/2002, *Greenpeace Research Laboratories*, Department of Biological Sciences, University of Exeter, Exeter, UK, 2002.
4. Chuntapa, B., Powtongsook, S. and Menasveta, P., Water quality control using Spirulina platens is in shrimp culture tanks. *Aquaculture*, 220(1), 355-366, 2003.
5. Treguer, P. and Pondaven, P., Global change: Silica control of carbon dioxide. *Nature*, 406 (6794), 358-359, 2000.
6. Reynolds, C.S., The ecology of phytoplankton. *Cambridge University Press*, New York, 2006.
7. FAO (Food and Agriculture Organization of the United Nations). The state of world fisheries and aquaculture. In: *Opportunities and challenges*. FAO Fisheries and Aquaculture Department. *Food and Agriculture Organization of the United Nations*, Rome, 2014.
8. DanSilva, R.L. and Barbosa, J.M., Seaweed meal as a protein source for the white shrimp *Lipopenaeusvannamei*. *Journal of Applied Phycology*. 20, 193-197, 2008.
9. Turcios, A. and Papenbrock, J., Sustainable treatment of aquaculture effluents-what can we learn from the past for the future? *Sustainability*, 6(2), 836-856, 2014.
10. Lananan, F., Hamid, S.H.A., Din, W.N.S., Khatoon, H., Jusoh, A. and Endut, A., Symbiotic bioremediation of aquaculture wastewater in reducing

ammonia and phosphorus utilizing effective microorganism (EM-1) and microalgae (*Chlorella* sp.). *Int. Biodeterior. Biodegradation*, 95, 127-134, 2014.

11. Luo, L., Zhao, Z., Huang, X., Du, X., Wang, C.A., Li, J., ….. Xu, Q., Isolation, identification, and optimization of culture conditions of a bioflocculant-producing bacterium *Baccilus megaterium* sp. 1 and its application in aquaculture wastewater treatment. *Biomed. Res. Int.*, 1-9, 2016.
12. Zhou, Q., Li, K., Xie, J., Bo, L., Role and functions of beneficial microorganisms in sustainable aquaculture. *Bioresour. Technol.* 100, 3780-3786, 2009.
13. Crab, R., Defoirdt, T., Bossier, P., Verstraete, W., Biofloc technology in aquaculture: beneficial effects and future challenges. *Aquaculture*, 356-357, 351-356, 2012.
14. Martinez-Cordova, I.R., Emerenciano, M., Miranda-Baeza, A., Martinez-Porchas, M., Microbial-based systems for aquaculture of fish and shrimp: an updated review. *Rev. Aquac.* 7, 131-148, 2015.
15. Ebeling, J.M., Timmons, M.B., Bisogni, J.J., Engineering analysis of the stoichiometry of photoautotrophic, autotrophic, and heterotrophic removal of ammonia-nitrogen in aquaculture systems. *Aquaculture*. 257, 346-358, 2006.
16. Muthukrishnan, A., Sabaratnam, S., Chong, V.C., Ammonia nitrogen removal by an aerobic heterotrophic bacterium, *Microbacterium* sp. VCN11. *Malays. J. Sci.* 31, 76-82, 2012.
17. Lin, C.Y. and Chen, J.C., Acute toxicity of nitrite on *Litopenaeusvannamei* (Boone) juveniles at different salinity levels. *Aquaculture*, 224, 193-201, 2003.
18. Xian, J.A., Wang, A.L., Chen, X.D., Gou, N.N., Miao, Y.T., Liao, S.A., *et al.*, Cytotoxicity of nitrite on haemocytes of the tiger shrimp, *Penaeus monodon*, using flow cytometric analysis. *Aquaculture*, 317, 240-244, 2011.
19. Gross, A., Abutbul, S., Zilberg, D., Acute and chronic effects of nitrite on white shrimp, *Litopenaeusvannamei*, cultured in low-salinity brackish water, *J. World Aquacult. Soc.* 35, 315-321, 2004.
20. Wang, W.N., Wang, A.L., Zhang, Y.J., Li, Z.H., Wang, J.X., Sun, R.Y., Effects of nitrite on lethal and immune response of *Macrobrachiumnipponense*. *Aquaculture*, 232, 679-686, 2004.
21. Raval, H.D., Rana, P.S., Maiti, S.A., Novel high-fux, thin-flm composite reverse osmosis membrane modified by chitosan for advanced water treatment. *RSC Adv*. 5, 6687–6694, 2015.
22. Kuokkanen, V., Kuokkanen, T., Ramo, J., Lassi, U., Roininen, J., Removal of phosphate from wastewaters for further utilization using electrocoagulation with hybrid electrodes-techno-economic studies. *J. Water Process Eng.*, 8, e50–e57, 2015.
23. Quan, X., Ye, C., Xiong, Y., Xiang, J., Wang, F., Simultaneous removal of ammonia, P and COD from anaerobically digested piggery wastewater using an integrated process of chemical precipitation and air stripping. *J. Hazard Mater*, 178, 326–332, 2010.

24. Lin, Y.F., Chen, H.W., Chen, Y.C., Chiou, C.S., Application of magnetic modified with polyacrylamide to adsorb phosphate in aqueous solution. *J. Taiwan Inst. Chem. Eng.* 44, 45–51, 2013.
25. Yang, H.L.G., Chemical reduction of nitrate by nanosized iron: Kinetics and pathways. *Water Res.*, 39, 884–894, 2005.
26. Van Voorthuizen, E.M., Zwijnenburg, A., Wessling, M. Nutrient removal by NF and RO membranes in a decentralized sanitation system. *Water Res.* 39, 3657–3667, 2005.
27. Appunni, S., Rajesh, M.P., Prabhakar, S. Nitrate decontamination through functionalized chitosan in brackish water. *Carbohydr. Polym.*, 147, 525–532, 2016.
28. Bassin, J.P., Kleerebezem, R., Dezotti, M., van Loosdrecht, M.C.M., Simultaneous nitrogen and phosphate removal in aerobic granular sludge reactors operated at different temperatures. *Water Res.* 46(12), 3805–3816, 2012.
29. Wang, H., Gao, Q., Liu, S., Chen, Q., Simultaneous nitrogen and carbon removal in a single biological aerated filter by the bioaugmentation with heterotrophic-aerobic nitrogen removal bacteria. *Environ. Technol.*, 2020. https://doi.org/10.1080/09593330.2020. 1739147.
30. Bhambri, A., Karn, S.K., Singh, R.K., *In-situ* remediation of nitrogen and phosphorus of beverage industry by potential strains *Bacillus* sp. (BK1) and *Aspergillus* sp. (BK2). *Nature Scientific Reports,* London, UK, 11, 12243, 2021. https://doi.org/10.1038/s41598-021-91539-y
31. Bhambri, A. and Karn, S.K., Bio-technique for Nitrogen and Phosphorus Removal: A Possible Insight. *Chemistry and Ecology*, London, 36, 785-809, 2020. https://doi.org/10.1080/02757540.2020.1777991
32. Spolaore, P., Joannis-Cassa, C., Duran, E., Isambert, A., Commercial applications of microalgae. *J. Biosci. Bioeng.* 101, 87–96, 2006.
33. Shah, M.R., Lutzu, G.A., Alam, A., Sarker, P., Chowdhury, M.A.K., Parsaeimehr, A., Liang, Y., Daroch, M., Microalgae in aquafeeds for a sustainable aquaculture industry. *J. Appl. Phycol.*, 30, 197–213, 2018.
34. Falaise, C., François, C., Travers, M.A., Morga, B., Haure, J., Tremblay, R., Turcotte, F., Pasetto, P., Gastineau, R., Hardivillier, Y., *et al.* Antimicrobial compounds from Eukaryotic Microalgae against human pathogens and diseases in aquaculture. *Mar. Drugs,* 14, 159, 2016.
35. Dang, V.T., Li, Y., Speck, P., Benkendorff, K., Effects of micro and macroalgal diet supplementations on growth and immunity of greenlip abalone, Haliotis laevigata. *Aquaculture,* 320, 91–98, 2011.
36. Yaakob, Z., Ali, E., Zainal, A., Mohamad, M., Takriff, M.S., An overview: Biomolecules from microalgae for animal feed and aquaculture. *J. Biol. Res.-Thessalon.*, 21, 6, 2014.
37. Austin, B. and Day, J., Inhibition of prawn pathogenic Vibrio spp. by a commercial spray-dried preparation of *Tetraselmissuecica. Aquaculture,* 90, 389–392, 1990.

38. Viso, A., Pesando, D., Baby, C., Antibacterial and antifungal properties of some marine diatoms in culture. *Bot. Mar.* 30, 41–46, 1987.
39. Das, B and Pradhan, J. Antibacterial properties of selected freshwater microalgae against pathogenic bacteria. *Indian J. Fish.*, 57, 61–66, 2010.
40. Nonwachai, T., Purivirojkul, W., Limsuwan, C., Chuchird, N., Velasco, M., Dhar, A.K., Growth, nonspecific immune characteristics, and survival upon challenge with Vibrio harveyi in Pacific white shrimp (*Litopenaeusvannamei*) raised on diets containing algal meal. *Fish Shellfish Immunol.* 29, 298–304, 2010.
41. Becker, E.W., Micro-algae as a source of protein. *Biotechnol Adv.*, 25, 207–210, 2007. https://doi.org/10.1016/j.biotechadv.2006.11.002.
42. Chu, W.L., Biotechnological applications of microalgae. *Int e-J Sci Med Educ.*, 6, S24–S37, 2012.
43. Priyadarshani, I. and Rath, B., Commercial and industrial applications of microalgae-a review. *J Algal Biomass Utln.*, 3, 89–100, 2012.
44. Sirakov, I., Velichkova, K., Stoyanova, S., Staykov, Y., The importance of microalgae for aquaculture industry. Review. *Int J Fish Aquat. Stud.*, 2, 81–84, 2015.
45. Bleakley, S. and Hayes, M., Algal proteins: extraction, application, and challenges concerning production. *Foods,* 6(5), 33, 2017. https://doi.org/10.3390/foods6050033.
46. Ruiz-Martinez, A., Martin Garcia, N., Romero, I., Seco, A., Ferrer, J., Microalgae cultivation in wastewater: nutrient removal from anaerobic membrane bioreactor effluent. *Bioresour. Technol.* 126, 247–253, 2012. https://doi.org/10.1016/j.biortech.2012.09.022
47. Apandi, N.M., Radin, M.S.R.M., Al-Gheethi, A., Kassim, A.H.M., Microalgal biomass production through phycoremediation of fresh market wastewater and potential applications as aquaculture feeds. *Environ Sci Pollut Res Environ Sci Pollut Res,* 26, 3226–3242, 2019. https://doi.org/10.1007/s11356-018-3937-3
48. Andreotti, V., Chindris, A., Brundu, G., Vallainc, D., Francavilla, M., García, J., Bioremediation of aquaculture wastewater from Mugil cephalus (Linnaeus, 1758) with different microalgae species. *Chem Ecol.*, 33, 750– 761, 2017. https://doi.org/10.1080/02757540.2017.1378351
49. Michels, M.H., Vaskoska, M., Vermu, M.H., Wijffels, R.H., Growth of *Tetraselmissuecica* in a tubular photobioreactor on wastewater from a fish farm. *Water Res,* 65, 290–296, 2014. https://doi.org/1`0.1016/j.watres.2014.07.017
50. Leng, L., Li, J., Wen, Z., Zhou, W., Use of microalgae to recycle nutrients in aqueous phase derived from hydrothermal liquefaction process. *Bioresour. Technol.* 256, 529–542, 2018.
51. Wang, J., Zhou, W., Yang, H., Wang, F., Ruan, R., Trophic mode conversion and nitrogen deprivation of microalgae for high ammonium removal from synthetic wastewater. *Bioresour. Technol.* 196, 668–676, 2015.

52. Lu, Q., Zhou, W., Min, M., Ma, X., Chandra, C., Doan, Y.T., Ma, Y., Zheng, H., Cheng, S., Griffith, R., Growing *Chlorella* sp. on meat processing wastewater for nutrient removal and biomass production. *Bioresour. Technol.* 198, 189–197, 2015.
53. Lu, Q., Li, J., Wang, J., Li, K., Li, J., Han, P., Chen, P., Zhou, W., Exploration of a mechanism for the production of highly unsaturated fatty acids in *Scenedesmus* sp. at low temperature grown on oil crop residue based medium. *Bioresour. Technol.*, 244, 542–551, 2017.
54. De-Bashan, L.E., Hernandez, J.P., Morey, T., Bashan, Y., Microalgae growth-promoting bacteria as "helpers" for microalgae: A novel approach for removing ammonium and phosphorus from municipal wastewater. *Water Res.*, 38, 466–474, 2004.
55. Enzing, C.M., Nooijen, A., Eggink, G., Springer, J., Wijffels, R., Algae and genetic modification. Research, production and risks. 2012. http://www.cogem.net/index.cfm/nl/publicaties/publicatie/onderzoeksrapportalgae-andgenetic-modification-research-production-and-risks.
56. Sarker, P., Gamble, M., Kelson, S., Kapuscinski, A., Nile tilapia (*Oreochromis niloticus*) show high digestibility of lipid and fatty acids from marine *Schizochytrium*sp. and of protein and essential amino acids from freshwater *Spirulina* sp. feed ingredients. *Aquac Nutr.* 22, 109–119, 2016a.
57. Rada-Ariza, A.M., Lopez-Vazquez, C.M., van der Steen, N.P. Lens, P.N.L., Nitrification by Microalgal-Bacterial Consortia for Ammonium Removal in Flat Panel Sequencing Batch Photobioreactors. *Bioresour. Technol.*, 245, 81–89, 2017.
58. Karya, N.G.A.I., van der Steen, N.P., Lens, P.N.L., Photo-Oxygenation to Support Nitrification in an Algal–Bacterial Consortium Treating Artificial Wastewater. *Bioresour. Technol.*, 134, 244–250, 2013.
59. Bilanovic, D., Holland, M., Starosvetsky, J., Armon, R., Co-Cultivation of Microalgae and Nitrifiers for Higher Biomass Production and Better Carbon Capture. *Bioresour. Technol.* 220, 282–288, 2016.
60. Fernández, E., Schnell, R., Ranum, L.P., Hussey, S.C., Silflow, C.D., Lefebvre, P.A. Isolation and characterization of the nitrate reductase structural gene of *Chlamydomonas reinhardtii.Proc Natl Acad Sci U S A*, 86, 6449–6453, 1989.
61. Berges, J., Miniview: algal nitrate reductases. *Eur J Phycol.*, 32, 3–8, 1997.
62. Florczyk, M., Łakomiak, A., Woêny, M., Brzuzan, P., Neurotoxicity of cyanobacterial toxins. *Environ Biotechnol.*, 10(1), 26-43, 2014.
63. El Midaoui, A., Elhannouni, F., Taky, M., *et al.* Optimization of nitrate removal operation from ground water by electrodialysis. *Sep Purif Technol.* 29(3), 235-44, 2002.
64. Glass, C. and Silverstein, J. Denitrification of high-nitrate, high-salinity wastewater. *Water Res.*, 33(1), 223-229, 1999.
65. Stewart, W.D.P. The effect of available nitrate and ammonium-nitrogen on the growth of two nitrogen-fixing blue-green algae. *J Exp Bot.*, 15(1), 138-145, 1964.

66. Azov, Y., Goldman, J.C., Free ammonia inhibition of algal photosynthesis in intensive cultures. *Appl Environ Microbiol.*, 43(4), 735-739, 1982.
67. Collos, Y. and Harrison, P.J., Acclimation and toxicity of high ammonium concentrations to unicellular algae. *Mar Pollut Bull.*, 80(1-2), 8-23, 2014.
68. Su, Y., Mennerich, A., Urban, B. Comparison of nutrient removal capacity and biomass settleability of four high-potential microalgal species. *Bioresour Technol.*, 124, 157-62, 2012.
69. Sydney, E.B., da Silva, T.E., Tokarski, A., *et al.* Screening of microalgae with potential for biodiesel production and nutrient removal from treated domestic sewage. *Appl Energy.*, 88 (10), 3291-94, 2011.
70. Kang, C.D., An, J.Y., Park, T.H., Sim, S.J., Astaxanthin biosynthesis from simultaneous N and P uptake by the green alga *Haematococcuspluvialis* in primary treated wastewater. *BiochemEng J.*, 31(3), 234-38, 2006.
71. Wang, L., Min, M., Li, Y., *et al.*, Cultivation of green algae *Chlorella* sp. in different wastewaters from municipal wastewater treatment plant. *Appl Biochem Biotechnol.*, 162(4), 1174-86, 2010.
72. Li, Y., Horsman, M., Wang, B., Wu, N., Lan, C.Q., Effects of nitrogen sources on cell growth and lipid accumulation of green alga *Neochloris oleoabundans*. *Appl Microbiol Biotechnol.*, 81 (4), 629-36, 2008.
73. Hulatt, C.J., Lakaniemi, A.M., Puhakka, J.A., Thomas, D.N., Energy demands of nitrogen supply in mass cultivation of two commercially important microalgal species, *Chlorella vulgaris* and *Dunaliellatertiolecta*. *BioEnergy Research,* 5(3), 669-84, 2012.

7

Heavy Metal Bioremediation and Toxicity Removal from Industrial Wastewater

Namrata Gupta[1]*, Monika Singh[2], Piyush Gupta[3], Preeti Mishra[2] and Vijeta Gupta[4]

[1]*Faculty of Engineering and Technology, RBS Engineering Technical Campus, Bichpuri, Agra, India*
[2]*Faculty of Pharmacy, RBS Engineering Technical Campus, Bichpuri, Agra, India*
[3]*Department of Chemistry, Faculty of Engineering and Technology, SRM Institute of Science and Technology, Modinagar, Ghaziabad, India*
[4]*Directorate of Plant Protection, Quarantine & Storage, Faridabad, India*

Abstract

Untreated industrial wastewater discharged directly into streams, rivers, and other water bodies pose serious environmental issues facing society right now. In order to protect nature, a cost-effective, eco-friendly approach is urgently required to treat inorganic metals, such as chromium (Cr), lead (Pb), cadmium (Cd), mercury (Hg), etc., that have been released into the environment. According to recent trends in studies related to the treatment of heavy metals, microbial bioremediation is being considered as an alternative to conventional methods. Excess amounts of heavy metals are unsafe to microorganisms because they are not biodegradable. In order to combat the toxic effects of these heavy metals, microorganisms have been investigated for the development of detoxification mechanisms. This chapter provides a detailed assessment of the bioremediation capabilities of microorganisms, particularly in regard to environmental protection. Additionally, the biosorption ability of algae, biofilms, bacteria, fungus, and other microorganisms for the purpose of heavy metal remediation are also investigated. Since studies have shown the synergetic effects resulting in heavy metal biofilm removal as a potential sustainable use of biofilms, the focus of this chapter is on microbial biofilms and algal films with respect to heavy metal sequestration. Biofilm includes adsorption and remediation of heavy metals as well as remediation using metal species, including

**Corresponding author*: ngp28@yahoo.com

Charles Oluwaseun Adetunji, Julius Kola Oloke, Naveen Dwivedi, Sabeela Beevi Ummalyma, Shubha Dwivedi, Daniel Ingo Hefft and Juliana Bunmi Adetunji (eds.) Next-Generation Algae: Volume I: Applications in Agriculture, Food and Environment, (163–194) © 2023 Scrivener Publishing LLC

metal uptake and biofilms. The bioremediation of heavy metals through adsorption and decontamination, including metal uptake and treatment in biofilms, is also presented.

Keywords: Algal biofilms, heavy metals, wastewater treatment and biosorption

7.1 Introduction

The excessive pollution of bodies of water with heavy metals is a serious environmental issue, which has a negative impact on not only human health but also on entire ecosystems as a result of food chain buildup. Various industries are responsible for pollution and environmental problems, including fertilizer manufacturing, metallurgy, leather production, aerospace, photography, and mining. Other industries responsible for pollution and environmental problems include iron and steel production, energy and fuel generation as well as appliances production [1]. Several hazardous metals, including chromium, lead, zinc, arsenic, copper, nickel, copper, cobalt, and mercury are highly damaging to the human body. Chemical precipitation includes the processes of flocculation, coagulation and hydroxide precipitation [2]. These approaches have the disadvantage of necessitating additional treatment to remove precipitated metals from the treated sludge, which poses additional environmental difficulties. Additional methods, such as electrochemical treatment, filtering, and reverse osmosis, are costly and ineffective at eliminating organic molecules as well as heavy metal pollutants. Alternative greener methods of heavy metal contaminants removal from waste streams include processes such as bioaccumulation, biosorption, phycoremediation, and bioaccumulation [3]. Because they are readily available and naturally occurring in large volumes of heavy metal waste industrial wastewater, they are considered economically and environmentally friendly [4]. Microorganism-based bioremediation has recently piqued the interest of academics because of its several advantages, including high efficiency, low cost, and environmental friendliness. Microbes, unlike plants and animals, have evolved to be able to withstand environmental stress through fast mutation processes. Toxic heavy metals have an effect on microbial diversity and metabolism when they are active, and organisms may develop resistance to heavy metallic ions to cope with the stress [5]. Heavy metals can be converted from active to passive state by some species. Microbes are promising biosorbents for inorganic substances such as heavy metals because of their high relativeness to kinetics and fast adsorption [6–8]. One of the primary reasons for the removal of heavy metals from wastewater treatment by biosorption is the existence of numerous functional groups in the cell walls of organisms, which assist in the collection, degradation, and neutralization of hazardous ions in water. Bacteria have been found to have a heavy metal absorption capability of up to 500 mg/g. Microalgae and bacteria are being utilized to clean wastewater, which is helping to decrease CO_2 emissions while also providing economic biomass for bioenergy and other biomass value addition applications [3]. Non-biodegradable metals accumulate in living organisms, leading to a number of diseases and disorders, some of which can be life-threatening [9]. As a result, certain toxins, such as bisphenol A (BPA), can disrupt a variety of biological processes, even at the level of concentrations parts per billion (ppb) [10]. While the conventional removal methods for heavy metal ions in industrial effluents do eventually lead to pollutants in waterways, biosorption is a newer method for eliminating pollutants. This chapter will discuss the possible applications of algae and bacteria for the treatment or removal of harmful pollutants from wastewater treatment facilities (WWTPs), with an overview of the toxicities of heavy metals, the benefits of biosorption, and various biosorbents utilized for the removal of metal ions. The impacts of immobilization and modification of biosorbents and other factors influence the process of biosorption. Several aspects, including the processes by which heavy metals are absorbed, factors affecting heavy metal remediation by microbes, as well as future possibilities, have been taken into account by the researchers.

7.2 Environmental Heavy Metal Sources

Like other toxic substances, metals are neither created nor destroyed by humans, metals are produced and consumed as an essential component of industry. Man-made as well as natural sources of environmental pollution can both cause these contaminants to accumulate in the environment.

The primary source of heavy metal pollution is industrial waste effluent [11]. Water supplies today include additional pollutants such as these, despite the fact that paper mills and fertilizer industries discharge different bases, cyanides, and heavy metals into the water supply (Table 7.1). From these industries, wastewater containing a significant concentration of heavy metal ions, such as metalworking wastewater, dyes and pigment wastewater, photographs and film wastewater, galvanometer wastewater, metal wastewater, electroplating wastewater, leather and mining wastewater, are common.

Table 7.1 Heavy metals drinking water standards.

Trace inorganic contaminants	**USEPA (mg/L)**	**Indian standards (mg/L)**
Arsenic, as As(V)	0.05	0.05
Cadmium, as Cd(II)	0.005	0.01
Chromium, as Cr(VI)	0.05	0.05
Iron, as Fe(II & III)	0.3	0.3
Lead, as Pb(II)	0.05	0.1
Manganese, as Mn(IV)	0.05	0.1
Mercury, as Hg(II)	0.002	0.001
Nickel, as Ni(II)	0.1	0.05

Pesticides and other toxicants break down but heavy metals do not. Metals like lead, mercury and cadmium enter the environment as a result of industrial processes like burning fossil fuels, mining, and metallurgy. People's activities and the location where contaminants are added have a significant impact on where and how contaminants are distributed. While the economy, transportation, and manufacturing account for approximately a third of the country's total environmental burden, roadways and automobiles each account for over a third of that burden due to the large amounts of particulate matter found in emissions from vehicles. Metals can enter the aquatic environment in one of three ways: they may be deposited from the atmosphere, erosion of the environment's geology or because of human activities such as industrial wastewater, sewage, and mining waste [12].

7.3 Heavy Metal Sources of Water Treatment Plants

In wastewater, common pollutants are the metallic ions, which in low concentrations can be dangerous to humans and animals. Lead is extremely harmful to the neurological, renal, and reproductive systems. Encephalopathic symptoms occur in those exposed to lead, which has harmful effects on the brain. Cadmium has been found in many industrial applications, such as electroplates, batteries, TVs, ceramics, photography, insecticides, electronics, metalwork and metallurgy. Rechargeable nickel-cadmium batteries are also sources of bone degeneration, liver and blood damage. Cadmium is now believed to be carcinogenic based on sufficient evidence.

As copper is an essential trace element for certain biological processes, such as photosynthesis, it must be ingested at specific levels to avoid any adverse effects on the human body. Nasal, eye, and mouth irritation from inhaling high concentrations of copper dust can occur. Vomiting and diarrhea may also occur. It is possible to damage your kidneys from excessive exposure. Even low concentrations of copper can harm a variety of aquatic organisms. Copper exposure in the environment may be found in a variety of settings, including mining, metallurgy, and industrial uses.

Zinc is also a major dietary element. Zinc is necessary, but consuming too much of it can be harmful to your health. If children consume large amounts of zinc, they may experience nausea and vomiting. Excessive zinc intake can lead to anemia and lipid problems in humans. The purification processing of zinc ores is one of the most significant sources of zinc in the atmosphere, soil, and water, and the mineral is also utilized in a variety of metal alloys. As a byproduct, coal combustion produces fly ash.

Nickel is an element that occurs naturally in volcanic soils and rocks. A range of industrial applications involves nickel and its salts, including electroplating units and aircraft. Ni–Cd batteries are made in large numbers on an industrial scale using nickel–cadmium sulphate, also known as nickel chloride. Inhabitants have gotten rid of most traces of it because it occurs naturally and breaks down into constituent elements in water bodies. Numerous problems can arise from water-soluble salts of nickel, which are the major contaminants in aquatic systems. Nail polish and enamel company effluents containing nickel are discharged into nearby bodies of water. Nickel is another name for the volatile compound nickel carbonyl, which is present in both cigarettes and food.

Arsenic is found in deposits all over the globe, which can include geologic deposits in various locations. Albertus Magnus was the first to isolate it as an element in 1250 AD. Arsenic is found in ores in powdered amorphous and crystalline forms. Some areas contain more arsenic than the amount present in a typical dose, and this results in significant public health problems, such as human poisoning and animal toxicity. Rock weathering and anthropogenic activities, such as mining and smelting, are responsible for natural materials entering the environment. Areas where groundwater is contaminated by arsenic are of great concern. In the process of adsorbing arsenic from water, metal oxides of iron, aluminium, and manganese have a role to play. Arsenic has been found naturally in water supplies in countries such as India, Bangladesh, Taiwan, Brazil, and Chile, which contain very high concentrations of arsenic. People and animals can be adversely affected by the high concentration of this chemical in water.

As in the case of Minamata Bay, Japan, mercury toxicity has become widely recognized across the globe. Fetal exposure to mercury as a result of eating contaminated fish results in babies being born with mental challenges and physical deformities. Natural mercury is defined as mercury obtained from natural sources such as volcanic eruptions, weathering of rock soils, pollution, and chemical processes, whereas anthropogenic mercury is defined as mercury derived from man-made sources such as pollution and chemical processes. The most toxic species of mercury is methyl mercury.

Large quantities of toxic chromium species, which occur as wastewater from industrial processes, have entered into water supplies as a result of extensive use of chromium compounds. The environment contains a source of chromium from natural inputs and anthropogenic emissions. Rock weathering in geology, soils, and sediments is the source of chromium, while human-induced processes (like burning fossil fuels and production of chromates) contribute to chromium in the environment. Hexavalent chromium is more toxic than trivalent chromium.

7.4 Heavy Metal Toxicity in Relation to Living Organisms

Heavy metals toxicity depends on how readily a metal can be absorbed by microbes, as well as the total amount absorbed. Toxicity of heavy metal involves a number of mechanisms that inhibit the function of essential enzymes, increase oxidative stress, and attack DNA and protein synthesis [13]. Heavy metals can affect the physiological and biochemical properties of microorganisms. The presence of Cr and Cd in microorganisms can be damaged and denatured, and the bioremediation capacity of microbes can be weakened. The organic compound chromium(III) can react with enzymes' carboxyl and thiol groups to change the structure and activity of the enzymes. The following might be connected: the introduction of cationic Cr(III) complexes into cells (an *in-vitro* assay) leads to cellular binding, cell electropotential, and thus epigenetic change (alteration in the gene pool).

Fenton and Haber-Weiss reactions can use metals, such as Cu^{+} and Cu^{++}, to produce ROS in a soluble electron carrier state. Damaging cytoplasmic molecules, DNA, lipids, and other proteins are the possible side effects. Superoxide radicals (or oxygen radicals) can be stabilized by aluminum, which causes fragmentation of DNA [14]. If metals (e.g., Hg, Pb, lead, Cd, etc.) interfere with the enzymatic functions of the enzymes' substrates, then the enzymes' structures will be changed, leading to inefficiency. The harmful results of ion imbalance can include attachment to the cell membrane and permeation into the cell via ion channels or transmembrane carriers.

Cadmium and lead harm cell membranes, DNA and bacteria. The harm occurs when an environment displaces metals from the binding sites [15]. Disturbed cell membranes limit enzyme activity and reduce phosphorylation. They also are involved in morphology, metabolism and development of microorganisms, which all have consequences [16] (Table 7.2).

Table 7.2 Heavy metal toxicity to microorganisms.

Heavy metals	Effects on microbes	Citations
Arsenic	Deactivates enzymes	[20]
Cadmium	Destroys nucleic acid; denatures protein; hinders transcription and cell division	[16]
Chromium	Elongates lag phase; inhibits growth; inhibits oxygen uptake	[21]
Copper	Disrupts cellular function; reduces enzyme activities	[16]
Selenium	Inhibits growth rate	[22]
Lead	Destroys nucleic acid and protein; inhibits enzyme actions and transcription	[16]
Mercury	Denatures protein; inhibits enzyme function; disrupts cell membrane	[16]
Nickel	Disrupts cell membrane; hinders enzyme activities and oxidative stress	[16]
Silver	Causes cell lyses; inhibits cell transduction and growth	[23]
Zinc	Decreases biomass; inhibits growth and death	[24]

7.5 Remediation Technologies for Heavy Metal Decontamination

7.5.1 Conventional Methods

Various metal contaminants flow into our reservoirs every day as a result of the discharge of wastewater from metal manufacturing companies. Although non-biodegradable, these substances have a tendency to accumulate in food chains, leading to illness. In the past, traditional treatment approaches for removal of heavy metals from contaminated water were performed. There are various ways to remove impurities, such as using chemicals to precipitate impurities, using ultrafiltration to remove impurities, ion exchange to remove impurities, electrowinning to remove impurities, and phytoremediation to remove impurities [17–19].

7.5.1.1 Chemical Precipitation

There are numerous ways to extract heavy metals from inorganic wastes; however, chemical precipitation is the most commonly used approach. The precipitation of dissolved metal ions results in the formation of insoluble metal hydroxides, sulfides, carbonates, and phosphates, according to the conceptual mechanism behind the phenomenon (filtering out the particles).

7.5.1.2 Ion Exchange

This technique is a reversible interchange of metal ions between a solid phase and a liquid phase. Anion exchangers are solid resins that can exchange both kinds of ions in an electrolyte, allowing for the equalization of the anion charge with the cation charge.

7.5.1.3 Membrane Filtration

Due to its capabilities, membrane filtration can remove both metal ions and suspended solids as well as organics. A membrane is a selective layer that allows two dissimilar phases to make contact while filtering contaminants of different sizes.

7.5.1.4 Reverse Osmosis

If the solution is separated by semi-permeable membrane (SPM), the heavy metal-containing liquid is forced to flow through a SPM, which is pushed by the pressure of the separated solution. To remove copper(II), nickel(II), and zinc(II) from water, the researchers employed polyamide thin-film composite membrane as SPM [25].

7.5.2 Ultrafiltration

This separation method uses pore diameters between 0.1 to 0.001 microns, and utilizes permeable membranes to permeate solvent and light substances while blocking bigger molecules, particles, and colloids after first removing the Cu^{2+}, Zn^{2+}, Ni^{2+}, and Mn^{2+} from water solutions with the help of ultrafiltration utilizing a polymer made of malic acid and acrylic acid [26].

7.5.3 Microfiltration

Microfiltration (MF) and ultrafiltration (UF) both use the same concept, known as differential sedimentation. MF (permeate rejection) separates out large solutes whereas UF (solutes rejected) rejects small solutes. Crossflow microfiltration was employed in the production of a bioaccumulation process in which yeast acts as a carrier for Cu^{2+}, Cd^{2+}, Pb^{2+}, Cr^{3+} and Cr^{+6}. Metal ions were effectively removed by the 31, 7, 63, and 71 percent efficiency methods.

7.5.4 Nanofiltration

In a nanofiltration process, molecular weight ranges from 300 DAs to 500 DAs are maintained with pores that have a diameter of 0.5 to 2 nm. Metal removal of 99%, 89%, and 74% were accomplished with commercially available nanofiltration membrane NF270.

7.5.5 Electrodialysis

The separation of ionic species in solution is accomplished by the use of electrodialysis, a new hybrid liquid membrane separation technology.

The removal of arsenic, lead, manganese, and nitrate nitrogen from groundwater in Korea was accomplished through the use of an electrodialysis system, which achieved removal rates of 73.9, 89.9, 98.9, and 95.1%, respectively.

7.6 Biological Approach in the Remediation of Heavy Metals

Microbes can be used to alter metal valences and extracellular precipitate compounds, and particular plants can be used to decontaminate soil or water by inactivating or moving metals to the rhizosphere and into their aerial portions. Plant remediation is the term used to describe the second technique. These approaches should be viewed as a novel and extremely promising rehabilitation technique for contaminated areas.

Biosorption is the use of economical biological materials to remove a solution of metals, metalloids, compounds and particles [27]. All biological materials can serve as excellent biosorbents for metal sequestration with the exception of mobile alkali metal cations such as sodium and potassium. Certain types are available in the following categories of cheap metal biosorbents, such as bacteria, fungus, algae and other polysaccharide materials. Wastewater, biosolids, industrial waste, agricultural waste and a range of other polysaccharide-containing materials. For the purposes of biosorption, all types of biomaterials tested (organic, inorganic, and hybrid) demonstrated comparable bioadsorption capabilities for all metal ions. a method for removing toxic heavy metals includes work done in the laboratory with microorganisms, as well as processing of diverse industries, wastewater treatment facilities, and various biomass sources [28, 29]. Heavy metals can be transformed and immobilised by microorganisms that can tolerate large concentrations of them. Bacteria have developed many strategies to survive exposure to metallic ions, including a number of processes to accommodate for uptake of such ions. One technique that can help reduce exposure to harmful metals is by the transfer of ions and the entrapment and buildup of the metal ions within the cell. This is accomplished through the use of binding and accumulation processes. Many creatures possess the ability to eliminate contaminants, and hence keep the ecosystem supplied. By using microorganisms, you may learn about biosorption of metal ions on cell surfaces, internal absorption of metal ions, and chemical alteration of metal ions [30]. While some methods need the use of several distinct techniques to remove metals from multi-element systems, a recent study indicated that biosorption is more selective.

Also, collecting metal-loving bacteria can be employed to process industrial effluents because of the same metabolic capabilities. A metal pollution heavy metal selectively prunes the microbial community and gives rise to strain-prone organisms with low extracellular enzyme activity. When examining the damaging effects of metal ions on cells, the beneficial use of microbes, including bacteria, algae, fungi, and yeast, is highly relevant [31]. This microbial consortium is also effective at collecting and removing hazardous metals from polluted locations because they have an exceptional ability to absorb the heavy metals. A large number of research studies on heavy metals remediation from industrial waste have been performed [32]. The reported high concentrations of several metal toxins among the strains reported by Soltan support his findings [33]. Workentine *et al.* found isolates of *Bacillus* and *Pseudomonas*, as well as other bacterial species, were resistant to mercury exposure [34]. Although only 43 isolates were found, the majority of them (almost 93%) were Gram-negative, and pigmentation rate was approx. 26 percent. They were *Moraxella, Pseudomonas, Bacillus Flavobacterium, Vibrio, Xanthomonas, Alcaligenes, Micrococcus, Aeromonas,* and *Acinetobacter.*

7.6.1 Bacteria as Heavy Metal Biosorbents

Biochemical bioremediation is the application of bacteria to remove contaminants, including metals ions and dyes, from wastewater streams that are not biodegradable. In practice, however, larger-scale isolation, screening, and harvesting could become quite difficult but are nevertheless some of the most effective methods of dealing with environmental pollutants. Various bacterial strains were employed to perform distinct metal ion removal jobs. Table 7.5 below illustrates the capability of different metal ions for biosorption by various types of bacterial biomass. Bacteria have evolved numerous efficient detoxification methods, most of which are meant to help the bacteria survive.

7.6.2 Algae as Heavy Metal Biosorbents

Algae are an inexpensive and efficient source of nutrients as they need only a small amount of nutrients. According to the findings of a research project on algae's biosorption capacity, it has been found that algae absorb about 15.3% to 84.6%, which is greater in comparison to other microbial biosorbents. Particularly, brown algae have been recognized to have strong absorbing capabilities. Ion exchange serves as a mechanism on the cell surface for sorption of metal ions by microorganisms. Carboxyl, sulfonate, amino, and sulfhydryl are the chemical groups on the brown sea algae's surface that are capable of absorbing metals like Cd, Ni, and Pb.

7.6.3 Fungi as Heavy Metal Biosorbents

It has been established that fungi are successfully applied as a biosorbent material and that they are an economic and eco-friendly option. There are many different kinds of fungi capable of producing extracellular enzymes, and these enzymes can be used to process complex polysaccharides and convert them into hydrolyzed products. In addition, they are also suitable for large-scale production because they are quite simple to cultivate in fermenters. It's also possible to separate fungal biomass with a filter since its structure is filamentous. Yeasts are much more sensitive to changes in nutrition, aeration, pH, and temperature than filamentous fungi [35, 36].

7.6.4 Phytoremediation

Botanical bioremediation is now being studied, which offers significant potential for cleanup of contaminated soils and water sources. Phytoremediation literally refers to several ways that plants remove pollutants from the environment. When considering whether to use a traditional strategy, such as engineering, or to use a novel way, like phytoremediation, a common argument is that phytoremediation is more efficient and less costly [37]. Ernst postulates that metal hyperaccumulation is a result of a plant's exposure to natural excesses of different metals, as well as plants' tolerance to those metals [38]. Almost all species in the genus *Ipomoea* have shown considerable potential for accumulating metals, such as iron, copper, or zinc [39]. Depending on species and cultivar, the capacity to accumulate heavy metals varies greatly. While there are various approaches for phytoremediation, there are also several mechanisms of remediation, meaning the phytoremediation process varies with each.

7.7 Mechanism Involved in Biosorption

Microorganisms are organisms that can tolerate adversity. Their abilities have been developed over the previous millions of years [40]. For many decades, various environmental scientists, engineers, and biotechnologists have been fascinated by the power of microbes to remove heavy metal ions as well as radionuclides and change them into less harmful forms. So, as a result, several ideas for bioremediation of waste streams and cleanup of heavy metal contaminants from the environment are being anticipated, including those that are being implemented for the first time [41]. Due to the variety of different mechanisms utilized in biosorption, many of them are not well understood. Incorporation into a cell's metabolism (i.e., non-metabolism-dependent) or to where the metal is sequestered after extraction from solution (i.e., intracellular or extracellular) is referred to as a metabolism-dependent condition [42].

When metal uptake occurs while the organism is in metabolic state, it is through the interactions between the microbe's reactive groups and the metal [43]. Metal binding groups in polysaccharides, proteins, and lipids can be found in vast quantities in microbial biomass's cell walls. Ahalya *et al.* 2003 found that biosorption in this specific manner, meaning it is not dependent on metabolism, is relatively quick and reversible, as explained in Figures 7.1 and 7.2 [44].

The mechanisms of detoxification may be categorized as:

(i) Intracellular sequestration;
(ii) Modifying the membrane transport system that occurs when harmful ions are present in the cell and preventing them from entering;

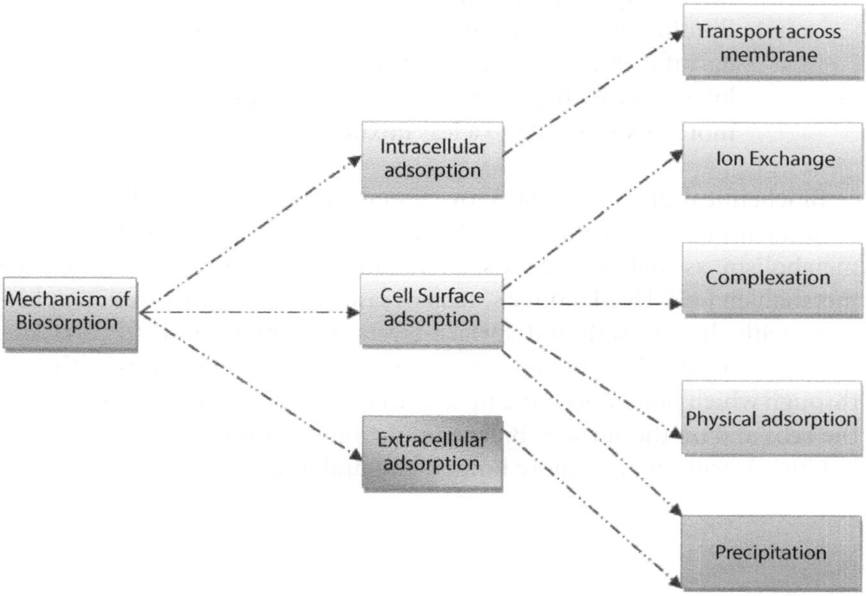

Figure 7.1 Mechanisms of biosorption.

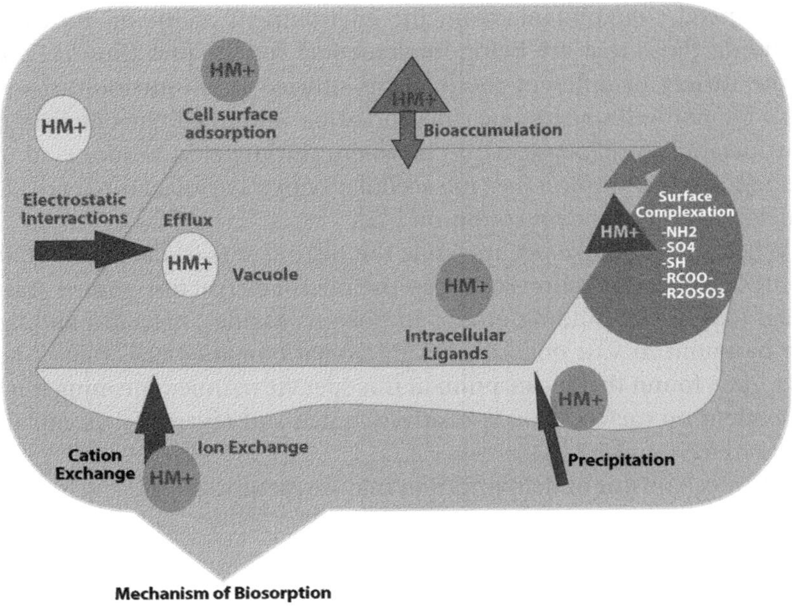

Figure 7.2 Biosorption mechanisms based on metal removal site.

(iii) Improving gas permeability;
(iv) Binding of mineral-ion in extracellular sequestration, the harmful cations or anions are removed from the extracellular space by the action of enzyme for the conversion of a more noxious state to a less noxious state (Tables 7.3–7.6).

Biochemical processes that move metals into microbial cells, like bioaccumulation, include processes that operate in a manner independent of metabolism, as well as processes where bioaccumulation occurs alongside metabolism [45]. Dead biomass, such as sawdust, which involves complexation with the cell wall and surface layers, may be utilized to carry out biosorption [46]. It is important to understand the numerous pathways through which pollutants build up in the body since they occur both within the cells and on the surface. Bioaccumulation relies on multiple pathways, but bioadsorption has a more limited and undefined effect [47].

Table 7.3 Common industrial units discharging toxic heavy metals [48].

S. no.	Heavy metals	Sources
1.	Arsenic	Semiconductors, oil refining, wood conservation products, feed additives, coal plants, automotive exhaust, dyeing and industrial dust.
2.	Cadmium	Metal smelting and refining, phosphate fertilizers, paint pigments, pesticides, plastics, polyvinyl and copper refineries.
3.	Chromium	Electroplating, leather, chrome plate, oil refining, tanning, manufacture of textiles and pulp processing equipment. Both hexavalent and trivalent forms exist.
4.	Copper	Electroplating industry, metal refining, plastic industry and industrial emissions.
5.	Iron	Metal refining, galvanization of engine parts.
6.	Lead	Automobile batteries, petrol-based materials, pesticides and paints.
7.	Mercury	Emissions from caustic soda, thermometers, adhesives, paints, light bulb industry, wood preservatives, leather industry, tanning industry and ointment industry.
8.	Nickel	Refining of metals, galvanization of paint and powder, processing batteries and fertilizers for superphosphates.
9.	Zinc	Rubber industry, paints, dyeing industry, wood preservatives and ointments.

Table 7.4 Algal species with biosorption capacity.

Metal removed	Algal species	Biosorption capacity (mg/g)	Citations
Cadmium	*Sargassum* sp.	84.7	[49]
Chromium	*Chlorella miniata*	34.60	[50]
Copper	*Spirulina platensis*	67.93	[51]
Cadmium	*Ulva lactuca* sp.	43.02	[52]
Lead		181.82	[52]
Lead	*Spirogyra* sp.	140	[53]
Zinc	*Sargassum muticum*	34.10	[54]

Table 7.5 Heavy metal-removing bacterial species.

Metal ion	Bacterial species	Biosorption capacity, (mg/g)	pH	Citations
Zinc Zn(II)	*Pseudomonas putida*	17.7	5	[55]
	Bacillus jeotgali	222.2	7	
Copper Cu(II)	*Enterobacter* sp. J1	32.5	5	[56]
	Arthrobacter sp.	17.87	5	[57]
Chromium Cr(VI)	*Pseudomonas fluorescens*	40.8	2	[58]
	Pseudomonas sp.	95	4	[59]
Nickel Ni(II)	*E. coli*	6.9	2.7–3.6	[60]
	Pseudomonas fluorescens	40.8	2	[58]
Cadmium Cd(II)	*Enterobacter* sp. J1	46.2	6	[60]

Table 7.6 Use of fungal species and their capacity for biosorption.

Fungal species	Metal ion	Biosorption capacity (mg/g)	Citations
Aspergillus niger	Pb	34.4	[61]
	Cu	28.7	[62]
Penicillium simplicissimum	Cd	52.50	[63, 64]
	Zn	65.60	[63, 65]
Saccharomyces cerevisiae	Pb	270	[66]
	Hg	64.2	
Penicillium chrysogenum	Ni	260	[67, 68]
	Pb	204	
	Cu	92.0	[69]
Penicillium purpurogenum	Cr(VI)	36.5	[69–71]

7.7.1 Intracellular Sequestration

Metal ions are sequestered within the cell cytoplasm by different chemicals, which complex the metal ions. Metals that are dissolved into microbial cells first bind with surface ligands and then enter via transport at a slow rate Metal sequestration within bacterial cells has found use in industries like wastewater treatment. Experiments showed that *P. putida* strain exhibited the capacity to sequester Cu, Cd, and Zi ions from the cytoplasm via the usage of cysteine-rich low molecular weight proteins that included cationic residues. Additionally, it was shown that glutathione inside *Rhizobium leguminosarum* cells was capable of sequestering cadmium ions within the cell [72].

There are various fungi that accumulate metals in their mycelium as well as spores. The cell wall of fungi has a ligand-like molecule that enables it to function as a labeling agent and assists the removal of inorganic metals [73–76]. Metal-binding ligands including –COOH, $-RCOO^-$, –OH, $-NH_2$, $-HPO_4^{2-}$, SO_4^{2-}, $R_2OSO_3^-$ and –SH are abundant in peptidoglycan, polysaccharide, and lipid cell wall components. Metal absorption, meanwhile, can be facilitated by cationic and anionic metal species as amine binds to anionic and cationic metal species by electrostatic and surface complexation.

7.7.2 Extracellular Sequestration

The extracellular sequestration of metallic ions by cell components occurs in the periplasm, and insoluble metal ion complexes form by metal ion association in the periplasm. In *Synechocystis* sp. PCC6803, zinc ions may enter the cytoplasm through an efflux mechanism and accumulate in the periplasm. The extracellular sequestration of metal is called metal precipitation. Iron-reducing bacteria, such as *Geobacter* sp. and sulfur-reducing bacteria, such as *Desulfuromonas* sp., can help get rid of dangerous metals, such as heavy metals, by either reducing the metals to benign metals or leaving them in a reduced state. *Geobacter* strict anaerobe is capable of converting manganese (Mn) to manganese (II) and uranium (U) to uranium (VI) (IV). White and Knowles reported that G. sulfurreducens and G. metallireducens have the potential to reduce the harmful chromium (VI) to safer Cr (III) [77]. Heavy metal ions precipitate when bacteria produce high amounts of hydrogen sulfide.

7.7.3 Extracellular Barrier of Metal Prevention in Microbial Cells

Using metal-binding membranes, cell walls, or capsules formed by microorganisms, it may be feasible to prevent metal ions from entering the cell. An ionizable region of the cell wall of bacteria can absorb metal ions (carboxyl, phosphate, hydroxyl and amino groups). Green-Ruiz and Taniguchi *et al.* stated that non-active cells of *Pseudomonas putida* and *Bacillus* sp. passively biosorbed heavy metals at high levels [78, 79]. While planktonic cells, as well as biofilm cells that are positioned at the periphery, were vulnerable to ionic forms of Cu, Pb, Zn, etc., cells placed at the perimeter of the biofilm showed more resilience. Metal ions collected in extracellular polymers of biofilm, and thus facilitated the protection of bacterial cells within the biofilm [80].

7.7.4 Metals Methylation

Higher metal lipophilicity and cell membrane penetration result in increased metal toxicity as a result of increased metal lipophilicity and cell membrane penetration, respectively. The elimination of hazardous metals from the environment is an important function of microbial methylation.

The conversion of arsenic and volatile dimethyl selenide to gaseous arsines was observed in contaminated soil, while the conversion of lead to dimethyl lead was discovered in contaminated soil.

7.7.5 Heavy Metal Ions Remediation by Microbes

By converting metal ions oxidation state, microbes can help reduce the overall toxicity of the metal ions [80, 81]. Metals as well as metalloids are frequently used by bacteria as electron donors or acceptors. Aerobic bacteria's anaerobic respiration occurs when electron acceptors, which are in the oxidized form, are used. By means of an enzymatic process, you can reduce the number of metal ions, resulting in less dangerous levels of mercury and chromium [82].

7.8 Alga-Mediated Mechanism

Different mechanisms for microalgae heavy metal remediation have been explained by various authors. Microalgae have a negatively charged cell wall surface as a result of the accumulation of functional groups, such as phosphate, carboxyl, sulfhydryl, hydroxyl and amino, on the surface of the cell wall. Functional groups in microalgae, much as in bacteria, are critical in the uptake of heavy metals by the organisms. Furthermore, the numerous metal transports in plasma containers are responsible for removing metals from the cytoplasm. Microalgae diaphragm and cell wall are made up of lipids, protein and polysaccharides. Membrane transporters are important for linking microalgae with heavy metals. The algae bioremediation mechanism includes the non-metabolic absorption and metabolic absorption. Cell surface adsorption is the primary mechanism of non-metabolic activities as ionic replacement, complexation and precipitation. Adsorption of metals within the cell was mediated by active metal transport across the membrane and into the metabolism-dependent metal removal system [83]. The cleanup mechanism of heavy metal by microalgae and bacteria is depicted in Figure 7.3.

7.9 Application of Biosorption for Waste Treatment Technology

Among the most popular applications for biofilm algal systems is wastewater treatment. This is because it provides a straightforward, energy-efficient

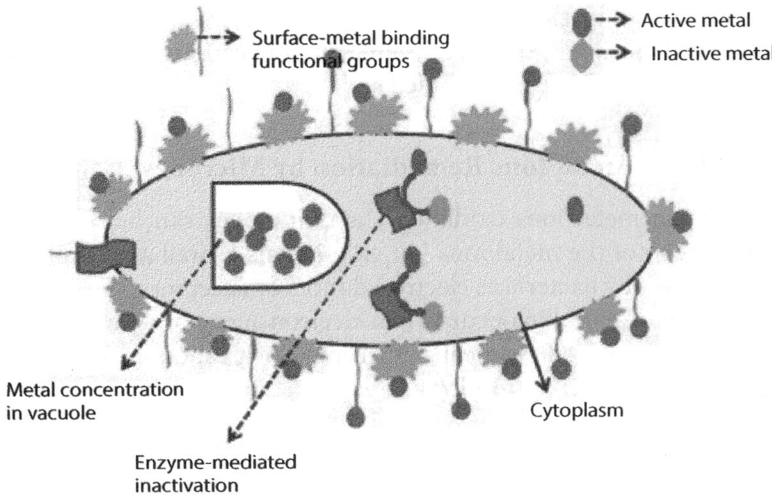

Figure 7.3 Mechanism of bacterial and algae heavy metal remediation.

technology that absorbs key nutrients such as N and P, followed by a straightforward separation between algal biomass and wastewater mass [84–88]. Microalgae biofilms can grow on any surface or carrier material with enough humidity and light, as evidenced by the several materials successfully evaluated for microalgic biofilms as carriers. Diatoms and green algae are commonly used to colonize the carrying material.

Over the course of many years, microalgae have been utilized in restricted algal ponds to cleanse municipal wastewater. Inorganic nitrogen and phosphorus microalgae remove ammonium (NH_4^+) and phosphate (PO_4^{3-}) from wastewater effluents, as well as (nitrified) nitrate (NO_3^-) and PO_4^{3-} from wastewater effluents, by absorbing them into the biomass of the algae. Protein and nucleic acid synthesis are accomplished via the use of absorbed nitrogen, whereas P is accomplished through the use of P. It is also possible that phospholipids are synthesized from phosphorus and potassium's abundant absorption, i.e., PO_4^{3-} internal storage, and that they may also arise as polyphosphates [89]. Precipitation can remove PO_4^{3-} from the environment in addition to absorption. The pH of the water rises when there is more NO_3^- and/or more carbon in the water that can be absorbed from the atmosphere. If the pH of the solution is too high, the ion PO_4^{3-} may surge into cations such as Ca^{2+} and Mg^{2+}, which are commonly found in municipal wastewater sources. In 2003, Metcalf and Eddy published their findings [90]. A lot of research has been done on the nutrient removal capabilities of the microalgae *Scenedesmus* and *Chlorella* [91, 92]. Although most wastewater treatment systems are based on microalgae monocultures, the microalgae mentioned above can be found in conjunction with a variety of other species in most wastewater treatment systems.

Microalgal biofilm systems can be comprised of huge biofilm panels that are positioned above the effluent and flow through it. An algal turf scrubber is an example of a level microalgae biofilming device [93] that cultivates microalgae films on a sloping flow way, suggesting that the effluent from a wastewater treatment facility is being polished efficiently [94]. Twin layer is an aerated biofilm system that isolates microalgae by immobilizing them on an atmospheric substratum layer, while enabling nutrients to pass through a second layer of enclosure to the surface of the water [92].

7.10 Microbial Heavy Metal Remediation Factors

The tendency to stimulate or inhibit heavy metals in microorganisms is determined by following factors:

- **Metal**

Metallic ion concentrations and metal's chemical forms and its redox potential.

- **Substrate**

Other parameters that influence bioremediation include the amount of biomass present, the amount of biosorbent used, and the length of time the biosorbent is exposed to the environment. The state of algae's growth and development can have an impact on their ability to absorb nutrients. In several studies, it was observed that microalgae and bacteria having a greater number of functional groups had better adsorption on the cell wall surface. This is dependent on the type of algae and bacterial biomass being used as biosorbents, due to the nature of the microbial biomass being utilized as biosorbents. After harvesting algae biomass, it is necessary to dry the biomass, and it is important to ensure that the algae's reactive groups are not damaged by the drying or pretreatment of the biomass. Higher levels of biosorbents increase costs, and although a lower dosage will have a significant impact on biosorption efficiency it will not be sufficient for a wide operation. The optimum dose can therefore be selected by carefully evaluating the experimental results for the extensive process design of heavy metal bioremediation.

- **Contact time**

It contributes to the overall efficiency of the absorption process. Heavy metal remediation has an effect on the height of the bed, the flow of effluent, and the initial concentrations of heavy metals during mass operation. The potential for metal absorption by microorganisms is boosted by increasing the height of the bed, which increases the amount of absorbent surface. Additionally, the higher flow rate has an effect on the volume of wastewater produced and the bioremediation process itself.

- **Environmental effects**

Concentration of hydrogen ions, ambient temperature, molecular weight, and humic acids are all important factors that can have an impact on

Table 7.7 Factors affecting heavy metal bioremediation.

Factors	Activities
Microbial	(i) Enrichment of capable microbial populations (ii) Enzyme induction (iii) Mutation and horizontal gene transfer (iv) Production of toxic metabolites
Substrate	(i) Chemical structure of contaminants (ii) Solubility of contaminants (iii) Too low concentration of contaminants (iv) Toxicity of contaminants
Mass transfer limitations	(i) Diffusion of nutrients (ii) Solubility and oxygen diffusion (iii) Solubility/miscibility in/with water
Growth substrate vs. co-metabolism	(i) Alternate carbon source present (ii) Concentration (iii) Microbial interaction
Biological aerobic vs. anaerobic process	(i) Availability of electron acceptors (ii) Microbial population present in the site (iii) Oxidation/reduction potential
Environmental	(i) Depletion of preferential substrates (ii) Inhibitory environmental conditions (iii) Lack of nutrients

the transformation, transport, and valence state of heavy metals and also heavy metal bioavailability to microorganisms. The free ionization of heavy metals occurs most frequently at acidic pH, where there are extra protons available to satisfy the binding sites of metals. Increased adsorbent surface loading occurs at higher pH values, resulting in a reduction in attraction between the adsorbent surface and metal cations. The ambient temperature has a critical effect on the adsorption of heavy metals. The rate of adsorption spread over the exterior boundary layer increases as the temperature of the surrounding environment rises. As temperature rises, it increases the metallic solubility, resulting in an increase in the bioavailability of metals for humans [95]. Elevated temperature causes microorganism activity to rise, as does microbial metabolism and enzyme activity, resulting in an acceleration of bioremediation and a reduction in the time it takes to complete it. Furthermore, temperature influences the metabolism and enzyme activity of microorganisms, thereby increasing the efficiency of bioremediation. As a result of the substrate and a variety of external influences, deterioration might occur [96]. Sorption sites, cell wall structure of microbes, and the chemical mobility of metal ionization on the cell walls all play a role in the complicated stability of microbe-metal interactions (Table 7.7).

A variety of factors influence the decomposition of bacteria and algae. These factors include the substrate, the type of biomass, and ambient conditions.

7.11 Conclusion

Presence of heavy metal ions, bacteria and microalgae are strong detoxification organisms that can remove toxins. As a consequence of these organisms, low-cost wastewater with high metal absorption and selectivity may be produced in large quantities. It is necessary to have a deeper knowledge of the physical and other factors mentioned in this chapter in order to achieve successful bioremediation. There have only been a few laboratory-scale bioremediation experiments recorded for wastewater; many field investigations are required to fully grasp the greatest bioremediation potential in order to make larger operations feasible and economically viable. One of the most appealing aspects of high-biomass algae growth in wastewater is the ability to utilize microalgae for both value-added products and low-cost biomass for the same process.

7.12 Future Prospects

The present state of bioremediating heavy metals using the various mechanisms described in this chapter, such as microalgae and bacteria, demonstrates that the metal absorption and detoxification technique is a viable option. Research has shown that there are still certain barriers to widespread use of those techniques, such as the difficulty in obtaining dependable, inexpensive biomass and the opposing effects on the biomass potential of organisms caused by the presence of a coexisting metal ion in the environment. There are significant variations in the ecological, physiological, and genetic expression of bioremediation microorganisms, as well as disparities in their preferred habitats. Research should be focused on the creation of engineered microorganisms that are suitable for the process of bioremediation in wastewater treatment plants (WWTPs) that handle complicated pollutant combinations. Another source of concern is that the biodegradation result, rather than the chemical parent, may be hazardous. The possibility of including this in future studies is being explored. Other organisms do not have major adsorption mechanisms for heavy metal ions, whereas bacteria and microalgae do. As established in a number of studies, they outperform numerous commercial and conventional physicochemical processing procedures, and their application in the field of wastewater remediation technology has shown promise. In the case of microorganisms, the efficacy is impacted by a variety of criteria such as strains used, tolerance to the environment, size of the organisms, growth phase, metal ion concentration, and the kind of metal ion. Temperature, pH, durability, salinity, and other abiotic elements all play a key part in the process. Biosorption studies of microalgae and bacteria from various effluents have shown that technology provides a sustainable, economically viable, and environmentally friendly route to a healthier environment.

References

1. Das N., Vimala R. andKarthika P., "Biosorption of heavy metals - An overview". *Indian J. Biotechnol.*, Volume Number 7, pp 159-169, 2008.
2. Aziz H.A., Adlan M.N. and Ariffin K.S., "Heavy metals (Cd, Pb, Zn, Ni, Cu and Cr (III)) removal from water in Malaysia: post treatment by high quality limestone". *Bioresour. Technol.*, Volume Number 99 (6), pp 1578-1583, 2008.
3. Alam M.A., Wan C., Zhao X.Q., Chen L.J. and Bai F.W., "Enhanced removal of Zn^{2+} or Cd^{2+} by self flocculating microalga Chlorella vulgaris". JSC7. *J. Hazard. Mater.*, Volume Number 298, pp 38-35, 2015.

4. Cai X., Zheng X., Zhang D., Iqbal W., Liu C., Yang B., Zhao X., Lu X. and Mao Y., "Microbial characterization of heavy metal resistant bacterial strains isolated from an electroplating wastewater treatment plant". *Ecotox. Environ. Saf.*, Volume Number 181, pp 472-480, 2019.
5. Bilal M., Rasheed T., Sosa-Hernandez J.E., Raza A., Nabeel F. and Iqbal H.M.N., "Biosorption: an interplay between marine algae and potentially toxic elements a review". *Mar. Drugs*,Volume Number 16, pp 1-16, 2018.
6. Banerjee G., Pandey S., Ray A.K. and Kumar R., "Bioremediation of heavy metals by a novel bacteria strain Enterobacter cloacae and its antioxidant enzyme activity, flocculant production and protein expression in presence of lead, cadmium, and nickel". *Water Air Soil Pollut.* Volume Number 226 (6), pp 1-9, 2015.
7. Banerjee, S., Kamila, B., Barman, S., Joshi, S.R., Mondal, T. and Halder, G., "Interlining Cr(IV) remediation mechanism by a novel bacterium Pseudomonas brenneri isolated from coalmine wastewater". *J. Environ. Manag.*, Volume Number 233, pp 271-282, 2019.
8. Chan A., Salsali H. and McBean E., "Heavy metal removal (copper and zinc) in secondary effluent from wastewater treatment plant by microalgae". *ACS Sust. Chem. Eng.*,Volume Number 2(2), pp 130-137, 2013.
9. Monachese M., Burton J.P. and Reid G., "Bioremediation and tolerance of humans to heavy metals through microbial processes: A potential role for probiotics?"*Appl. Environ. Microbiol.*, Volume Number 78, pp 6397-6404, 2012.
10. Bulgariu, D. and Bulgariu, L., "Potential use of alkaline treated algae waste biomass as suitable biosorbent for clean recovery of cadmium (II) from aqueous media: batch and column studies". *J. Clean. Prod.*, Volume Number 112, pp 4525-4533, 2016.
11. Srivastava N. K. and Majumder C. B., "Novel biofiltration methods for the treatment of heavy metals from industrial wastewater", *J. Hazard. Mater.*,Volume Number 151, pp 1–8, 2008.
12. Lee K.Y. and Kim K.W., "Heavy metal removal from shooting range soil by hybrid electrokinetics with bacteria and enhancing agents", *Environ. Sci. Technol.*,Volume Number 44, pp 9482-9487, 2010.
13. Gauthier P. T., Norwood W. P., Prepas E. E. and Pyle G. G., "Metal-PAH mixtures in the aquatic environment: A review of co-toxic mechanisms leading to more-than-additive outcomes", *Aquat. Toxicol.*,Volume Number 154, pp. 253–269, 2014.
14. Booth S. C., Weljie A. M. and Turner R. J., "Metabolomics reveals differences of metal toxicity in cultures of Pseudomonas pseudoalcaligenes KF707 grown on different carbon sources", *Front. Microbiol.*, Volume Number 6, pp 1-17, 2015.
15. O. Olaniran, A. Balgobind, and B. Pillay, "Bioavailability of heavy metals in soil: Impact on microbial biodegradation of organic compounds and possible

improvement strategies", *Int. J. Mol. Sci.*,Volume Number 14 (5), pp 10197-10228, 2013.

16. Fashola M. O., Ngole-Jeme V. M. and Babalola O. O., "Heavy metal pollution from gold mines: Environmental effects and bacterial strategies for resistance", *Int. J. Environ. Res. Public Health*, Volume Number 13(11), pp 1047, 2016.
17. Joshi N.C. Heavy metals, conventional methods for heavy metal removal, biosorption and the development of low cost adsorbent. *European Journal of Pharmacy and Medical Research.*, Volume Number 4, pp 388-393, 2017.
18. Lakherwal D., "Adsorption of heavy metals: A review". *Int J Environ Res. & Develop.*,Volume Number 4, pp 41-48, 2014.
19. Gunatilake S., "Methods of removing heavy metals from industrial wastewater". *J. Multidiscip. Eng. Sci. Technol.*, Volume Number 1, pp 12-18, 2015.
20. Sankarammal M., Thatheyus A. and Ramya D., "Bioremoval of Cadmium Using Pseudomonas fluorescens", *J Water Pollut Control Fed.*, Volume Number 2, pp 92–100, 2014.
21. Cervantes C., Campos-Garcia J. and Devars S., "Interactions of chromium with microorganisms and plants". *FEMS Microbiol. Rev.*, Volume Number 25(3), pp 335–347, 2001.
22. Dixit R., Wasiullah, Malaviya D., Pandiyan K., Singh U.B., Sahu A., Shukla R., Singh B.P., Rai J.P., Sharma P.K., Lade H. and Paul D., "Bioremediation of Heavy Metals from Soil and Aquatic Environment: An Overview of Principles and Criteria of Fundamental Processes". *Sustainability*, Volume Number 7, pp 2189-2212, 2015.
23. Prabhu S. and Poulose E. K., "Silver nanoparticles: mechanism of antimicrobial action, synthesis, medical applications, and toxicity effects". *Int. Nano Lett.*, Volume Number 2(1), Article no. 32, pp 1-10, 2012.
24. Malik A., "Metal bioremediation through growing cells". *Environ. Int.*, Volume Number 30(2), pp 261–278, 2004.
25. Bakalar T., Bugel M., Gajdosova L., "Heavy metal removal using reverse osmosis". *Acta Montan. Slovaca*, Volume Number 14, pp 250-253, 2009.
26. Qiu Y.R. and Mao L.J., "Removal of heavy metal ions from aqueous solution by ultrafiltration assisted with copolymer of maleic acid and acrylic acid". *Desalination*, Volume Number 329, pp 78-85, 2013.
27. Wang J. and Chen C., "Biosorbents for heavy metals removal and their future". *Biotechnol. Adv.*, Volume Number 27, pp 195-226, 2009.
28. Kumar R., Sharma A.K., Singh P., Dhir B. and Mehta D., "Potential of Some Fungal and Bacterial Species in Bioremediation of Heavy Metals". *J. Nuclear Phys. Mater. Sci. Radiat. Appl.*, Volume Number 1, pp 213-223, 2014.
29. Macek T. and Mackova M., "Potential of biosorption technology", In Book: Microbial Biosorption of Metals, *Edited by Pavel Kotrba, Martina Mackova, & Tomas Macek*, Springer, Dordrecht, pp 7-17, 2011.

30. Zimmerman R., "Abiotic and microbial oxidation of laboratory-produced black carbon (Biochar)". *Environ. Sci. Technol.*, Volume Number 44, pp 1295-1301, 2010.
31. Wiatrowski H. A., Ward P. M. and Barkay T., "Novel reduction of mercury (II) by mercury-sensitive dissimilatory metal reducing bacteria", *Environ. Sci. Technol.*, Volume Number 40, pp 6690–6696, 2006.
32. Renella G., Landi L., Ascher J., Ceccherini M. T., Pietramellara G., Mench M. and Nannipieri P., "Long-term effects of aided phytostabilisation of trace elements on microbial biomass and activity enzyme activities, and composition of microbial community in the Jales contaminated mine spoils". *Environ. Pollu.*, Volume Number 152, pp 702-712, 2008.
33. Soltan E.S., "Isolation and Charaterization of antibiotic and heave metal resistance P. Stutzeri". *BioMetals*, Volume Number 7, pp 30-40, 2001.
34. Workentine M. L., Harrison J. J.,.Stenroos P. U., Ceri H. and Turner R. J., "Pseudomonas fluorescence view of the periodic table". *Environ. Microbiol.*, Volume Number 10, pp 238-250, 2008.
35. Sardrood B.P., Goltapeh E.M. and Varma A., "An Introduction to Bioremediation Fungi as Bioremediators". Springer, Berlin, pp 3-27, 2013.
36. Leitao A.L., "Potential of Penicillium species in the bioremediation field". *Int. J. Environ. Res. Public Health*, Volume Number 6, 1393-1417, 2009.
37. Gabos M. B., Abreu C. A. and Coscione A. R., "EDTA assisted phytoremediation of a Pb contaminated soil: metal leaching and uptake by jack beans". *Sci. Agric.*, Volume Number 66, 506–514, 2009.
38. Ernst W. H. O., "Bioavailability of heavy metals and decontamination of soils by plants". *Appl. Geochem.*, Volume Number 11, 163-167, 1996.
39. Bartoli F., Coinchelin D., Robin C. and Echevarria G., "Impact of active transport and transpiration on nickel and cadmium accumulation in the leaves of the Ni hyperaccumulator Leptoplaxemarginata: a biophysical approach". *Plant Soil*, Volume Number 350, 99-115, 2012.
40. Shuhong Y., Meiping Z., Hong Y., Han W. and Shan X., "Biosorption of Cu^{2+}, Pb^{2+} and Cr^{6+} by a novel exopolysaccharide from Arthrobacter ps-5", *Carbohydr. Polym*, Volume Number 101, pp 50-56, 2014.
41. Hrynkiewicz K. and Baum C., "Application of microorganisms in bioremediation of environment from heavy metals". In book: *Environmental Deterioration and Human Health, Editors:* A. Malik *et al.*, Springer, Dordrecht, pp 215-227, 2014.
42. Yu F.B., Shan S.D., Luo L.P., Guan L.B. and Qin H., "Isolation and characterization of a Sphingomonas sp. strain F-7 degrading fenvalerate and its use in bioremediation of contaminated soil". *J. Environ. Sci. Health – B,* Volume Number 48, pp 198-207, 2013.
43. Neethu C.S., Mujeeb Rahiman K.M., Saramma A.V. and Mohamed Hatha A.A., "Heavy-metal resistance in Gram-negative bacteria isolated from Kongsfjord, Arctic". *Can. J. Microbiol.* Volume Number 61, pp 429-435, 2015.

44. Ahalya N., Ramachandra T. and Kanamadi R., "Biosorption of heavy metals". *Res. J. Chem. Environ.* Volume Number 7, pp 71-79, 2003.
45. Vijayaraghavan K. and Yun Y.S., "Bacterial biosorbents and biosorption". *Biotechnol. Adv.,* Volume Number 26(3), pp 266-291, 2008.
46. Fomina M. and Gadd G. M., Biosorption: current perspectives on concept, definition and application, *Bioresour. Technol.,* Volume Number 160, pp. 3–14, 2014.
47. Ayangbenro S. and Babalola O.O., "A new strategy for heavy metal polluted environments: A review of microbial biosorbents". *Int. J. Environ. Res. Public Health,* Volume Number 14(94), pp 1-16, 2017.
48. Lone M.I., He Z.L., Stoffella P.J. and Yang X.E., "Phytoremediation of heavy metal polluted soils and water: progresses and perspectives". *J. Zhejiang Univ. Sci. B,* Volume Number 9, pp 210-220, 2008.
49. Hajar M., "Biosorption of cadmium from aqueous solution using dead biomass of brown alga Sargassum Sp." *Chem. Eng. Trans.,* Volume Number 17, pp 1173-1178, 2009.
50. Li J., Lin Q. and Zhang X., "Mechanism of electron transfer in the bioadsorption of hexavalent chromium within Leersiahexandra Swartz granules by X-ray photoelectron spectroscopy". *J. Hazard. Mater.,* Volume Number 182, pp 598-602, 2010.
51. Celekli A., Yavuzatmaca M. and Bozkurt H., "An eco-friendly process: predictive modelling of copper adsorption from aqueous solution on Spirulina platensis". *J. Hazard. Mater.,* Volume Number 173, pp 123-129, 2010.
52. Bulgariu L., Lupea M., Bulgariu D., Rusu C. and Macoveanu M., "Equilibrium study of Pb (II) and Cd (II) biosorption from aqueous solution on marine green algae biomass". *Environ. Eng. Manag. J.,* Volume Number 12, pp 183-190, 2013.
53. Gupta V.K. and Rastogi A., "Biosorption of lead from aqueous solutions by green algae Spirogyra species: kinetics and equilibrium studies". *J. Hazard. Mater.,* Volume Number 152, pp 407-414, 2008.
54. Freitas O.M., Martins R.J., Delerue-Matos C.M. and Boaventura R.A., "Removal of Cd (II), Zn (II) and Pb (II) from aqueous solutions by brown marine macro algae: Kinetic modelling". *J. Hazard. Mater.,* Volume Number 153, pp 493-501, 2008.
55. Green-Ruiz C., Rodriguez-Tirado V. and Gomez-Gil B., "Cadmium and zinc removal from aqueous solutions by Bacillus jeotgali: pH, salinity and temperature effects". *Bioresour. Technol.,* Volume Number 99, pp 3864-3870, 2008.
56. Lu W.B., Shi J.J., Wang C.H. and Chang J.S., "Biosorption of lead, copper and cadmium by an indigenous isolate Enterobacter sp. J1 possessing high heavy-metal resistance". *J. Hazard. Mater.,* Volume Number 134, pp 80-86, 2006.

57. Hasan S.H. and Srivastava P., "Batch and continuous biosorption of Cu(2+) by immobilized biomass of Arthrobacter sp.". *J. Environ. Manage.* Volume Number 90, pp 3313-3321, 2009.
58. Uzel A. and Ozdemir G., "Metal biosorption capacity of the organic solvent tolerant Pseudomonas fluorescens TEM08". *Bioresour. Technol.,* Volume Number 100, pp 542-548, 2009.
59. Ziagova M., Dimitriadis G., Aslanidou D., Papaioannou X. and Litopoulou Tzannetaki E., "Comparative study of Cd (II) and Cr (VI) biosorption on Staphylococcus xylosus and Pseudomonas sp. in single and binary mixtures". *Bioresour. Technol.,* Volume Number 98, pp 2859-2865, 2007.
60. Quintelas C., Rocha Z., Silva B., Fonseca B. and Figueiredo H., "Removal of Cd (II), Cr (VI), Fe (III) and Ni (II) from aqueous solutions by an E. coli biofilm supported on kaolin". *Chem. Eng. J.,* Volume Number 149, pp 319-324, 2009.
61. Zeng X., Wei S., Sun L., Jacques D.A. and Tang J., "Bioleaching of heavy metals from contaminated sediments by the Aspergillus niger strain SY1". *J. Soils Sediments,* Volume Number 15, pp 1029-1038, 2015.
62. Dursun A.Y., "A comparative study on determination of the equilibrium, kinetic and thermodynamic parameters of biosorption of copper (II) and lead (II) ions onto pretreated Aspergillus niger". *Biochem. Eng. J.,* Volume Number 28, pp 187-195, 2006.
63. Fan T., Liu Y., Feng B., Zeng G. and Yang C., "Biosorption of cadmium(II), zinc(II) and lead(II) by Penicillium simplicissimum: Isotherms, kinetics and thermodynamics". *J. Hazard. Mater.,* Volume Number 160, pp 655-661, 2008.
64. Martins L.R., Lyra F.H., Rugani M.M. and Takahashi J.A. "Bioremediation of Metallic Ions by Eight Penicillium Species". *J. Environ. Eng.,* Volume Number 142(9), pp 1061-1068, 2015.
65. Ghosh A., Ghosh Dastidar M., and Sreekrishnan T., "Recent Advances in Bioremediation of Heavy Metals and Metal Complex Dyes: Review". *J. Environ. Eng.,* Volume Number 142(9), pp 4099-4100, 2016.
66. Ozer A., and Ozer D., "Comparative study of the biosorption of Pb(II), Ni(II) and Cr(VI) ions onto S. cerevisiae: determination of biosorption heats". *J. Hazard. Mater.,* Volume Number 100, pp 219-229, 2003.
67. Kumar R., "Potential of Some Fungal and Bacterial Species in Bioremediation of Heavy Metals". *J. Nuclear Phys. Mater. Sci. Radiat. Appl.,* Volume Number 1, pp 213-223, 2014.
68. Tan T.W., Hu B. and Su H., "Adsorption of Ni^{2+} on amine-modified mycelium of Penicillium chrysogenum". *Enzyme Microb. Technol.,* 35, 508-513, 2004.
69. Infante J., De Arco R., Angulo M., "Removal of lead, mercury and nickel using the yeast Saccharomyces cerevisiae". *Rev. MVZ Cordoba,* Volume Number 19, pp 4141-4149, 2014.

70. Tian J., Peng X.W., Li X., Sun Y.J. and Feng H.M., "Isolation and characterization of two bacteria with heavy metal resistance and phosphate solubilizing capability". *Huan Jing Ke Xue*, Volume Number 35, pp 2334-2340, 2014.
71. Wang L.K., Hung Y.T. and Shammas N.K., "Advanced physicochemical treatment technologies", Springer, Volume Number 5, pp 174-175, 2007.
72. Lima I. G., Corticeiro S. C. and de Almeida Paula Figueira E. M. "Glutathione-mediated cadmium sequestration in Rhizobium leguminosarum". *Enzyme Microb. Technol.*, Volume Number 39(4), pp. 763–769, 2006.
73. Jha S., Dikshit S. and Pandy G., "Comparative study of agitation rate and stationary phase for the removal of Cu2+ by A. lentulus", *Int. J. Pharm. Biol.*, Volume Number 2, pp 208–211, 2011.
74. Selvam K., Arungandhi B., Vishnupriya B., Shanmugapriya T. and Yamuna M., "Biosorption of chromium (VI) from industrial effluent by wild and mutant type strain of saccharomycescere-visiae and its immobilized form", *Biosci. Discov.*, Volume Number 4(1), pp 72–77, 2013.
75. Gupta V. K., Nayak A. and Agarwal S., "Bioadsorbents for remediation of heavy metals: Current status and their future prospects", *Environ. Eng. Res.*, Volume Number 20(1), pp. 1-18, 2015.
76. Xie Y., Fan J. and Zhu W., "Effect of heavy metals pollution on soil microbial diversity and bermudagrass genetic variation". *Front. Plant Sci.*, Volume Number 7 Article 775, pp 1-12, 2016.
77. White V. E. and Knowles C. J., "Effect of metal complexion on the bioavailability of nitriloacetic acid to Chelatobacterheintzil ATCC 2900". *Arch. Microbiol.*, Volume Number 173(5-6), pp 373-382, 2000.
78. Green-Ruiz C., "Mercury(II) removal from aqueous solutions by nonviable Bacillus sp. from a tropical estuary". *Bioresour. Technol.*, Volume Number 97(15), pp 1907–1911, 2006.
79. Taniguchi J., Hemmi H., Tanahashi K., Amano N., Nakayama T. and Nishino T., "Zinc biosorption by a zinc-resistant bacterium, Brevibacterium sp. strain HZM-1". *Appl. Microbiol. Biotechnol.*, Volume Number 54(4), pp 581-588, 2000.
80. Teitzel G. M. and Parsek M. R., "Heavy metal resistance of biofilm and planktonic Pseudomonas aeruginosa". *Appl. Environ. Microbiol.*, Volume Number 69(4), pp 2313-2320, 2003.
81. Jyoti B. and Harsh K. S. N., "Utilizing Aspergillus niger for bioremediation of tannery effluent", *Octa J. Environ. Res.*, Volume Number 2(1), pp 77–81, 2014.
82. Viti C., Pace A. and Giovannetti L., "Characterization of Cr(VI)-resistant bacteria isolated from chromium-contaminated soil by tannery activity", *Curr. Microbiol.*, Volume Number 46(1), pp 1-5, 2003.
83. Ummalyma S. B. and Singh A., "Importance of algae and bacteria in the bioremediation of heavy metals from wastewater treatment plants". Book: New Trends in Removal of Heavy Metals from Industrial Wastewater, Elsevier, pp 343-357, 2021.

84. Shi J., Podola B. and Melkonian M. "Application of a prototype-scale twin-layer photobioreactor for efective N and P removal from diferent process stages of municipal wastewater by immobilized microalgae". *Bioresour Technol.*, Volume Number 154, pp 260-266, 2014.
85. Boelee N.C., Janssen M., Temmink H., Taparaviciute L., Khiewwijit R., Janoska A., Buisman C.J.N. and Wijfels R.H., "The effect of harvesting on biomass production and nutrient removal in phototrophic bioflm reactors for effluent polishing". *J. Appl. Phycol.*, Volume Number 26(3), pp 1439–52, 2014.
86. Boelee N.C., Temmink H., Janssen M., Buisman C.J.N., Wijfels R.H., "Balancing the organic load and light supply in symbiotic microalgal-bacterial bioflm reactors treating synthetic municipal wastewater". *Ecol. Eng.*, Volume Number 64, pp 213–221, 2014.
87. Kesaano M. and Sims R.C., "Algal bioflm based technology for wastewater treatment". *Algal Res.* Volume Number 5, pp 231-40, 2014.
88. Li T., Lin G.Y., Podola B. and Melkonian M., "Continuous removal of zinc from wastewater and mine dump leachate by a microalgal bioflm PSBR". *J. Hazard. Mater.*, Volume Number 297, pp 112-118, 2015.
89. Powell N., Shilton A.N., Pratt S. and Chisti Y., "Factors Influencing Luxury Uptake of Phosphorus by Microalgae in Waste Stabilization Ponds". *Environ. Sci. Technol.*, Volume Number 42(16), pp 5958-5962, 2008.
90. Metcalf and Eddy I., "Wastewater Engineering: Treatment and Reuse". 4th edn. McGraw-Hill, 2003.
91. Johnson M.B. and Wen Z., "Development of an Attached Microalgal Growth System for Biofuel Production". *Appl. Microbiol. Biotechnol.*, Volume Number 85(3), pp 525-534, 2009.
92. Shi J., Podola B. and Melkonian M., "Removal of Nitrogen and Phosphorus from Wastewater Using Microalgae Immobilized on Twin Layers: An Experimental Study". *J. Appl. Phycol.*, Volume Number 19, pp 417-423, 2007.
93. Craggs R.J., Adey W.H., Jenson K.R., John M.S., Green F.B. and Oswald W.J., "Phosphorous Removal from Wastewater Using an Algal Turf Scrubber". *Water Sci. Technol.*, Volume Number 33 (7), pp 191-198, 1996.
94. Schumacher G., Blume T. and Sekoulov I., "Bacteria Reduction and Nutrient Removal in Small Wastewater Treatment Plants by an Algal Biofilm". *Water Sci. Technol.*, Volume Number 47 (1), pp 195-202, 2003.
95. A. M. Bandowe, Bigalke M., Boamah L., Nyarko E., Saalia F. K. and Wilcke W., "Polycyclic aromatic compounds (PAHs and oxygenated PAHs) and trace metals in fish species from Ghana (West Africa): Bioaccumulation and health risk assessment". *Environ. Int.*, Volume Number 65, pp 135-146, 2014.
96. Boopathy R., "Factors limiting bioremediation technologies". *Bioresour. Technol.*, Volume Number 74(1), pp 63-67, 2000.

8

The Application of DNA Transfer Techniques That Have Been Used in Algae

Thilini Jayaprada and Jayani J. Wewalwela*

Department of Agricultural Technology, Faculty of Technology, University of Colombo, Pitipana, Homagama, Sri Lanka

Abstract

Microalgae are considered captivating microbes for a robust future bioeconomy via biotechnological production of biofuels and a wide range of high-value natural components. For industrial scalability, hardy non-model algae species that can survive any environment are being used for targeted development in robust and diverse biotechnology platforms. The DNA delivery into cells is the first step which depends on the host organism (cell membrane/wall permeabilization) and targeted cell component (organellar or nuclear). Traditional methods of DNA delivery such as mechanical agitation, electroporation, particle bombardment, and bacterial DNA transfer; and emerging technologies such as cell-penetrating peptides/polymers, metal-organic frameworks, liposome-mediated and CRISPR/Cas-mediated transformation, have been used in algae.

Keywords: Microalgae, DNA delivery, cell wall, transformation, biotechnology

8.1 Introduction

Microalgae are unicellular, photosynthetic, and fast-growing organisms that usually exist in aquatic ecosystems [124]. Their ability to grow in minimal nutrient systems to produce high biomass yield with the aid of photosynthesis permits cultivation in very low-cost culture media [41].

**Corresponding author*: jayani@at.cmb.ac.lk

Charles Oluwaseun Adetunji, Julius Kola Oloke, Naveen Dwivedi, Sabeela Beevi Ummalyma, Shubha Dwivedi, Daniel Ingo Hefft and Juliana Bunmi Adetunji (eds.) Next-Generation Algae: Volume I: Applications in Agriculture, Food and Environment, (195–224)

Production of biomass for biofuel, cosmetics, food and biopharmaceuticals by microalgae has recently received significant attention (Table 8.1) [32, 34, 49]. Microalgae consist of a diversified group of organisms. Microalgae, such as *Chlorella* spp., *Acutodesmus obliquus, Neochloris oleoabundans, Arthrospira platensis, Chlamydomonas reinhardtii* and *Euglena gracilis* spp., are intensively used in commercial biomanufacturing and have a plethora of potential biotechnological applications, nanotechnology and environmental technologies [34, 86]. For industrial scalability, hardy non-model algae species that can survive any environment are in use, and their targeted development is needed for robust and diverse biotechnology platforms. However, in order to achieve a commercial-scale production, microalgae need to be advanced to express potential high-quality recombinant proteins using synthetic biology. Genetic engineering has been transformed to become a strong approach to succeed in this goal [34, 108].

The first step for a pipeline of the production of targeted molecules is determining the gene that will express the biopharmaceutical followed by cloning it into a vector [47, 74]. The vector DNA delivery into algae cells is paramount for the successful expression of the functional biopharmaceuticals [47]. Even in an era of recent cutting-edge inventions of gene editing tools like CRISPR/cas9, in order to deliver a ribonucleoprotein (Cas9/sgRNA complex) an efficient DNA delivery technique is a must. DNA delivery depends on the algae host biology in terms of cell membrane/wall permeabilization and also on target cellular compartment (organelles or nuclear) [72, 93]. Moreover, during DNA delivery cells have the ability to come through chemical or mechanical handling. Therefore, the versatility of tools needed depends on the species. Nevertheless, to engineer algae to optimize productivity, the tool sets accessible now have significantly contributed rapidly and accurately [93, 106].

Likewise, as of now, some conventional DNA transfer techniques are in use for microalgae, such as the particle bombardment technique [20, 40], using PEG-$CaCl_2$ [62], *Agrobacterium*-mediated transformation [18, 100], and electroporation [6, 115]. However, in addition to being time-consuming and tedious, these conventional methods face issues like low efficiency and stability in integrating transgenes. Usually, a steady transformation is not achieved. Thus, genetic transformation for biotechnological manipulation of algal seemed to be lag behind. Due to these constraints, novel transformation techniques are garnering attention. Liposome-mediated transformation [45, 112], metal-organic frameworks [104], cell-penetrating polymers [11], cell-penetrating peptides [36, 44], nanoparticles/nanobiolistics [18] and silicon carbide whisker-mediated transformation [4] are such emerging and novel techniques that can be used for DNA delivery in

Table 8.1 List of applications of genetic engineering in microalgae.

	Species of microalgae	Expression product	Application	Reference
1	*Dunaliella salina*	White spot syndrome virus (WSSV), viral envelope protein 28 (VP28) in the nucleus	Transgenic commercial crayfish feed alga with WSSV resistance	[33]
2	*N. oceanica*	Sterol biosynthesis	Dietary nutrient	[76]
3	*P. tricornutum*	Lupeol biosynthesis	Potential medicinal properties, like anticancer and anti-inflammatory activity	[22]
4	*C. reinhardtii*	Anti-CD22-gelonin	Immunotoxin against B-cell lymphoma	[120]
5	*Nannochloropsis* sp.	Triacylglycerol (TAG)	Enhanced lipid production	[52]
6	*Dunaliella salina*	β-carotene hydroxylase	Carotene production	[113]
7	*Chlorella vulgaris*	Endogenous omega-3 fatty acid desaturase	Fatty acid biosynthesis	[90]
8	*Phaeodactylum tricornutum*	Bacterial-type acyl-CoA dehydrogenase (PtMACAD1)	Determine PtMACAD1's role	[53]
9	*Picochlorum renovo*		Development of a high-productivity, halophilic, thermotolerant microalga	[23]

(*Continued*)

Table 8.1 List of applications of genetic engineering in microalgae. (*Continued*)

	Species of microalgae	Expression product	Application	Reference
10	*Haematococcus pluvialis*	Astaxanthin	Overaccumulation	[37]
11	*Chlorella vulgaris*	Tea tree essential oil	Dermic and cosmetic applications	[85]

microalgae. In addition, many other techniques are currently in discussion to increase the DNA transformation efficiency in microalgae. Hence, the conventional and emerging set of DNA transformation tools will be discussed throughout this chapter with the hope of extending our knowledge to advance microalgae biotechnology related to the future bioeconomy.

8.2 Conventional DNA Transfer Techniques in Algae

Versatile DNA transferring techniques are needed to genetically manipulate a broad range of algae species due to the fact that efficient DNA transfer depends on the species. To date, in many strains of marine algae, foreign DNA delivery has been achieved by conventionally used transformation methods. This part of the chapter examines the progress in conventional gene transfer techniques and their technical aspects, advantages, limitations, and their potential with respect to algal biotechnology are further discussed.

8.2.1 Electroporation

In electroporation technique, the transient pores in the cell membrane are produced by high-voltage pulses, results in the cell entry of exogenous DNA. Among the transformation methods that are in use currently, electroporation is the frequently used method. Any electroporation protocol is considered to be efficient if it can provide the highest cell membrane permeability that keeps cell viability intact. This depends on factors like makeup of the cell membrane/cell wall, size of the cell and its growth stage, properties of exogenous molecules, and electroporation medium [69].

Single cell *Synechococcus* sp. was the first organism that electroporation was performed on [80]. Most recently, several modifications have been done on conventional electroporation transformation protocols in algae. Furthermore, novel microalgae species have been employed in genetic engineering for many purposes due to the discovery of novel transformation protocols through electroporation. For instance, *Nannochloropsis limnetica* is the only known freshwater species in the genus *Nannochloropsis* which is thought to be one of the most likely organisms for biofuels production. An electrotransformation protocol has been established for *N. limnetica.* In this protocol, 100-folds times increase in transformation efficiency was achieved by pretreatment of *N. limnetica* with lithium acetate and dithiothreitol prior to electroporation [16]. In another attempt, a fast and efficient gene transfer pipeline for *Chaetoceros muelleri*, which is considered as a feed microalga in aquaculture, has been attained by electroporation recently. Yin and Hu used an optimum voltage intensity of 500 V with the presence of salmon sperm DNA (ssDNA) and achieved >1000 per 10^8 cells of transformation efficiency [127]. Wang *et al.* also reported an efficient DNA transfer technique using the square wave electroporation system *for Chlamydomonas reinhardtii* as strains with cell walls [121].

However, the electroporation method has serious obstacles relating to cell viability, delivery efficiency, and productivity. Several new techniques have been integrated with conventional protocols to overcome them. For example, with a single platform, Kang *et al.* devised a polyimide (PI) film based on chip electroporation system [56]. It can shield the cells from the electrodes with four sheath flows, enabling a 3D flow focusing. A long spiral channel for efficient delivery associated with high-weight molecules is damaging to the cell viability with longer exposure to the electric field. Even under those conditions this setting has enhanced the cell viability. This novel movement in the field of microfluidics is anticipated to contribute greatly to algal research as well as to broader applications [57]. In addition to that, the microalgal cell wall is an intrinsic obstacle that limits the gene delivery efficiency in algae genetic engineering. In such cases, in order to increase the intake of genes, Chen and Lee proposed hard-uptake nanoparticles (huNPs) for use in *Chlamydomonas reinhardtii during electroporation.* They observed the precipitation of NPs due to gravity gave gene-huNPs (530 nm) complexes to stably deliver more genes to the cells. They suggest these revelations describe the role of gene-NP complexes in microalgae electroporation [17]. In order to optimize the transformation parameters of *Nannochloropsis salina* (Eustigmatophyceae), another microalgae species with cell wall, Jeon *et al.* discovered a hybrid approach. Secondary metabolites from myxobacteria were utilized to weaken the cell wall and also to maximize the transformation efficiency. Taken together, they proposed that an integrative approach using high field strength square wave pulses and conditioned cells with myxobacterial extract is an efficient method for the productive genetic transformation of microalgae [54]. Furthermore, in electroporation technique, the DNA transfer into *C. reinhardtii* is considerably affected by the cell growth stage. Taking this into consideration, Kim could successfully control cell growth and DNA transfer by manipulating initial cell concentration and light conditions. The diurnal light had positive effects on exogeneous DNA transfer in electroporation. In addition to that, he found that late log phase cell harvesting is ideal to achieve more efficient transformation in *C. reinhardtii* [66]. Likewise, electroporation integrated with novel techniques has rapidly advanced the application of microalgae in biotechnology applications.

8.2.2 *Agrobacterium*-Mediated Transformation

In 1988, *Agrobacterium tumefaciens* was permitted to transfer DNA into higher plants; and since then, it has been exploited by scientists in genetic engineering. Mixing of target cell cultures and *Agrobacterium* cells together in the presence of virulence agent acetosyringone is the first step. *Agrobacterium* cells consist of transgene vectors. Bacterium is signaled by acetosyringone to infect wounded targeted cells. Lastly, an antibiotic is used to screen transformants. *C. reinhardtii* and *Haematococcus pluvialis* are a few microalgae species which respond efficiently to *A. tumefaciens*-mediated transformations. In addition, efficient *A. tumefaciens* protocols have been developed for several other microalgae species. However, *A. tumefaciens* transformation efficiency in microalgae is determined and affected by several conditions. The T-DNA of pCAMBIA1304 plasmid was efficiently transferred into *Dunaliella pseudosalina* microalgal cells within two months by *A. tumefaciens* GV3301, EHA101 and GV3850 strains [26]. Microalgae, *Chlorella sorokiniana*, has gained attention as a potent species for biofuel and recombinant proteins production. Sharma *et al.* originally establish *A. tumefaciens*-mediated transformation protocol productively for *C. sorokiniana* as a solution to lacking an efficient transformation procedure for genetic engineering of this species. According to his findings, cocultivation of *C. sorokiniana* and *Agrobacterium* at OD_{600} 1.0 and 0.6, respectively, in the simultaneous presence of 100 μM of acetosyringone, gave a significant transformation efficiency (220 ± 5 hygromycin-resistant colonies per 10^6 cells) [111]. In an another attempt in microalgae *Dunaliella tertiolecta*, researchers concluded that *A. tumefaciens* transformation on solid nutrient medium at $OD_{600} = 0.5$ for 72 h co-culture duration displayed the highest amount of stable *gusA* expression [91].

Bashir *et al.* were successful in expressing "erythropoietin," which is an important therapeutic protein, in a non-model green microalga, *Dictyosphaerium pulchellum* via *A. tumefaciens* transformation. After 2 to 3 weeks, successful transformants were achieved by co-cultivation of *D. pulchellum* with *A. tumefaciens* binary vector [9]. Genetic engineering studies on red algae, which is a valuable source of phycocolloids, have been rarely attempted, and *Agrobacterium*-mediated gene transfer techniques are yet to be explored. Ramessur *et al.* developed a step-by-step methodology of *Agrobacterium*-mediated transformation for red algae *Chondrus crispus* [103]. *Chlamydomonas reinhardtii* is a unicellular green alga commonly used for a variety of genetic studies. Insignificant levels and non-stable expression of targeted genes have been seen as the major obstacles to further *Chlamydomonas reinhardtii* transformation studies. To remedy this, Mini *et al.* suggested the benefit of *Agrobacterium* transformation in insertional mutagenesis, due to its higher proportion of within-gene, single-locus insertions [82]. However, *Agrobacterium*-mediated transformation has its difficulties for a good transformation efficiency, which is attributed to its labor-intensive nature.

8.2.3 Bacterial Conjugation

Except for *Agrobacterium*-mediated transformation, *E. coli.* is being used to transfer genetic material into microalgae by conjugation [10]. To date, particle bombardment and electroporation have been found to be successful in microalgae DNA delivery, yet the inserted DNA seems to be arbitrarily integrated into the nuclear genome. Bacterial conjugation allows the integrated vector to exist as an episome in the targeted cell. This is considered as an important difference of bacterial conjugation from other methods. In a recent study, in order to increase the triacylglycerol (TAG) production in microalgae *Acutodesmus obliquus* and *Neochloris oleoabundans, E. coli* conjugation-based episome delivery has been achieved. The transformation protocol could also be optimized by changing the incubation time at 30 °C in dark conditions (60, 90, and 120 min) and keeping microalgae at a bacteria ratio of 1:1 or 1:2 [86].

Karas *et al.* reported the plasmid integration in diatoms *Phaeodactylum tricornutum* and *Thalassiosira pseudonana* through *E. coli* conjugation. It is reported as the original attempt to transfer nuclear episomal vector to diatoms. They also reported the steady episome replication in these diatoms even without antibiotic selection and further demonstrated that the episomes are retained in close nature to native chromosomes [59].

In another study, both conjugation and biolistic transformation methods are used to deliver CRISPR/Cas9 system into marine diatom *Phaeodactylum tricornutum*. The aim of this attempt was to introduce mutations into a target gene and compare the two transfer methods with respect to mutation efficiency and evaluate issues associated with constitutive expression of Cas9. They revealed that the Cas9-induced mutant percentage is similar between the two transfer techniques. But a longer time duration was needed to induce mutation when the CRISPR/Cas9 system was episomal via conjugation. They further suggested that the common issues related to other DNA delivery techniques, like random insertions and unstable mutants by constitutive expression of Cas9, can be prevented via conjugative CRISPR/Cas9 delivery system. Here, both biolistic and conjugative transformations were carried out with *P. tricornutum*. Conjugative delivery was 20–100 folds more efficient [110].

8.2.4 Biolistic Particle Bombardment

In this method, DNA-coated microparticles will be released at high speed, so exogeneous DNA will be integrated into the targeted genome after entering the cell. Most examples are found to be targeting microalgae chloroplasts. The factors affecting the successful delivery of this method mainly depend on how far the targeted cell and particle released device are situated, the type of microparticles used and the released pressure. Gold microparticles are considered to be a better choice than tungsten with respect to low damage caused to DNA, low rate of cell growth inhibition, and favorable DNA binding efficiency [95].

A recent study reported the *Phaeodactylum tricornutum* chloroplast transformation via particle bombardment technique for simultaneous transformation of combined CRISPR/Cas9 plasmid and pAF6 plasmid zeocin resistance selection marker [111]. Furthermore, based on a CRISPR-Cas9 ribonucleoprotein (RNP) delivery system, Chang *et al.* were successful in enhancing lipid synthesis in marine green microalga, *Tetraselmis* spp. After 2 to 3 weeks of particle transformation they observed cells grew on the f/2 medium, confirming the successful delivery and integration [14]. In another recent study, successful production of phytase, which is a necessary nutrient compound in animal feed, was achieved in the chloroplast of green microalga *C. reinhardtii* by particle bombardment transformation [98]. The insertion of plant chlorophyllase (CLH) gene, which can enhance the resistance to predators, is another application of microparticle bombardment to engineer microalgae [60].

However, there have been several occurrences where transformation through bombardment did not give the expected outcome. Khatiwada *et al.* explored *Agrobacterium*-mediated transformation, particle bombardment and electroporation to achieve *E. gracilis* nuclear transformants. Out of the three transformation techniques, successful integrations of 10, 7 and 1 per 10,000 plated cells were given by *Agrobacterium*-mediated transformation, biolistic bombardment and electroporation respectively. Transformants from biolistic bombardment and electroporation lost the *hptII* gene expression upon being cultivated several times. This implies the deficiency in steady DNA integration under biolistic bombardment and electroporation. However, even after 12 rounds of subculture, *Agrobacterium*-mediated transformation gave stable nuclear transformants growing on plates supplemented with hygromycin [61].

8.2.5 Agitation with Glass Beads

Agitation with glass beads is similar to particle bombardment. This refers to agitating a mixture of targeted cell suspension, exogeneous DNA and glass beads together. To enter DNA into the targeted cells, chemical/enzymatic removal of cell walls becomes a paramount prerequisite [31]. Therefore, this has also been successful in cell wall deficient cells. The model green microalgae *C. reinhardtii* and *Cyanidioschyzon merolae*, *Chlorella vulgaris*, *Dunaliella salina* have been involved in DNA delivery into both chloroplast and nucleus via agitation with glass beads [31, 63]. Apparently, deficiency in cell wall, size of the cell, agitation speed and circular or linear nature of DNA are the main factors that determine the transformation ratio [105].

In a recent study, *Chlamydomonas reinhardtii* was transformed for synthetic phytoene desaturase-encoding gene (CRTIop) as a selectable marker through glass bead agitation method with delivery ratio of 550 colonies μg^{-1} DNA [83]. Chloroplast-mediated transformation of microalgae is an efficient strategy which has the potential for so many other microalgae species. Wannathong *et al.* were successful in targeted chloroplast transformation of *Chlamydomonas reinhardtii* via agitation with glass beads. Chloroplast transformation was carried out through agitation of a mixture of 5–10 μg circular plasmid DNA and glass beads of 400–625 μm diameter together [123].

Agitation with glass beads technique is considered to be more beneficial than others because it does not need special equipment and is cheap and fast. However, low transformation efficiency and the requirement of cell wall removal are considerable barriers of this method.

8.3 Novel Emerging DNA Transfer Techniques in Algae

Most of the microalgae species show less efficiency to previously discussed transformation methods due to cell membrane barriers such as cell wall. Owing to the revolution in gene editing, many carriers that can carry large molecules like nucleic acids and can efficiently enter tissues and cells are advancing rapidly. Some of these novel carriers that will eliminate existing barriers are expected to be commercially available in the near future.

8.3.1 Protoplast Fusion

Protoplasts are somatic plant cells which lack cell walls and are produced through enzymatic hydrolysis using enzymes like cellulase, hemicellulose, macerozyme and pectinases [2, 107]. To prevent protoplasts from rupturing in the process, osmotic regulators like mannitol and sorbitol are added with hydrolyzing enzymes. This has resulted in significant cell viability and better protoplast yields in *Chlamydomonas* sp. [75]. Polyethylene glycol (PEG) is the common fusogen agent used. Protoplast fusion allows a combination of genomes between intra- and interspecies. Therefore, this has emerged as a significant tool to improve microalgae for a broad range of industrially important applications. Several recent examples of producing protoplasts, fusion techniques as well as applications of this method in context to microalgae are mentioned below.

Recently, *D. salina* and *C. vulgaris* protoplasts were derived and were combined with PEG and $CaCl_2$. After several rounds of subculture, growth stability and significant carotenoid production were achieved in the hybrid. The purpose of this attempt was to improve the β-carotene content for sustainable aquaculture [71]. In 2018, the same group of scientists were successful in significantly increasing the lipids, proteins and carbohydrates contents in a microalgae hybrid obtained by interspecies fusion of *Chlorella* protoplasts [70]. Another recent attempt focused on improving *Chlorella,* which is considered to be an important feed source in aquaculture. Two mutant *Chlorella* strains were derived via UV and chemical mutagenesis and wall-less cell protoplasts were obtained after cellulase and driselase treatment. In the hybrid, they observed a significantly high content of biomass and protein compared to the original algae [126]. In addition, Abomohra *et al.* carried out successful interphylum genetic recombination between *Ochromonas danica* of phylum Ochrophyta and *Haematococcus pluvialis* of phylum Chlorophyta using PEG as the fusing agent. The heterofusants were verified through fatty acid analysis where each particular characteristic of fatty acids were traced in the novel hybrid. The findings shed light on successful genetic recombination between two microalgal phyla [1].

Electrofusion and chemofusion are other novel alternative fusion techniques that can be employed in microalgae. Chemofusion is used to induce recombination between two or more protoplasts. The technique is used to expose targeted microalgae cells to an alkaline solution of pH 9.0–10.5 in the presence of a fusogen agent, an osmotic regulator and divalent cations [96]. The electrofusion technique uses an electrical pulse to increase the cell membrane permeability by reversible rupturing of cell membranes. This is followed by a sudden disruption in the applied electric field, enabling the fusion between closest protoplasts. The cell membrane natural composition will be restored after a few seconds or minutes [42]. Furthermore, nuclear fusion is another technique which uses glass capillary microinjection to insert exogeneous DNA into cells [128]. Protoplast fusion is a technique which has not been used to its full potential yet in terms of improving microalgae. This method can be used for interspecies, interphyla and beyond genome fusion, leading to super-producing strains.

8.3.2 Liposome-Mediated Transformation

Lipofection, or liposome-mediated transfection (LMT), is based on introducing exogenous DNA into cells with the help of lipophilic reagents, which are considered to heighten the cellular intake of polynucleotides. Cationic liposomes (liposomes bearing a net positive charge) have been shown to bind with DNA, followed by efficient delivery of cloned genes both *in vitro* and *in vivo*. These DNA-cationic liposome complexes have been shown to transfer and express cloned genes in a wide variety of cultured cells. The LMT technique also has the potential to carry gene-editing-associated proteins into cells [112]. A recent LMT-based strategy is liposome-mediated mRNA transfection, which is shown to be ideal in post-mitotic cells due to its ability to prevent genes entering into the nucleus, and therefore is considered to be independent of cell proliferation [114].

The efficiency and versatility of liposome-mediated DNA delivery, as well as the wide range of cationic DNA carrier molecules which have been synthesized and tested for gene delivery, have generated substantial interest in this approach. The use of cationic liposomes as a DNA carrier system for transfecting cultured cells offers several potential advantages over other cellular transfection systems [24]. For example, LMT carries a significant DNA packaging capacity which allows it to be highly suitable for DNA transfer in genetic engineering [55]. The strategy followed by this technique is to attract and engulf negatively charged DNA by lipophilic reagents [13]. Therefore, DNA will be further protected by liposomes from nuclease degradation [7]. All of these result in producing a high number of putative transformants and simultaneously it omits the need for special equipment. Taken all together, LMT makes a highly potential tool for genetic manipulation of algae strains of *C. reinhardtii, D. salina,* and *C. merolae*, which are cell wall deficient or produced protoplasts of microalgae. To date, no attempts have been made of LMT on algae.

8.3.3 Metal-Organic Frameworks

Metal organic frameworks (MOFs) is a category of crystalline materials with pores which are molecularly defined. This makes them ideally applicable for introducing DNA inside these pores and use as delivery vehicles. The scope of their interaction with large molecules is still emerging. In a recent study, Peng *et al.* used three MOFs with accurately harnessed pore environments combined with organic linkers for the insertion of exogenous DNA molecules from the solution [97].

Zeolitic imidazolate framework (ZIF-8) is a recently developed MOF which has been used for CRISPR/Cas9 delivery. The large surface area, finely tuned pore size, biocompatibility and controlled degradability are attributes that contribute to serve MOF as successful DNA carriers. For example, ZIF-8 has given three times higher GFP expression compared to conventional transformation methods. MOFs provide a potential solution for the deficiency of CRISPR/Cas9 gene editing elements that are effective and cell-type-specific [3]. Aminoclay has been used to effectively deliver a broad range of molecules like DNA, polysaccharides, enzymes, and proteins into mammalian, plant, and *C. reinhardtii* microalgae cells. Aminoclay is considered as yet another successful MOF [64]. The potential of MOF in microalgae biotechnology is yet to be revealed.

8.3.4 Cell-Penetrating Polymers

This class of carriers can deliver genetic material by making nano-sized polyplexes via electrostatic interactions. The examples for cell-penetrating polymers include polyethyleneimine (PEI), polyamidoamine (PAMAM) and polyethylene acrylic acid (PEAA) [77, 78]. These have been frequently used in mammalian cell types for DNA delivery. These carriers bear unique attributes such as excellent encapsulation capacity, non-immunogenicity, the ability to prevent integration of exogeneous genes into host chromosomes, are able to produce cheap transgenic lines efficiently, are a simple technique to handle, and also have a modifying capacity for a broad range of DNA delivery functions. At the same time, some drawbacks of this technique also have been reported such as low transformation efficiency, toxicity to cells, or unstable gene integration [65].

Due to a lot of unique attributes, cell-penetrating polymers are promising genetic transformation vehicles in algal biotechnology. No attempts have been made to use cell-penetrating polymers in algae genetic engineering. Algae cell wall seems to be a major barrier and research is needed to improve the available polymers with respect to algae.

8.3.5 Cell-Penetrating Peptides

Cell-penetrating peptides (CPPs) are short chains of amino acid residues which give them the ability to move through membrane lipid bilayers in targeted cells. These CPPs, which carry positive charges and are derived from viral proteins [15], deliver a broad range of genetic materials even in their biologically active form and are widely used for animal, bacterial and plant cells [48, 119]. The mechanism behind delivery are the amino acid residues of CPPs, which interact with different biomolecules to form biomolecule-CPP complexes, facilitating an efficient delivery [88].

Although CPP delivery of exogeneous DNA into microalgae is rarely found, this technique carries huge potential for advanced algal research. In a recent study, a novel CPP pVEC-mediated protein delivery tool for microalgae was introduced. This tool has been successful in carrying proteins in a range of 6 kDa to 150 kDa into wild-type *C. reinhardtii.* Delivery of 66 kDa-sized protein was also achieved in *Nannochloropsis salina* and *Chlorella vulgaris* [58]. For the first time, Kang *et al.* reported pVEC-mediated delivery of ribonucleoprotein (Cas9/sgRNA complex) into microalgae *Chlamydomonas reinhardtii.* This attempt is expected to provide new avenues for algae genetic engineering in the context of CRISPR/Cas9 system [57, 58]. Tat-Sar-EED, a CPP delivery system, is another tool which came into use recently in microalgae *Chlamydomonas reinhardtii* and *Euglena gracilis.* These species are considered to be difficult in terms of DNA delivery yet this novel tool was capable of successfully introducing citrine protein into the cell lines [19]. The applications so far propose CPPs as prompt protein delivery tools which can open novel avenues in algae genetic engineering in the future.

8.3.6 Nanoparticle-Mediated Transformation

Nanotechnology is a field which is in the forefront of science today which also has paved the way to more efficient DNA delivery into many types of cells, including microalgae. For example, a recent study illustrated that quantum dots are capable carriers of siRNA and many more molecules into human and mice cells [63]. Key factors determining successful nanoparticle-mediated transformation are concentration of plasmid DNA, polymer coating, composition of nanoparticle and reaction medium pH. A recent study clearly describes the strategy behind nanoparticle-mediated transformation. According to the study, nanoparticles enter cells either through cell membrane external pores or by folding back the cell membrane on itself to form a cavity. In and out transport of molecules by nanoparticles was observed based on invagination in the algae *Penium* [35]. Invagination mode of action was further confirmed by Giraldo *et al.* [43]. According to them, carbon nanotubes can move through the double-membranes of chloroplasts through diffusion, in the process getting wrapped by glycerolipids in green microalgae *Desmodesmus subspicatus* [79]. Many avenues in nanoparticle-mediated transformation are yet to be revealed with respect to microalgae.

8.4 Limitations to Genetic Transformation in Algae

Although plenty of transformation methods are available, genetic material transfer still remains a challenge for some algae. Less efficient transformation methods is thought to be because of less effective gene delivery, unstable transgene expression, random genome integration and genetic divergence. Several such significant limitations are explained.

8.4.1 Cell Wall as a Significant Barrier

Complex and rigid cell wall is reasoned to be the most evident barrier to microalgae genetic transformation [117]. In almost every transformation protocol that has been described, the success of the transformation depends on the extracellular DNA entering the cells through the cell wall barrier. Researchers keep experimenting on ways to increase the cell membrane permeability via changing other attributes, but the interference still remains a challenge. For example, in order to increase the cell permeability, researchers have been manipulating electroporation parameters, but recovery of treated cells has been a great challenge after such adjustments [87].

CRISPR/Cas-based gene editing is the tool which is in the forefront of genetic engineering research nowadays [116]. This tool achieves gene editing upon delivery of a protein-gRNA complex where cell removal becomes a paramount prerequisite [116]. Therefore, to be on par with current genetic engineering technologies, cell wall is a barrier that needs to be overcome.

Chlamydomonas reinhardtii is the model algae, which has been mostly used in exogeneous DNA transformation attempts. Yet, only wall-less cell mutants or strains without cell walls report high efficiency transformation in this species [51, 102, 121]. Therefore, in most algae, including *C. Reinhardtii,* protoplasts production has become a necessary step to facilitate successful genetic transformation [2, 107]. Several attempts have been taken to discover enzymes that degrade cell walls. Recently, Hwang *et al.* tested seven enzymes that are commercially available on *Chlamydomonas reinhardtii*. Out of the tested enzymes, subtilisin (Alcalase) at a low concentration of 0.3 Anson units/mL gave the most efficient results in cell wall lysis. According to them, the developed subtilisin cell wall lysis-based transformation approach was more efficient than the conventional methods [50].

Agrobacterium or other conjugative bacteria-mediated transformation is another strategy that can be used to avoid the cell wall barriers [26, 109]. However, in most of the transformation methods, creating cell wall permeability to allow transferring enough DNA and cell death in doing so requires fine-tuning. Currently, strict protocol optimization is the strategy for successful transformation with enough cell wall permeability and cell recovery.

8.4.2 Native Antibiotics Resistance

Using antibiotic is the most abundantly used approach in screening successful transformants. Generally, transformation is performed on cells of a high density (10^5 to 10^9). Out of them, only a few cells will be successful in exogenous DNA integration and expression. Successful screening of the transformants is the next key step in an efficient transformation approach. The common strategy that has been used is to integrate a particular antibiotic resistance gene through the transformation vector and screen out survivors in the presence of that antibiotic or herbicide. But there are some species of non-model microalgae which carry inherent antibiotic resistance, which makes it challenging for efficient screening [28, 118]. This enables cells which have not been transformed by the transgenes to grow well in antibiotic medium. These kinds of incidences have been reported in several cases. According to some reports, this also has led to the use of high concentrations of antibiotics during screening due to already existing intrinsic resistance at lower concentrations [30].

Therefore, for a productive algae transformation outcome, the careful selection of the screening agent is paramount and is determined by the microalgae species and its intrinsic resistance. This follows a thorough examination to make sure that at a usable and reasonable concentration a minimal number of given microalgae cells will spontaneously resist the screening agent.

8.4.3 Low Genetic Stability of Transgenes

Frequently, especially microalgae nuclear transformation led to low levels of stably integrated transgenes along with high variable expression levels among confirmed transformants. Strong stability and expression of transgenes after transformation play an important role in a productive transformant selection. According to some reports, this is species-dependent. For example, algae, such as *Ulva lactuca, Kappaphycus alvarezii, Poryhyra miniata,* and *Gracilaria changii,* are reported to not be transgene stable, especially after nuclear transformation [39]. To date, the transgene stability after transformation is extremely indefinite in microalgae. Several attempts have been made to assess this concern [29, 86]. For example, lately Molina-Márquez *et al.* invented a novel multicistronic expression plasmid for microalgae called Phyco69. The key advantage of this plasmid is that it facilitates the large phenotypic screening normally needed for the selection of high-expression stable clones. The polylinker region of this plasmid allows any gene of interest to be inserted. The gene of interest will be linked with the amino terminus of aminoglycoside 3′-phosphotransferase (AphVIII) through a short viral self-cleaving peptide. This AphVIII from *Streptomyces rimosus* gives resistance to the antibiotic paramomycin. The fusion between the gene of interest and selection agent gene AphVIII renders an efficient approach in screening transformants with the highest level of expression [84]. Likewise, during the development of transgenic algae lines for commercialization, it is important to select reliable and stable gene expression levels by laboratory testing.

8.5 Future Prospects of Algae Transformation

To date, considerable progress has made in genetic manipulations of algae, yet many barriers need to be overcome and so much potential to be advanced. Novel genetic elements and tools related to genetic engineering are advancing rapidly and in this scenario, and microalgae are gaining momentum to produce a wide range of commercially important products. Lack of efficient genetic elements for transformation is a feature which exists pertaining to the low efficient transformation in microalgae. In addition, advanced genetic engineering attempts often need the simultaneous expression of several transgenes (gene stacking).

Synthetic biology is a field that is in the forefront of science today. In order to achieve a desired function or a product, designing an efficient expression vector with unique genetic elements is a must. Synthetic biology uses novel and rapid tools, like recombining, to assemble different transgenes into efficient expression vectors [72]. In a recent attempt, 100 gene parts standing for 67 unique genetic elements were assembled and the first Golden Gate Modular Cloning (MoClo) toolkit for a microalga, *Chlamydomonas reinhardtii,* was developed. This toolkit has been validated among researchers specially to advance genetic engineering of *Chlamydomonas* [21]. In another attempt, gene stacking was carried out to integrate several expression genes to a single vector to facilitate transformation of microalgae *Nannochloropsis oceanica.* The combined transgenes included multiple resistance marker genes, bidirectional promoters, and multiple expression cassettes [99]. Dehghani *et al.* genetically engineered three species of microalgae, *Chlamydomonas reinhardtii, Chlorella vulgaris,* and *Dunaliella salina, to produce* human interleukin-2 protein. This was achieved by designing an efficient, novel "Gained *Agrobacterium*-2A plasmid for microalgae expression" (named GAME plasmid). Further research validated that this GAME plasmid can efficiently transform designated microalgae with steady production of the hIL-2 [25]. With the rapid advance in synthetic biology, novel and efficient toolkits are expected to be developed and be available to the microalgae research community in the future.

Viral vectors like type IV secretion system (T4SS-VirB/D4 T4SS) is an emerging tool to deliver exogenous DNA into cells [68]. There is so much potential to integrate this tool into microalgae, although few recent attempts have been reserved in this regard. The DNA geminiviral vector is an efficient viral vector being used and recent attempts have been made through *Agrobacterium*-mediated transformation to examine the ability of this vector to produce recombinant proteins in microalgae *Chlamydomonas reinhardtii* and *Chlorella vulgaris.* In this study, nuclear transformation was carried out to transiently express the recombinant proteins of SARS-CoV-2 receptor binding domain (RBD) and basic fibroblast growth factor (bFGF) [111]. Likewise, viral vectors are simple and efficient transformation elements that have a high potential to overcome barriers with microalgae genetic transformation. Sequence optimization algorithms are another novel technique that is being used to optimize transformation vectors recently in genetic engineering. This computational method has already been attempted in *C. reinhardtii* [46, 81].

When it comes to microalgae transformation, it is clear that which method should be applied depends on availability of molecular tools and the incidence of cell wall. Furthermore, there is a huge potential to advance and update the existing transformation methods to enhance the transformation efficiency [73]. Recently, more attention has been given to microalgal chloroplasts instead of nuclear transformation, which is considered an efficient approach to the classical transformation platforms used in microalgae biotechnology. A recent successful attempt at chloroplast transformation was achieved in Nannochloropsis by inserting a nuclear-targeted construct into the chloroplast genome via electroporation. Genetic transformation of Nannochloropsis was considered challenging due to its small cell size and unclarified cell wall [38]. The advantage of chloroplast transformation is its ability to integrate transgenes precisely into targeted loci via homologous recombination. This avoids two barriers related to conventional transformation methods. The first one is the ability to avoid screening of a large number of transformant lines, and the second is to avoid unstable transgene expression by overcoming the barrier of positional effect. In addition to that, polycistronic transcription units are a novel and remarkable tool associated with chloroplast transformation which facilitate the expression of multiple genes under single promoter simultaneously. This is an important prerequisite to produce protein complexes by allowing synchronized production of enzymes from complex metabolic pathways. Furthermore, chloroplast transformation allows significantly high expression levels without the need for gene silencing because of the high number of transgene copies and endogenous strong promoters. Chloroplast transformation has not made significant progress over the last decade in microalgae species like *Nannochloropsis* sp., *Chlorella* sp., and *C. reinhardtii* [122]. The above-mentioned advantages and recent advances in chloroplast transformation confirm chloroplast as a viable and sustainable tool for future microalgae genetic engineering.

Microfluidics is a novel technology which needs to be associated with microalgae biotechnology for its dilute cultures and microscale sizes. One of the time- and cost-effective techniques associated with microfluidics is the microfluidic lab-on-a-chip (LOC) systems, which carries immense potential to advance microalgae biotechnology research by furnishing high-precision and high-efficiency cell transformation in a fully automated fashion. For instance, to overcome the limitations of microalgae electroporation, Qu *et al.* discovered a droplet microfluidics-based continuous-flow electroporation method for *C. reinhardtii.* The proposed technique could give a significantly higher efficiency of transformation than the classical process. The strategy was to confine both microalgae cells and transgene DNA as encapsulated droplets followed by exposing droplets to constant voltage to facilitate electroporation [102]. Another application of microfluidics technology in microalgae transformation is associated with performing delivery of transgenes via a combination of nanowire array and a pneumatically actuated microvalve system [5]. This strategy was employed to transform *C. reinhardtii* with hygromycin B resistance gene, and the transformation efficiency was 10,000 times higher than conventional methods. The emerging applications of microfluidic technologies on microalgal DNA delivery has a plethora of future directions to defeat current barriers associated with microalgae genetic transformation [94].

Development of novel bacteria strains as novel hosts that have the ability to store and deliver DNA to plants has already begun and is an emerging tool in genetic transformation. For example, Brumwell *et al.* developed *Sinorhizobium meliloti* to be a universal host as multi-host and multi-functional shuttle vectors (MHS). This universal vector could store DNA with high G+C content and efficiently deliver DNA into eukaryotic microalgae *Phaeodactylum tricornutum.* Crucially, they suggest this newly developed *S. meliloti* is going to be an efficient universal donor for MHS plasmids with high G+C content in conjugation-based transformation attempts [12]. This tool has a huge potential towards advancing microalgae transformation.

Over the last decade, microalgae genetic engineering has progressed from DNA transmission, transformation, selection, and now the advanced tool CRISPR/Cas9. The emerging of CRISPR/Cas9-based gene editing technique started during the 2000s with the discovery of tools such as RNA interference, ZFNs, and TALENs [8]. Especially since 2014, frequent attempts have been made to genetically engineer *C. reinhardtii* using CRISPR/Cas9, marking a new period of microalgae genome editing [101]. However, most attempts demonstrated a lack of efficiency and have a poor survival ratio because of vector-driven Cas9 toxicity [92]. Therefore, in recent years, relatively few studies have been documented related to CRISPR/Cas9 genome editing attempts in microalgae, which needs further extensive research and applications. For example, critical barriers related to microalgae CRISPR/Cas9 gene editing, such as DNA load delivery, species independence, germline transformation and transformation/editing efficiency, can be overcome by nanomaterials [27].

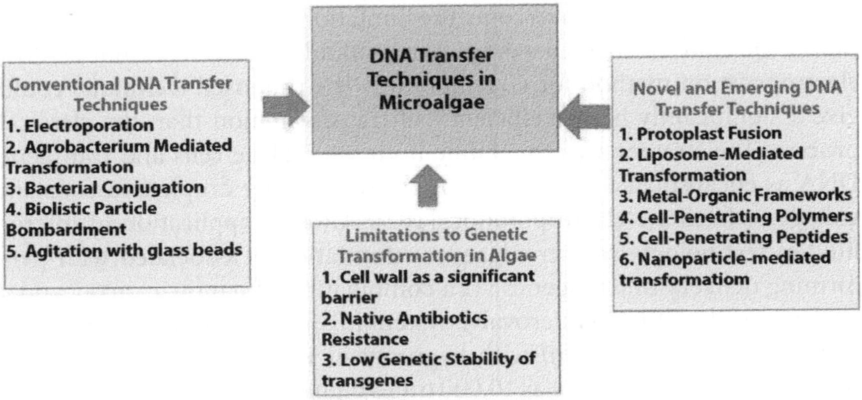

Figure 8.1 Graphical abstract representing the conventional and emerging DNA transfer techniques used in algae and their associated barriers.

This chapter particularly sheds light on identifying conventional and emerging DNA transformation techniques associated with algae. This is especially important to advance a broad range of other avenues of microalgae genetic engineering to their full potential. A graphical summary of all the techniques discussed in this chapter is given in Figure 8.1.

References

1. Abomohra, A.E.F., El-Sheekh, M. and Hanelt, D., Protoplast fusion and genetic recombination between Ochromonas danica (Chrysophyta) and Haematococcus pluvialis (Chlorophyta). *Phycologia, 55*(1), pp.65-71, 2016.
2. Al-Someidae, T.A., Al-Nema, Q.S. and Al-Shahery, Y.J., Isolation and culture of protoplasts from *Chloricoccum Humicola. Plant Archives, 20*(1), pp.2866-2868, 2020.
3. Alyami, M.Z., Alsaiari, S.K., Li, Y., Qutub, S.S., Aleisa, F.A., Sougrat, R., Merzaban, J.S. and Khashab, N.M., Cell-type-specific CRISPR/Cas9 delivery by biomimetic metal organic frameworks. *Journal of the American Chemical Society, 142*(4), pp.1715-1720, 2020.
4. Arshad, M. and Asad, S., Embryogenic Calli Explants and Silicon Carbide Whisker-Mediated Transformation of Cotton (Gossypium hirsutum L.). In *Transgenic Cotton* (pp. 75-91). Humana Press, New York, NY, 2019.
5. Bae, S., Park, S., Kim, J., Choi, J.S., Kim, K.H., Kwon, D., Jin, E., Park, I., Kim do, H. and Seo, T.S., Exogenous Gene Integration for Microalgal Cell Transformation Using a Nanowire-Incorporated Microdevice, *ACS Applied Materials & Interfaces, 7*(49), pp.27554-27561, 2015.

6. Bahi, M.M., Tsaloglou, M.N., Mowlem, M. and Morgan, H., Electroporation and lysis of marine microalga Karenia brevis for RNA extraction and amplification. *Journal of The Royal Society Interface*, *8*(57), pp.601-608, 2011.
7. Barampuram, S. and Zhang, Z.J., Recent advances in plant transformation. *Plant Chromosome Engineering*, 701, pp.1–35, 2011.
8. Barman, H.K., Rasal, K.D., Chakrapani, V., Ninawe, A.S., Vengayil, D.T., Asrafuzzaman, S., Sundaray, J.K. and Jayasankar, P., Gene editing tools: state-of-the-art and the road ahead for the model and non-model fishes. *Transgenic Research*, *26*(5), pp.577-589, 2017.
9. Bashir, K.M.I., Kim, M.S., Stahl, U. and Cho, M.G., Agrobacterium-mediated genetic transformation of Dictyosphaerium pulchellum for the expression of erythropoietin. *Journal of Applied Phycology*, *30*(6), pp.3503-3518, 2018.
10. Beratto-Ramos, R.P., Castillo-Felices, N.A., Troncoso-Leon, A., Agurto-Munoz, C. and Agurto-Munoz., Selection criteria for high-value biomass: seasonal and morphological variation of polyphenolic content and antioxidant capacity in two brown macroalgae, *Journal of Applied Phycology*, *31*(1). pp.653–664, 2019. https://doi.org/ 10.1007/s10811-018-1528-9.
11. Bilichak, A., Sastry-Dent, L., Sriram, S., Simpson, M., Samuel, P., Webb, S., Jiang, F. and Eudes, F., Genome editing in wheat microspores and haploid embryos mediated by delivery of ZFN proteins and cell-penetrating peptide complexes. *Plant Biotechnology Journal*, *18*(5), pp.1307-1316, 2020.
12. Brumwell, S.L., MacLeod, M.R., Huang, T., Cochrane, R.R., Meaney, R.S., Zamani, M., Matysiakiewicz, O., Dan, K.N., Janakirama, P., Edgell, D.R. and Charles, T.C., Designer Sinorhizobium meliloti strains and multi-functional vectors enable direct inter-kingdom DNA transfer. *PloS One*, *14*(6), p.e0206781, 2019.
13. Chai, R., Zhang, G., Sun, Q., Zhang, M. and Zhao, S., Liposome-mediated mycelial transformation of filamentous fungi. *Fungal Biology*, *117*(9), pp.577–583, 2013.
14. Chang, K.S., Kim, J., Park, H., Hong, S.J., Lee, C.G. and Jin, E., Enhanced lipid productivity in AGP knockout marine microalga Tetraselmis sp. using a DNA-free CRISPR-Cas9 RNP method. *Bioresource Technology*, *303*, pp.122932, 2020.
15. Chang, M., Huang, Y.W.S., Aronstam, R. and Lee, H.J., Cellular delivery of noncovalently-associated macromolecules by cell-penetrating peptides. *Current Pharmaceutical Biotechnology*, *15*(3), pp.267-275, 2014.
16. Chen, Y. and Hu, H., High efficiency transformation by electroporation of the freshwater alga Nannochloropsis limnetica. *World Journal of Microbiology and Biotechnology*, *35*(8), pp.1-10, 2019.
17. Chen, Z. and Lee, W.G., A switching role of hard-uptake nanoparticles in microalgae cell electroporation. *Analyst*, *144*(11), pp.3581-3589, 2019.
18. Cheng, R., Ma, R., Li, K., Rong, H., Lin, X., Wang, Z., Yang, S. and Ma, Y., Agrobacterium tumefaciens mediated transformation of marine microalgae Schizochytrium. *Microbiological Research*, *167*(3), pp.179-186, 2012.

19. Chuah, J.A., Odahara, M., Kodama, Y., Miyamoto, T., Tsuchiya, K., Motoda, Y., Kigawa, T. and Numata, K., Synthetic peptide-induced internalization of biomolecules into various plant and algal cells via micropinocytosis. *BioRxiv*, pp.630301, 2019.
20. Coll, J.M., Methodologies for transferring DNA into eukaryotic microalgae. *Spanish Journal of Agricultural Research*, *4*(4), pp.316-330, 2006.
21. Crozet, P., Navarro, F.J., Willmund, F., Mehrshahi, P., Bakowski, K., Lauersen, K.J., Pérez-Pérez, M.E., Auroy, P., Gorchs Rovira, A., Sauret-Gueto, S. and Niemeyer, J., Birth of a photosynthetic chassis: a MoClo toolkit enabling synthetic biology in the microalga Chlamydomonas reinhardtii. *ACS Synthetic Biology*, *7*(9), pp.2074-2086, 2018.
22. D'Adamo, S., Schiano di Visconte, G., Lowe, G., Szaub-Newton, J., Beacham, T., Landels, A., Allen, M.J., Spicer, A. and Matthijs, M., Engineering the unicellular alga Phaeodactylum tricornutum for high-value plant triterpenoid production. *Plant Biotechnology Journal*, *17*(1), pp.75-87, 2019.
23. Dahlin, L.R., Gerritsen, A.T., Henard, C.A., Van Wychen, S., Linger, J.G., Kunde, Y., Hovde, B.T., Starkenburg, S.R., Posewitz, M.C. and Guarnieri, M.T., Development of a high-productivity, halophilic, thermotolerant microalga Picochlorum renovo. *Communications Biology*, *2*(1), pp.1-9, 2019.
24. Debs, R.J., Cationic liposome-mediated gene transfer *in vitro* and *in vivo*. In *Liposomes as Tools in Basic Research and Industry*, pp. 171-176, CRC Press, 2017.
25. Dehghani, J., Adibkia, K., Movafeghi, A., Pourseif, M.M. and Omidi, Y., Designing a new generation of expression toolkits for engineering of green microalgae; robust production of human interleukin-2. *BioImpacts: BI*, *10*(4), pp.259, 2020.
26. Dehghani, J., Movafeghi, A., Barzegari, A. and Barar, J., Efficient and stable transformation of *Dunaliella Pseudosalina* by 3 strains of Agrobacterium tumefaciens. *BioImpacts: BI*, *7*(4), p.247, 2017.
27. Demirer, G.S., Silva, T.N., Jackson, C.T., Thomas, J.B., Ehrhardt, D.W., Rhee, S.Y., Mortimer, J.C. and Landry, M.P., Nanotechnology to advance CRISPR–Cas genetic engineering of plants. *Nature Nanotechnology*, *16*(3), pp.243-250, 2021.
28. Dias, E., Oliveira, M., Manageiro, V., Vasconcelos, V. and Caniça, M., Deciphering the role of cyanobacteria in water resistome: Hypothesis justifying the antibiotic resistance (phenotype and genotype) in Planktothrix genus. *Science of The Total Environment*, *652*, pp.447-454, 2019.
29. Doron, L., Segal, N.A. and Shapira, M., Transgene expression in microalgae—from tools to applications. *Frontiers in Plant Science*, *7*, pp.505, 2016.
30. Douchi, D., Mosey, M., Astling, D.P., Knoshaug, E.P., Nag, A., McGowen, J. and Laurens, L.M., Nuclear and chloroplast genome engineering of a productive non-model alga Desmodesmus armatus: Insights into unusual and selective acquisition mechanisms for foreign DNA. *Algal Research*, *53*, pp.102152, 2021.

31. Economou, C., Wannathong, T., Szaub, J. and Purton, S., A simple, low-cost method for chloroplast transformation of the green alga Chlamydomonas reinhardtii. In *Chloroplast Biotechnology*, pp. 401-411. Humana Press, Totowa, NJ, 2014.
32. Elshobary, M.E., Zabed, H.M., Yun, J., Zhang, G. and Qi, X., Recent insights into microalgae-assisted microbial fuel cells for generating sustainable bio-electricity. *International Journal of Hydrogen Energy, 46*(4), pp.3135-3159, 2021.
33. Feng, S., Feng, W., Zhao, L., Gu, H., Li, Q., Shi, K., Guo, S. and Zhang, N., Preparation of transgenic Dunaliella salina for immunization against white spot syndrome virus in crayfish. *Archives of Virology*, 159(3), pp.519-525, 2014.
34. Fernández, F.G.A., Reis, A., Wijffels, R.H., Barbosa, M., Verdelho, V. and Llamas, B., The role of microalgae in the bioeconomy. *New Biotechnology, 61*, pp.99-107, 2021.
35. French, K.E., Nanoparticle-mediated transformation expands horizon of organism engineering. *BioRxiv*, pp.559252, 2019.
36. Gadamchetty, P., Mullapudi, P.L.V., Sanagala, R., Markandan, M. and Polumetla, A.K., Genetic transformation of Chlorella vulgaris mediated by HIV-TAT peptide. *3 Biotech*, *9*(4), pp.1-7, 2019.
37. Galarza, J.I., Gimpel, J.A., Rojas, V., Arredondo-Vega, B.O. and Henríquez, V., Over-accumulation of astaxanthin in Haematococcus pluvialis through chloroplast genetic engineering. *Algal Research, 31*, pp.291-297, 2018.
38. Gan Q, Jiang, J., Han, X., Wang, S. and Lu, Y., Engineering the chloroplast genome of oleaginous marine microalga *Nannochloropsis Oceanica*, *Frontiers of Plant Science*, *9*,1–6, 2018.
39. Gan, S.Y., Qin, S., Othman, R.Y., Yu, D. and Phang, S.M., Transient expression of lacZ in particle bombarded Gracilaria changii (Gracilariales, Rhodophyta). *Journal of Applied Phycology, 15*(4), pp.345-349, 2003.
40. Gangl, D., Zedler, J. A., Rajakumar, P. D., Martinez, E.M.R., Riseley, A., Włodarczyk, A. and Robinson, C., Biotechnological exploitation of microalgae. *Journal of Experimental Botany*, *66*(22), pp. 6975-6990, 2015.
41. Gao, F., Sá, M., Teles, I., Wijffels, R.H. and Barbosa, M.J., Production and monitoring of biomass and fucoxanthin with brown microalgae under outdoor conditions. *Biotechnology and Bioengineering, 118*(3), pp.1355-1365, 2021.
42. Geerts, P., Druart, P., Ochatt, S. and Baudoin, J.P., Protoplast fusion technology for somatic hybridisation in Phaseolus. Biotechnologie, Agronomie, Société et Environnement/Biotechnology, Agronomy, *Society and Environment, 12*(1), pp.41-46, 2008.
43. Giraldo, J.P., Landry, M.P., Faltermeier, S.M., McNicholas, T.P., Iverson, N.M., Boghossian, A.A., Reuel, N.F., Hilmer, A.J., Sen, F., Brew, J.A. and Strano,

M.S., Plant nanobionics approach to augment photosynthesis and biochemical sensing. *Nature Materials, 13*(4), pp.400-408, 2014.
44. Golestanipour, A., Nikkhah, M., Aalami, A. and Hosseinkhani, S., Gene delivery to tobacco root cells with single-walled carbon nanotubes and cell-penetrating fusogenic peptides. *Molecular Biotechnology, 60*(12), pp.863-878, 2018.
45. Gutiérrez, S. and Lauersen, K.J., Gene Delivery Technologies with Applications in Microalgal Genetic Engineering. *Biology, 10*(4), p.265, 2021.
46. Hajieghrari, B., Farrokhi, N., Goliaei, B. and Kavousi, K., Identification and characterization of novel miRNAs in *Chlamydomonas Reinhardtii* by computational methods. *MicroRNA, 5*(1), pp.66-77, 2016.
47. Hallmann, A., Advances in Genetic Engineering of Microalgae. In *Grand Challenges in Algae Biotechnology,* pp. 159-221. Springer, Cham, 2019.
48. Higuchi-Takeuchi, M., Miyamoto, T., Foong, C.P., Goto, M., Morisaki, K. and Numata, K., Peptide-Mediated Gene Transfer into Marine Purple Photosynthetic Bacteria. *International Journal of Molecular Sciences, 21*(22), pp.8625, 2020.
49. Hussain, F., Shah, S.Z., Ahmad, H., Abubshait, S.A., Abubshait, H.A., Laref, A., Manikandan, A., Kusuma, H.S. and Iqbal, M., Microalgae an ecofriendly and sustainable wastewater treatment option: Biomass application in biofuel and bio-fertilizer production. A review. *Renewable and Sustainable Energy Reviews, 137,* pp.110603, 2021.
50. Hwang, H.J., Kim, Y.T., Kang, N.S. and Han, J.W., A simple method for removal of the Chlamydomonas reinhardtii cell wall using a commercially available Subtilisin (Alcalase). *Journal of Molecular Microbiology and Biotechnology, 28*(4), pp.169-178, 2018.
51. Hyman, J.M., Geihe, E.I., Trantow, B.M., Parvin, B. and Wender, P.A., A molecular method for the delivery of small molecules and proteins across the cell wall of algae using molecular transporters. *Proceedings of the National Academy of Sciences, 109*(33), pp.13225-13230, 2012.
52. Iwai, M., Hori, K., Sasakisekimoto, Y., Shimojima, M. and Ohta, H., Manipulation of oil synthesis in Nannochloropsis strain NIES-2145 with a phosphorus starvation-inducible promoter from Chlamydomonas reinhardtii. *Frontiers in Microbiology, 6,* pp. 912, 2015.
53. Jallet, D., Xing, D., Hughes, A., Moosburner, M., Simmons, M.P., Allen, A.E. and Peers, G., Mitochondrial fatty acid β-oxidation is required for storage-lipid catabolism in a marine diatom. *New Phytologist, 228*(3), pp.946-958, 2020.
54. Jeon, S., Kang, N.K., Suh, W.I., Koh, H.G., Lee, B. and Chang, Y.K., Optimization of electroporation-based multiple pulses and further improvement of transformation efficiency using bacterial conditioned medium for Nannochloropsis salina. *Journal of Applied Phycology, 31*(2), pp.1153-1161, 2019.

55. Joshi, M.D. and Müller, R.H., Lipid nanoparticles for parenteral delivery of actives. *European Journal of Pharmaceutics and Biopharmaceutics, 71*(2), pp.161-172, 2009.
56. Kang, S., Jeon, S., Kim, S., Chang, Y.K. and Kim, Y.C., Development of a pVEC peptide-based ribonucleoprotein (RNP) delivery system for genome editing using CRISPR/Cas9 in *Chlamydomonas Reinhardtii. Scientific Reports, 10*(1), pp.1-11, 2020a.
57. Kang, S., Kim, B., Yim, S.J., Kim, J.O., Kim, D.P. and Kim, Y.C., On-chip electroporation system of Polyimide film with sheath flow design for efficient delivery of molecules into microalgae. *Journal of Industrial and Engineering Chemistry, 88*, pp.159-166, 2020b.
58. Kang, S., Suresh, A. and Kim, Y.C., A highly efficient cell penetrating peptide pVEC-mediated protein delivery system into microalgae. *Algal Research, 24*, pp.360-367, 2017.
59. Karas, B.J., Diner, R.E., Lefebvre, S.C., McQuaid, J., Phillips, A.P., Noddings, C.M., Brunson, J.K., Valas, R.E., Deerinck, T.J., Jablanovic, J. and Gillard, J.T., Designer diatom episomes delivered by bacterial conjugation. *Nature Communications, 6*(1), pp.1-10, 2015.
60. Kashiyama, Y., Ishizuka, Y., Terauchi, I., Matsuda, T., Maeda, Y., Yoshino, T., Matsumoto, M., Yabuki, A., Bowler, C. and Tanaka, T., Engineered chlorophyll catabolism conferring predator resistance for microalgal biomass production. *Metabolic Engineering, 66*, pp.79-86, 2021.
61. Khatiwada, B., Kautto, L., Sunna, A., Sun, A. and Nevalainen, H., Nuclear transformation of the versatile microalga Euglena gracilis. *Algal Research, 37*, pp.178-185, 2019.
62. Kim, D.H., Kim, Y.T., Cho, J.J., Bae, J.H., Hur, S.B., Hwang, I. and Choi, T.J., Stable integration and functional expression of flounder growth hormone gene in transformed microalga, *Chlorella Ellipsoidea. Marine Biotechnology, 4*(1), pp.63-73, 2002.
63. Kim, S., Choi, Y., Park, G., Won, C., Park, Y. and Lee. Y., Highly efficient gene silencing and bioimaging 132 based fluorescent carbon dots *in vitro* and *in vivo. Nano Research, 10* (2), pp. 503–51 9, 2017.
64. Kim, S., Lee, Y.C., Cho, D.H., Lee, H.U., Huh, Y.S., Kim, G.J. and Kim, H.S., A Simple and Non-Invasive Method for Nucleus Transformation of Intact-walled *Chlamydomonas Reinhardtii. PLoS ONE, 9*, pp. e101018, 2014.
65. Kim, T.I., Rothmund, T., Kissel, T. and Kim, S.W., Bioreducible polymers with cell penetrating and endosome buffering functionality for gene delivery systems. *Journal of Controlled Release, 152*(1), pp.110-119, 2011.
66. Kim, Y.H., Control of the culture conditions of Chlamydomonas reinhardtii for efficient delivery of exogenous materials in electroporation. *Algal Research, 35*, pp.388-394, 2018.
67. Kindle, K.L.,. High-frequency nuclear transformation of Chlamydomonas reinhardtii. *Proceedings of the National Academy of Sciences, 87*(3), pp.1228-1232, 1990.

68. Kiyokawa, K., Ohmine, Y., Yunoki, K., Yamamoto, S., Moriguchi, K. and Suzuki, K., Enhanced Agrobacterium-mediated transformation revealed attenuation of exogenous plasmid DNA installation in recipient bacteria by exonuclease VII and SbcCD. *Genes to Cells, 25*(10), pp.663-674, 2020.
69. Kotnik, T., Frey,W., Sack, M., Haberl Meglič, S., Peterka, M. and Miklavčic, D., Electroporation-based applications in biotechnology. *Trends Biotechnol.* 33, 480–488, 2015.
70. Kusumaningrum, H. and Zainuri, M., Improvement of Nutrition Production by Protoplast Fusion Techniques in Chlorella vulgaris. *Journal of Food Processing & Technology, 9*(1), pp. 1–5, 2018. https://doi.org/10.4172/2157-7110.1000711.
71. Kusumaningrum, H.P. and Zainuri, M., Optimization and stability of total pigments production of fusan from protoplast fusion of microalgae Dunaliella and Chlorella *in vivo*: Attempts on production of sustainable aquaculture natural food. *International Journal of Marine and Aquatic Resource Conservation and Co-existence, 1*(1), pp.1-5, 2014.
72. Kwon, Y.M., Kim, K.W., Choi, T.Y., Kim, S.Y. and Kim, J.Y.H., Manipulation of the microalgal chloroplast by genetic engineering for biotechnological utilization as a green biofactory. *World Journal of Microbiology and Biotechnology, 34*(12), pp.1-11, 2018.
73. Lee, J.W., Lee, M.W., Ha, J.S., Kim, D.S., Jin, E., Lee, H.G. and Oh, H.M., Development of a species-specific transformation system using the novel endogenous promoter calreticulin from oleaginous microalgae Ettlia sp. *Scientific Reports, 10*(1), pp.1-12, 2020.
74. Lin, B., Cui, Y., Yan, M., Wang, Y., Gao, Z., Meng, C. and Qin, S., Construction of astaxanthin metabolic pathway in the green microalga Dunaliella viridis. *Algal Research, 44*, pp.101697, 2019.
75. Liu, S., Liu, C., Huang, X., Chai, Y. and Cong, B., Optimization of parameters for isolation of protoplasts from the Antarctic sea ice alga Chlamydomonas sp. ICE-L. *Journal of Applied Phycology, 18*(6), 783–786, 2006. https://doi.org/10.1007/s10811-006-9093-z.
76. Lu, Y., Zhou, W., Wei, L., Li, J., Jia, J., Li, F., Smith, S.M. and Xu, J., Regulation of the cholesterol biosynthetic pathway and its integration with fatty acid biosynthesis in the oleaginous microalga Nannochloropsis oceanica. *Biotechnology for Biofuels, 7*(1), pp.1-15, 2014.
77. Lv, J., Tan, E., Wang, Y., Fan, Q., Yu, J. and Cheng, Y., Tailoring guanidyl-rich polymers for efficient cytosolic protein delivery. *Journal of Controlled Release, 320*, pp. 412-420, 2020.
78. Lv, J., Fan, Q., Wang, H. and Cheng, Y., Polymers for cytosolic protein delivery. *Biomaterials, 218*, pp.119358, 2019.
79. Malejko, J., Szymańska, N., Bajguz, A. and Godlewska-Żyłkiewicz, B., Studies on the uptake and transformation of gold (III) and gold nanoparticles in a water–green algae environment using mass spectrometry techniques. *Journal of Analytical Atomic Spectrometry, 34*(7), pp.1485-1496, 2019.

80. Matsunaga, T., Takeyama, H. and Nakamura, N., Characterization of cryptic plasmids from marine cyanobacteria and construction of a hybrid plasmid potentially capable of transformation of marine cyanobacterium, Synechococcus sp., and its transformation. *Applied Biochemistry and Biotechnology, 24*(1), pp.151-160, 1990.
81. McQuillan, J., *Developing Metabolic Engineering Tools for Enhanced Synthesis of High-Value Products in Chlamydomonas Reinhardtii* (Doctoral dissertation, University of Sheffield), 2021.
82. Mini, P., Demurtas, O.C., Valentini, S., Pallara, P., Aprea, G., Ferrante, P. and Giuliano, G., Agrobacterium-mediated and electroporation-mediated transformation of Chlamydomonas reinhardtii: a comparative study. *BMC Biotechnology, 18*(1), pp.1-12, 2018.
83. Molina-Márquez, A., Vila, M., Rengel, R., Fernández, E., García-Maroto, F., Vigara, J. and León, R., Validation of a New Multicistronic Plasmid for the Efficient and Stable Expression of Transgenes in Microalgae. *International Journal of Molecular Sciences, 21*(3), p.718, 2020.
84. Molina-Márquez, A., Vila, M., Vigara, J., Borrero, A. and León, R., The Bacterial Phytoene Desaturase-Encoding Gene (CRTI) is an Efficient Selectable Marker for the Genetic Transformation of Eukaryotic Microalgae. *Metabolites, 9*(3), p.49, 2019.
85. Morais, F.P., Simões, R. and Curto, J.M., Biopolymeric Delivery Systems for Cosmetic Applications Using Chlorella vulgaris Algae and Tea Tree Essential Oil. *Polymers, 12*(11), p.2689, 2020.
86. Muñoz, C.F., Sturme, M.H., D'Adamo, S., Weusthuis, R.A. and Wijffels, R.H., Stable transformation of the green algae *Acutodesmus obliquus* and *Neochloris oleoabundans* based on E. coli conjugation. *Algal Research, 39*, p.101453, 2019.
87. Muñoz, C.F., de Jaeger, L., Sturme, M.H., Lip, K.Y., Olijslager, J.W. and Springer, J., Improved DNA/protein delivery in microalgae-A simple and reliable method for the prediction of optimal electroporation settings. *Algal Research, 33*, 448–455, 2018.
88. Ng, K.K., Motoda, Y., Watanabe, S., Sofiman Othman, A., Kigawa, T., Kodama, Y. and Numata, K., Intracellular delivery of proteins via fusion peptides in intact plants. *PLoS One, 11*(4), p.e0154081, 2016.
89. Niu, Y.F., Zhang, M.H., Xie, W.H., Li, J.N., Gao, Y.F., Yang, W.D., Liu, J.S. and Li, H.Y., A new inducible expression system in a transformed green alga, Chlorella vulgaris. *Genet Mol Res, 10*(4), pp.3427-34, 2011.
90. Norashikin, M.N., Loh, S.H., Aziz, A. and San Cha, T., Metabolic engineering of fatty acid biosynthesis in Chlorella vulgaris using an endogenous omega-3 fatty acid desaturase gene with its promoter. *Algal Research, 31*, pp.262-275, 2018.
91. Norzagaray-Valenzuela, C.D., Germán-Báez, L.J., Valdez-Flores, M.A., Hernández-Verdugo, S., Shelton, L.M. and Valdez-Ortiz, A., Establishment of an efficient genetic transformation method in Dunaliella tertiolecta

mediated by *Agrobacterium Tumefaciens. Journal of Microbiological Methods, 150*, pp.9-17, 2018.

92. Nymark, M., Sharma, A.K., Sparstad, T., Bones, A.M. and Winge, P., A CRISPR/Cas9 system adapted for gene editing in marine algae. *Scientific Reports, 6*(1), pp.1-6, 2016.
93. Ortiz-Matamoros, M.F., Villanueva, M.A. and Islas-Flores, T., Genetic transformation of cell-walled plant and algae cells: delivering DNA through the cell wall. *Briefings in Functional Genomics, 17*(1), pp.26-33, 2018.
94. Ozdalgic, B., Ustun, M., Dabbagh, S.R., Haznedaroglu, B.Z., Kiraz, A. and Tasoglu, S., Microfluidics for microalgal biotechnology. *Biotechnology and Bioengineering, 118*(4), pp.1716-1734, 2021.
95. Ozyigit, I.I. and Yucebilgili Kurtoglu, K., Particle bombardment technology and its applications in plants. *Molecular Biology Reports, 47*(12), pp.9831-9847, 2020.
96. Pasha, C., Kuhad, R.C. and Rao, L.V., Strain improvement of thermotolerant Saccharomyces cerevisiae VS3 strain for better utilization of lignocellulosic substrates. *Journal of Applied Microbiology, 103*(5), pp.1480-1489, 2007.
97. Peng, S., Bie, B., Jia, H., Tang, H., Zhang, X., Sun, Y., Wei, Q., Wu, F., Yuan, Y., Deng, H. and Zhou, X., Efficient Separation of Nucleic Acids with Different Secondary Structures by Metal–Organic Frameworks. *Journal of the American Chemical Society, 142*(11), pp.5049-5059, 2020.
98. Peraza-Echeverria, S., Bernardo-Candelero, S., Baas-Espinola, F.M., Puch-Hau, C., Rivera-Solís, R.A., Echevarría-Machado, I., Borges-Argáez, I.C. and Herrera-Valencia, V.A., Production of a ruminal bacterial phytase in the green microalga Chlamydomonas reinhardtii with potential applications in monogastric animal feed. *Bioresource Technology Reports, 14*, pp.100660, 2021.
99. Poliner, E., Clark, E., Cummings, C., Benning, C. and Farre, E.M., A high-capacity gene stacking toolkit for the oleaginous microalga, Nannochloropsis oceanica CCMP1779. *Algal Research, 45*, pp.101664, 2020.
100. Pratheesh, P.T., Vineetha, M. and Kurup, G.M., An efficient protocol for the Agrobacterium-mediated genetic transformation of microalga Chlamydomonas reinhardtii. *Molecular Biotechnology, 56*(6), pp.507-515, 2014.
101. Qi, L.S., Larson, M. H., Gilbert, L. A., Doudna, J. A., Weissman, J. S., Arkin, A. P. and Lim, W.A., Repurposing CRISPR as an RNA-guided platform for sequence-specific control of gene expression. *Cell, 152*(5), pp. 1173-1183, 2013.
102. Qu, B., Eu, Y.J., Jeong, W.J. and Kim, D.P., Droplet electroporation in microfluidics for efficient cell transformation with or without cell wall removal. *Lab on a Chip, 12*(21), pp.4483-4488, 2012.
103. Ramessur, A.D., Bothwell, J.H., Maggs, C.A., Gan, S.Y. and Phang, S.M., Agrobacterium-mediated gene delivery and transient expression in the red macroalga Chondrus crispus. *Botanica Marina, 61*(5), pp.499-510, 2018.

104. Ringaci, A., Yaremenko, A.V., Shevchenko, K.G., Zvereva, S.D. and Nikitin, M.P., Metal-organic frameworks for simultaneous gene and small molecule delivery *in vitro* and *in vivo*. *Chemical Engineering Journal, 418*, pp.129386, 2021.

105. Rivera, A.L., Magana-Ortiz, D., Gomez-Lim, M., Fernandez, F. and Loske, A.M., Physical methods for genetic transformation of fungi and yeast. *Physics of Life Reviews, 11*(2), pp.184-203, 2014.

106. Rosales-Mendoza, S., Solís-Andrade, K.I., Márquez-Escobar, V.A., González-Ortega, O. and Bañuelos-Hernandez, B., Current advances in the algae-made biopharmaceuticals field. *Expert Opinion on Biological Therapy, 20*(7), pp.751-766, 2020.

107. Savvidou, M.G., Ferraro, A., Hristoforou, E., Mamma, D., Kekos, D. and Kolisis, F.N., Incorporation of Magnetic Nanoparticles into Protoplasts of Microalgae Haematococcus pluvialis: A Tool for Biotechnological Applications. *Molecules, 25*(21), p.5068, 2020.

108. Shanmugam, S., Hari, A., Kumar, D., Rajendran, K., Mathimani, T., Atabani, A.E., Brindhadevi, K. and Pugazhendhi, A., Recent developments and strategies in genome engineering and integrated fermentation approaches for biobutanol production from microalgae. *Fuel, 285*, pp.119052, 2021.

109. Sharma, A.K., Nymark, M., Flo, S., Sparstad, T., Bones, A.M. and Winge, P., Simultaneous knockout of multiple LHCF genes using single sgRNAs and engineering of a high-fidelity Cas9 for precise genome editing in marine algae. *Plant Biotechnology Journal, 19*(8), pp.1658-1669, 2021.

110. Sharma, A.K., Nymark, M., Sparstad, T., Bones, A.M. and Winge, P., Transgene-free genome editing in marine algae by bacterial conjugation–comparison with biolistic CRISPR/Cas9 transformation. *Scientific Reports, 8*(1), pp.1-11, 2018.

111. Sharma, P.K., Goud, V.V., Yamamoto, Y. and Sahoo, L., Efficient *Agrobacterium tumefaciens*-mediated stable genetic transformation of green microalgae, Chlorella sorokiniana. *3 Biotech, 11*(4), pp.1-11, 2021.

112. Shi, L., Chen, D., Xu, C., Ren, A., Yu, H. and Zhao, M., Highly-efficient liposome-mediated transformation system for the basidiomycetous fungus Flammulina velutipes. *The Journal of General and Applied Microbiology*, pp.2016-10, 2017.

113. Simon, D.P., Anila, N., Gayathri, K. and Sarada, R., Heterologous expression of β-carotene hydroxylase in Dunaliella salina by Agrobacterium-mediated genetic transformation. *Algal Research, 18*, pp.257-265, 2016.

114. Tan, P., Xiang, X., Guo, H. and Ai, Q., Liposome-mediated messenger RNA: An alternative for fish cell transfection in culture. *Aquaculture Research, 51*(7), pp.2745-2757, 2020.

115. Tang, D.K., Qiao, S.Y. and Wu, M., Insertion mutagenesis of Chlamydomonas reinhardtii by electroporation and heterologous DNA. *Biochemistry and Molecular Biology International, 36*(5), pp.1025-1035, 1995.

116. Tanwar, A. and Kumar, S., Genome editing of algal species by CRISPR Cas9 for biofuels. In *Genome Engineering via CRISPR-Cas9 System,* (pp. 163-176), Academic Press, 2020.
117. Tanwar, A., Sharma, S. and Kumar, S., Targeted genome editing in algae using CRISPR/Cas9. *Indian Journal of Plant Physiology, 23* (4), pp. 653–669, 2018.
118. Telke, A.A. and Rolain, J.M., Functional genomics to discover antibiotic resistance genes: the paradigm of resistance to colistin mediated by ethanolamine phosphotransferase in Shewanella algae MARS 14. *International Journal of Antimicrobial Agents, 46*(6), pp.648-652, 2015.
119. Thagun, C., Motoda, Y., Kigawa, T., Kodama, Y. and Numata, K., Simultaneous introduction of multiple biomacromolecules into plant cells using a cell-penetrating peptide nanocarrier. *Nanoscale, 12*(36), pp.18844-18856, 2020.
120. Tran, M., Henry, R.E., Siefker, D., Van, C., Newkirk, G., Kim, J., Bui, J. and Mayfield, S.P., Production of anti-cancer immunotoxins in algae: ribosome inactivating proteins as fusion partners. *Biotechnology and Bioengineering, 110*(11), pp.2826-2835, 2013.
121. Wang, L., Yang, L., Wen, X., Chen, Z., Liang, Q., Li, J. and Wang, W., Rapid and high efficiency transformation of *Chlamydomonas reinhardtii* by square-wave electroporation. *Bioscience Reports, 39*(1), 2019.
122. Wang, S., Liang, H., Xu, Y., Li, L., Wang, H., Sahu, D.N., Petersen, M., Melkonian, M., Sahu, S.K. and Liu, H., Genome-wide analyses across Viridiplantae reveal the origin and diversification of small RNA pathway-related genes. *Communications Biology, 4*(1), pp.1-11, 2021.
123. Wannathong, T., Waterhouse, J.C. and Young, R.E.B., New tools for chloroplast genetic engineering allow the synthesis of human growth hormone in the green alga *Chlamydomonas Reinhardtii. Appl Microbiol Biotechnol, 100*, pp. 5467–5477, 2016.
124. Wu, N., Dong, X., Liu, Y., Wang, C., Baattrup-Pedersen, A. and Riis, T., Using river microalgae as indicators for freshwater biomonitoring: Review of published research and future directions. *Ecological Indicators, 81*, pp.124-131, 2017.
125. Wu, N., Dong, X., Liu, Y., Wang, C., Baattrup-Pedersen, A. and Riis, T., Using river microalgae as indicators for freshwater biomonitoring: Review of published research and future directions. *Ecological Indicators, 81*, pp.124-131, 2017.
126. Xie, F., Zhang, F., Zhou, K., Zhao, Q., Sun, H., Wang, S., Zhao, Y. and Fu, J., Breeding of high protein Chlorella sorokiniana using protoplast fusion. *Bioresource Technology, 313*, pp.123624, 2020.
127. Yin, W. and Hu, H., High-efficiency transformation of a centric diatom Chaetoceros muelleri by electroporation with a variety of selectable markers. *Algal Research, 55*, p.102274, 2021.
128. Zhou, X., Zhang, X., Boualavong, J., Durney, A. R., Wang, T., Kirschner, S. and Mukaibo, H., Electrokinetically controlled fluid injection into unicellular microalgae. *Electrophoresis, 38*(20), 2587-2591, 2017.

9

Algae Utilization as Food and in Food Production: Ascorbic Acid, Health Food, Food Supplement and Food Surrogate

Abiola Folakemi Olaniran[1*], Bolanle Adenike Akinsanola[1], Abiola Ezekiel Taiwo[2], Joshua Opeyemi Folorunsho[1], Yetunde Mary Iranloye[1], Clinton Emeka Okonkwo[3] and Omorefosa Osarenkhoe Osemwegie[1]

[1]*Responsible Consumption and Production Group, Landmark University SDG 12; and Department of Food Science and Microbiology, Landmark University, Omu-Aran, Kwara State, Nigeria*
[2]*Faculty of Engineering, Mangosuthu University of Technology, Durban, South Africa*
[3]*Department of Food Science and Technology, College of Agriculture and Veterinary Medicine, United Arab Emirates University, AI ain, United Arab Emirates*

Abstract
Algae, besides being edible, are valuable in other areas of human endeavors that span agriculture, pharmaceuticals, medicine, dietetics, and industries. Advancements on their consumption benefits are expanding to their functionality, reaffirming them as a sustainable bioresource with future promise in global bioeconomic development and amelioration of hunger. The potential benefits of algae as a food and in food processing are the subject of this chapter. Many edible algae which comprise micro- and macroalgae are found to be rich in proteins, carbohydrates, polyunsaturated fatty acids, fiber, vitamins and minerals. This is in addition to their applications in the production of algal-based bioplastic, biopesticides, biofertilizers, and bioproducts like flavoring, thickeners, sweeteners and confectioneries. The innate vitamin B complexes, bioactive ingredients and phytochemicals expressed by algae make them attractive to humans as delicacies and for culinary purposes. Furthermore, these components, when consumed in the food matrix, were found to increase human wellbeing by rejuvenating the

**Corresponding author*: abiolaolaniran@gmail.com; olaniran.abiola@lmu.edu.ng

Charles Oluwaseun Adetunji, Julius Kola Oloke, Naveen Dwivedi, Sabeela Beevi Ummalyma, Shubha Dwivedi, Daniel Ingo Hefft and Juliana Bunmi Adetunji (eds.) Next-Generation Algae: Volume I: Applications in Agriculture, Food and Environment, (225–240) © 2023 Scrivener Publishing LLC

physiological mechanisms that underpin antioxidants' activities, photosynthesis, hormone biosynthesis and enzymological functions. Also, application of algae in the food industry may be as supplements or surrogates, suggesting their combination with other food products and molecules for optimum beneficial outcomes. Therefore, promoting the sustainable application of algae in diverse ways for future human benefits or harnessing the opportunities they offer humanity is propelled by research intensity, technological advancement, a cultural shift towards algae acceptance, and pro-environmental behavior.

***Keywords*:** Food, algae, nutrients, bio-economy, consumption, ascorbic acid, health food, food supplement, food surrogate

9.1 Introduction

Algae are a heterotrophic group of predominantly photoautotrophic, aquatic and non-vascular eukaryotic organisms without identifiable stems, leaves, or roots, but with chlorophylls as well as other pigments. They are as common as other microbes even though they are widely found in water bodies, semi-aquatic and moist habitats with relatively few occurring as non-parasites on or within plant cells [1, 2]. Algae are fundamental to the trophic economy of many aquatic and semi-aquatic ecosystems since they are the primary producers upon which other consumers depend on for energy. Furthermore, their role as the main source of oxygen in aquatic environments and in the food webs cannot be overemphasized. Algal ecological lifestyle is diverse, may be invasive, and has the propensity to transform simple chemical precursor into value-adding metabolites coupled with polymers [3]. Hence, in more recent years, algae have been attracting attention as a research biogenetic resource used purposefully to screen for low-cost alternative bioactive, biofunctional, and other innocuous compounds valuable to improving human's standard of living. This inclination may also be attributed to efforts by the scientific, agricultural, and ruling communities to tackle global food gaps.

They exhibit a variety of pathways that allow them to remove carbon dioxide from the atmosphere while also efficiently utilizing nutrients and creating biomass. Furthermore, many algae have evolved a means to store large amounts of lipids and polysaccharides which may serve as valuable intermediates for the production of omega-3 [4]. Suffice it to say that their metabolites, polymers and biomass have been harnessed through various technological models (biotechnology, nanotechnology, digital technology) over the last couple of decades for use as biofuel, preservatives, nutraceuticals, pharmaceutics, foods, feeds, biofertilizers, stabilizing agents, vitamins, gelling agent, flavor enhancer, and as capping agents in green synthesis [5–8]. Algae, particularly kelps (seaweeds) and microalgae, have attracted more attention as a bioresource for the generation of usable products, which has escalated interest in their cultivation as well as their business prospects. The procedures required to harness their economical values were found to be environmentally friendly, cheap, and putatively cleaner than other known technologies [9]. While it is logical to assume that the wealth of algal diversity is underexploited for different and novel applications or product developments, the knowledge of their potential uses in the food industries is lacking. More so, the reports on the consumption of algae as foods, food supplements, superfoods or even to fill the protein gaps in global diets are less popular in the scientific literature compared to their applicability in pollution control, energy generation, water treatment, pharmaceutics, and other circular bioeconomic beneficiations [10]. The European Food Safety Authority (EFSA) under the regulation of the European Commission No 2015/2283 has approved numerous microalgae, such as *Odontella aurita, Anabaena flos-aquae, Chlorella luteoviridis, Arthrospira platensis, Tetraselmis chui, Chlorella vulgaris* and *Chlorella pyrenoidosa*, as algal foods and attractive alternatives for vegan diets. Hence, this chapter focuses on the broad application potentials of algae as food and food modulators.

9.2 The Utilization of Algae

9.2.1 Use of Algae in the Food Industry

Microalgae and seaweeds are among the promising biogenetic resources applicable for several biotechnological inventions and harnessing eco-friendly natural bioactive metabolites. Their cells naturally have a chemically different range of vitamins which can be explored for harnessing vitamins C, B12 and E for large production as food supplements. While not all algae are edible, those that are edible are rich in high-value nutritional and nutraceutical compounds which constitute natural foods in many Western countries. Even though edible algae are low in calories and fat, they are reported by Wang *et al.* [11] to be in proteins, carbohydrates and dietary fiber besides their mineral, macromineral and trace element contents [12]. Several species of algae are used in the food, dairy, pharmaceutical, and cosmetic industries even though most edible algae have been reported to possess many medical values, including therapeutic properties such as antiviral, antioxidant, antifungal, anticancer, anti-allergic, anti-inflammatory, anticoagulant, and antifouling [13]. Marine algae were reported by Surendhiran *et al.* [7] to have high concentrations of antimicrobial compounds that are valuably exploited for food preservation in the food industries. Their ability to exert antimicrobial protection when consumed has also endeared them to humans as dietary options. Del Mondo *et al.* [14] observed a rich concentration of ascorbic acids (0.06–18.79 mg/g) when compared to kiwis, lemons or strawberries (0.52 0.42, and 0.54 mg/g). While this further reinforces their application in the food industry, their consumption as food or derived products has been linked to Alzheimer's disease and the attenuation of isoproterenol-induced oxidative damage [15, 16]. Consequently, the inherent health benefit differs depending on the species and their derived products have been subjected to processing methods (harvesting, drying, sterilization and extraction) that maintain their bioactive characteristics in either dehydrated or powdered form. Whole edible algae in diets have a higher pharmacokinetics advantage that translates into optimal nutritional and biofunctional benefits compared to their derived products that may have their potency altered by technology. Their consumption also enhances the biological release of protein, dietary fiber, minerals, carotenoids, vitamins, polyphenols, polysaccharides and other bioactive compounds from the food matrix [17]. So, the wide implementation of algae in the food industry still requires further study of their diversity, chemical compositions, edibility, ethnophycology, and screening of their innumerable secondary metabolites or biopolymers for bioactivity when consumed in the food matrix.

Himanthalia elongata, Enteromorpha, Porphyra umbilicalis and *Undaria pinnatifida* are used as a food supplement to enhance the nutritional function as well as bioactive quality of meat (sausages, steaks, pasties, and frankfurters), fish, oil, and cereal-based products. A combination of microalgae is applied in the dairy products industry to processed cheese, yogurt, cream, cottage cheese and milk deserts. This is premised on the knowledge-based perception of their use in therapeutic diets for patients that suffer from hyperglycemia, oxidative stress, and hyperlipidemia [18]. Carrageenans, which are sulphated polysaccharides derived from an extracellular matrix of red algae (Irish moss), have found use in food production as an emulsifier, thickener and stabilizer in making ice cream, soymilk, and beer [19]. According to Guo *et al.* [20], their superior biocompactibility, biodegradative, solubility, rheological, thermal, mechanical, immunogenicity, nontoxic and low-cost attributes have made them more attractive in bio-based delivery systems, bionanotechnology, bioplastic and biopacking technologies respectively. Carrageenans such as k-carrageenan and l-carrageenan are identified with a thickening property which is useful in making deserts, jams, and meat products [21]. *Porphyra*, a red alga, is also used to make sushi, a traditional Japanese protein-rich meal, while *Macrocystis*, a brown alga, is a sustainable source of alginic acid, which is an important food thickener [3]. Sharma and Sharma [9] reported the use of *Chlorella vulgaris* as a supplement in food and carotene from *Dunaliella* sp. Consequently, algal use in the food industry has extended beyond consumption for hunger satiety to include functional roles delivered by bioactive substances released in the food matrix and income generation for many household communities through their scalar cultivation. This is in addition to their being valuable to dietetics, therapeutics, environment and conservation strategies. Their growing acceptance as functional foods has directly provoked scientific and business ventures for newer substances with more desirable physical and chemical qualities that could enrich food items. The anti-oxidant, anti-cancer, and anti-inflammatory properties of algal ingredients in dietary products are tremendous [22]. Some algae induce dual function because they possess chemical substances that improve food nutritional quality and increase their rate of bioactivity on consumption. Hence, many edible algae have been dubbed "superfoods" over the years because they have been modulated to function as per supplements in food as capsules, powder extracts, and liquid suspension. Their edible biomass was reported to be rich in a variety of active compounds containing vitamins. Vitamins E, C, A, and B1, B2, B6, which contain niacin, biotin, nicotinate, folic, and pantothenic acids respectively, are examples of some of the vitamins reported in algae [4, 23]. Concentration of proteins containing essential amino acids, polysaccharide hydrocolloids, antioxidant compounds, including pigments like minerals, carotenoids, active enzymes, chlorophylls, mono- and polyunsaturated fatty acids in algal biomass, enhance feed systems, food aesthetics, pharmacological activities, and promote dietary consumption of algae as well as their use as animal feed [24]. More still needs to be learned about the composition of biofunctional ingredients of edible algae and their therapeutic and pharmacological dynamics. Since their sustainable edibility is apparently not universal and may fundamentally be based on their availability, accessibility and closeness to human or maritime communities relative to other food alternatives, efforts are being made to popularize their cultivation [25]. Despite this limitation, algal-based snack meals, pasta, candy bars, and gums coupled with drinks and beverages flavored with some microalgal products are commonly encountered on the international market.

Microalgae serve as food substitutes which are a valuable additional source of bulk proteins, as they can be greatly harnessed to fill the gap in the demand for protein, especially in light of a growing population. When compared to other common protein sources, microalgae protein (MP) may be preferred, especially because it could be sustainably produced using far less cultivable land. One kilogram (1 kg) of MP takes less than 2.5 m^2 in comparison to chicken (42–52 m^2), pork (47–64 m^2), and beef (144–258 m^2) respectively. Surprisingly, certain plant protein sources that were extensively advertised as suitable replacements for animal proteins, particularly those intensively produced (soybean, pea, potato) for food and feed, require larger quantities of land than MP for cultivation [26]. Furthermore, the popularity of microalgae utilization is predicated on adequate considerations such as expertise, arable and non-arable requirements, availability of freshwater, and suitability of seawater, particularly for algal farmers living in non-coastal communities. However, many maritime countries have a long history of using macroalgae as a food source and more recently for various industrial applications.

9.2.2 Macroalgae with Application Prospects in Food

Seaweed consumption is popular in Asia, predominantly in Japan, India, China and Korea, where commercial seaweed production is a major industry. The cultural food heritage in Western countries coupled with the macroalgal-sourced habitats distance may be accountable for the slow cultivation and industrial use of seaweeds. Notwithstanding, some macroalgae species have been identified to either be edible or have other value-adding potentials that gave impetus to their cultivation in the European oceans. Protein, fiber, minerals, and vitamins like folic acid, vitamin C, A, and B_1 are all abundant in this seaweed [27]. *Porphyra* grows in shallow waters and is a popular delicacy rich in proteins (30–35%), carbohydrates (40–45%), and diverse vitamins. Also, the sea vegetable product known as "amanori" is produced in Japan, while *Porphyra* is commercially cultivated and widely consumed. On the northern coasts of the Atlantic and Pacific oceans, *Palmaria palmata,* also known as red dulse or sea lettuce flakes, is a naturally predominant seaweed. It is rich in vitamins and proteins with less essential amino acids compared with other oriental seaweeds. Salty and dehydrated *Palmaria palmata* is typically served in bars as cocktail snacks in Maine and Nova Scotia as well as being eaten in raw or minimally cooked form in Ireland [28]. *Laminaria* (kelp) is a brown alga that contains about 2% fat, 10% protein, and a considerable number of vital minerals like iron, iodine, magnesium, calcium, and potassium. The popularity of their values has resulted in an increase in macroalgae harvesting, especially in the wild, with value-adding implications for the burgeoning agricultural sector of some nations in Asia. *Laminaria* species are typically grown on ropes, cylinders, and seashore stones, and has a sustainable market demand due to its rich glutamic acid. This has made it desirable as one of the key elements in "kombu" or "konbu" food products made with dashi, which is a widely used flavoring additive in Japanese cuisines, particularly in their soup stock [29].

The *Alaria* genus has roughly seventeen edible species that are extensively consumed in dried form in South America, Western Europe, Japan, China and Korea. Other potential non-culinary uses for the *Alaria* species are documented such as bioindicator of environmental pollutants, ingredients of ruminant animal silage, and sequestration of dissolved inorganic nutrients coupled with oxygen in aquaculture systems [30, 31] Japanese "Sarumen" is made from *Alaria esculenta* that has rich supplies of high-quality protein and iodine compounds, while a Cochayuyo diet made from *Durvillaea antarctica* is often consumed in Chile. The *Durvillaea* species dominates the coastlines of southern New Zealand and Chile with their stem (hulte) applied in a variety of Chilean cuisine preparations, including salads and stews.

9.2.3 Microalgae Application Prospects in Foods

Monostroma is a green microalga that is industrially grown in bay zones of Korea, Taiwan, and Japan alongside *Ulva*. Its rich magnesium, lithium and calcium contents coupled with vitamins as well as methionine made it a culturally acceptable healthy food material [27]. Even though it is composed of a single-celled thick leaf, its biomass is usually mass-produced into sheets and dried before boiling in broth containing ingredients such sugar, soy sauce, etc., to produce nori jam. Sea lettuce (*Ulva*) is the second most well-known green macroalga with application in foods. *Ulva* is gathered and blended with *Monostroma* and *Enteromorpha* sp. to make "aonori," a condiment for making simple warm foods such as salads, rice and soups. *Ulva lactuca* is popular in Europe, particularly Scotland, where it serves as an important ingredient in salads and soups due to their high protein and iron contents. The *Enteromorpha* genus, a once free species now classed as *Ulva*, thrives in Japan's river and bay highlands and is colloquially known as "green nori." In Europe and North America, this green alga is equally well-known where it is commonly used as a garnish after drying of their biomass or roasted minimally prior to being used as flavoring [32].

Caulerpa racemosa, often known as sea grapes, is a green alga that occurs in shallow waters and used in the Australasia regions (Thailand, Fiji, Philippines, Japan) primarily for making salads. Permatasari *et al.* [33] observed that sea grapes application in foods helps lower cholesterol, inflammation, blood glucose, and may potentially induce anti-aging qualifying it as a functional food. This position was corroborated by Xia *et al.* [34], who reported the importance of its polysaccharides in biomarking immunomodulatory activities. *Nostoc commune* is a member of the blue-green algae (cyanobacteria) group which, though prokaryotic, is popular as a vegetable with high nutritive values. It is farmed and consumed as a foodstuff in the Philippines as well as the Peruvian highlands because it is high in protein and vitamins, particularly ascorbic acid. In addition, *Nostoc commune* along other prokaryotic microalgae, like *Oscillatoria*, *Chlorella*, *Chroococcus*, and *Spirulina*, have been implicated as foods and reported to possess bioactive ingredients that exert anti-ulcerative colitis, antimicrobial, anticancer, antioxidant, anti-allergic and anti-inflammatory properties when consumed in food matrix [35, 36]). Although, *Nostoc* exist in marine waters, it can also be found in soil, on wet rocks, and at the bottom of lakes and springs.

9.3 Pharmacological Potential of Algae in Foods

Several edible algae species, either consumed as food or implemented in diverse forms in food preparation and food product development, possess bioactive compounds that exhibit a wide range of pharmacological as well as therapeutic properties when consumed. The diversity of innate bioactive compounds of algae remains chemically and pharmacologically underexplored. Even though most edible algae function as food to satisfy hunger, they are also preferred for their health benefits. Different natural polysaccharide compounds found in algae (fucoidan, hexadecanoic acid, and fucoxanthin) inhibit cancer cell growth and reduce oxidative stress by scavenging active oxygen and superoxide [13, 37]. As a result, the high antioxidant levels present in algae have been linked to the reduction of cardiovascular illnesses [38]. This supports their promising role in the manufacture of antibacterial, antifungal, and antiviral medicinal chemicals [7]. In several experiments, the range of organic chemicals released by *Chondrus crispus* were found to inhibit the growth of surrounding diatoms, while studies also showed that carrageenans have antitumor and anticoagulant properties [9].

9.3.1 Algae Produced Vitamins

Vitamins are an indispensable part of global diets and an essential requirement for attaining good health. This is because they are precursors to antioxidative defense cofactors required for metabolic processes in biological systems [39]. Vitamins are ingested orally from food matrixes or in capsulated forms. Since the human body is unable to manufacture vitamin-rich diets, its dependence on biogenic sources of vitamins is inevitable. Pro-vitamin E (tocopherols and tocotrienols), vitamin C (ascorbic acid), vitamin A (carotene, apocarotenoids), and many B vitamin variants, such as B12 (cobalamin), B3 (niacin), B2 (riboflavin), and B1 (thiamine), are found in algae (red, cyanobacteria, brown, green, etc.). Jäpelt and Jakobsen [40] confirmed the presence of vitamin D in many algae; ascorbic acid, on the other hand, is a water-soluble potent reducing and oxidant agent that guards the human body against infection, intoxication, and contributes to collagen production for tissue coupled with organ repairs [41]. Hypoxia-inducible factor 1 (HIF-1) is a critical microenvironmental driver of tumor angiogenesis and carcinogenesis with ties to vitamin C as a regulator. Ascorbic acid is an important food additive that is associated with a variety of health benefits like preventing atherosclerosis and acting as an immunomodulatory agent [42]. Smerilli *et al.* [43] detected a significant quantity of ascorbic acid in *Skeletonema marinoi*, a centric coastal diatom, as well as their capacity to self-modulate in response to light intensity and spectrum. This qualifies it as potential bio-indicator of environmental change, and a sustainable source of oxylipin [44].

Vitamins derived from the consumption of algal diets by humans mediate in various physiological activities such as protein synthesis, enzymes' cofactors, nutrient absorption, tissue regeneration, repairs (healing) and preservation, and overall healthiness. Some of them can also be used to reduce and cap metal nanoparticles [45]. Pharmacologic amounts of some of these vitamins, particularly vitamin C, which may have promising roles in cancer treatment, operate as a pro-oxidant, thereby producing hydrogen peroxide. Furthermore, vitamin C-rich algal diets reduce monocyte adherence to the endothelium, increase endothelium-dependent nitric oxide generation, vasodilation, and vascular smooth muscle cell death [46, 47]. The pharmacokinetics of bioactive compounds locked up in edible algal-based foods and food products coupled with the underlying mechanisms release of beneficial nutrients after consumption, is, however, not yet fully understood.

9.4 Future and Prospect of Edible Algae

Microalgae and macroalgae are progressively gaining traction as a potential food source even though they have promise in contributing to the amelioration of the global challenges of hunger and food insecurity. Furthermore, the popularization of algal farming to meet the demand stretches beyond edibility and food (processing, preservation, biofortification) products, and eventual commodification might offer a partial resolution to the negative environmental impact of microalgae (phytoplanktons). Microalgae are abundant in fat, protein and fiber, and are a high-quality raw material for developing innovative culinary products. Torres-Tiji *et al.* [48] affirmed the potential future prospects of algae in the food industry, noting that the assessment of the worldwide market value is difficult owing to supply volatility and geographic limitation of algal-based foods. Carrageenans and alginates are versatile in applications, with an estimated global market value of 6.7 billion dollars per annum according to Wells *et al.* [49]. However, many aspects of the manufacturing process involving algal biogenics, like medium composition, growth systems, genotype, and product yield, need to be improved for optimal benefits in food items. Moreover, the fallacy that algae do more harm than good should be dismissed since some of the negativities associated with them were caused by anthropogenic activities. The rising body of research on the ecological importance of algae, and their existential capacity for a low-carbon economy is already being reported. This shows that these organisms have the potential to reduce greenhouse gas emissions, reduce carbon footprints and palliate the menace of climate change. Consequently, the popularization of algae-based foods and algal-derived products for use in agriculture, medicine and cosmetics, as well as their biomass use for electricity, are benign. In the development of low-carbon agriculture, the utilization of chemical content and algae culture as a manure alternative is still underexplored.

The scientific, business and political communities are slowly embracing algae biogenics and may need to put in more effort if algae-based initiatives are to become ranked among the future foreign exchange earners. Suffice it to say that the commercial cultivation of algae, biological mining of their inherently beneficial compounds and their preparations for the global market have enormous prospects. Individuals and industries have the foresight to capitalize on the commercial potential of algae in tandem with research that focuses on improving algae production efficiency, bioactive agents' isolation technologies, extracts fractionation levels, and product acceptance. The dynamic expansion of organic farming is one factor that may have an influence on the growing demand for algal biomass in agriculture, in addition to farmers' increased knowledge of benign engagement practices, and pro-ecological influence on the changes in food consumers' taste preferences. Furthermore, the expansion of algae cultivation in rural regions (in both open and closed systems) could be a critical component of agriculture's long-term environment-friendly sustainability and multi-faceted development. The development of algae production and its contribution to the agricultural economy can be bidimensional. This may be in terms of algae's contribution to low-carbon development on the socio-economic and spatial levels. The integration, popularization, and commercialization of algal biomass through the activities of biotech companies, agricultural holdings, and agricultural organizations as well as agencies for agricultural purposes is the other dimension. These should be driven by institutional and political supports, particularly agricultural policy programs to enhance the development and commercialization of algae-based agricultural production technologies. Furthermore, taking into consideration the impact of the supply chain on the natural environment, research and scientific effort in the field of optimization models is essential for a sustainable supply of algal biomass.

9.5 Conclusion

The recent paradigm shift in the agricultural sector involving the gradual replacement of chemically-synthesized pesticidal products with biotechnology products has promoted interest in screening different algae for pesticidal possibilities. This may reinvent a significant area of agribusiness development by the manufacture of bio-products with biotechnology. The pro-health features of algae coupled with the underlying technologies used for their sustainable development and ecology validates them as a future necessity in bioeconomic growth and development. Algae as a biological bioresource have latent potentials which are still untapped for economic, domestic, environmental, pharmaceutical and industrial benefits besides agricultural purposes.

References

1. Sahoo, D and Baweja. General Characteristics of Algae. General characteristics of algae. In *The Algae World*. Springer, Dordrecht pp. 3-29, 2015.
2. Sheath, R. G and Wehr, J. D. Introduction to the Freshwater Algae. *Freshwater Algae of North America*, second ed. Academic Press, Boston, pp. 1-11, 2015
3. Chapman, R. L. Algae: the world's most important "plants"-an introduction. *Mitigation and Adaptation Strategies for Global Change*, pp.5-12, 2013.
4. Piwowar, A. and Harasym, J. The Importance and Prospects of the Use of Algae in Agribusiness. *Sustainability*, 12 pp.1-13, 2020.

5. Chaudhary R, Nawaz K, Khan AK, Hano C, Abbasi BH, Anjum S. An overview of the algae-mediated biosynthesis of nanoparticles and their biomedical applications. *Biomolecules* 10,pp. 1498, 2020.
6. Chugh, D., Viswamalya, V.S. & Das, B. Green synthesis of silver nanoparticles with algae and the importance of capping agents in the process. *J. Genet Eng Biotechnol* 19, pp. 126, 2021. https://doi.org/10.1186/s43141-021-00228-w
7. Surendhiran D, Changzhu Li, Cui H, Lin L. Marine algae as efficacious bioresources housing antimicrobial compounds for preserving foods – a review. *Int. J. Food Microbiol.*, 358, pp. 109416, 2021.
8. Terriente-Palacios G, Castellari M. Levels of taurine, hypotaurine and homotaurine, and amino acids profiles in selected commercial seaweeds, microalgae, and algae-enriched food products. *Food Chem*, 368, pp. 130770, 2022. https://doi.org/10.1016/j.foodchem.2021.130770
9. Sharma, P. and Sharma, N. Industrial and Biotechnological Applications of Algae: A Review. *JAPB*, 1, pp.1-25, 2017.
10. Koyande, A.K., Chew, K.W., Manickam, S., Chang, J.S. and Show, P.L., Emerging algal nanotechnology for high-value compounds: A direction to future food production. *Trends Food Sci. Technol.*, 116, pp.290-302, 2021.
11. Wang, T., Jonsdottir, R., Olafsdottir, G., & Kristinsson, H. G. Antioxidant properties of marine macroalgae. Antioxidants and Functional Components in Aquatic Foods; Kristinsson, HG, Ed.; John Wiley & Sons, Ltd.: Chichester, UK, 283-317, 2014.
12. Rupérez P. Mineral content of edible marine seaweeds. *Food Chem*, 79 no. 1, pp. 23-26, 2002.
13. Etman SM, Elnaggar YSR, Abdallah OY. Fucoidan, a natural biopolymer in cancer combating: from edible algae to nanocarrier tailoring. *Int. J. Biol. Macromol.*, 147, pp. 799-808, 2020.
14. Del Mondo, A., Smerilli, A., Sané, E., Sansone, C., & Brunet, C. Challenging microalgal vitamins for human health. *Microb.* Cell Factories, 19 no, 1, pp. 1-23, 2020.
15. Olaniran, A. F., Taiwo, A. E., Bamidele, O. P., Iranloye, Y. M., Malomo, A. A., & Olaniran, O. D. The role of nutraceutical fruit drink on neurodegenerative diseases: a review. *Int. J. Food Sci.* 57, 3, pp. 1442-1450, 2022.
16. Xu, N., Wu, X., Zhu, Y., Miao, J., Gao, Y., Cheng, C., Peng, S., Zou, L., McClements, D.J. and Liu, W., Enhancing the oxidative stability of algal oil emulsions by adding sweet orange oil: Effect of essential oil concentration. *Food Chem.*, *355*, pp.129508, 2021.
17. Singh, D.V., Upadhyay, A.K., Singh, R. and Singh, D.P. Health benefits of bioactive compounds from microalgae. In *Phytomedicine* (pp. 291-319). Academic Press, 2021.
18. Ścieszka, S. and Klewicka, E., Algae in food: A general review. *Crit Rev Food Sci Nutr*, 59 no. 21, pp.3538-3547, 2019.

19. Dong Y, Wei Z, Xue C. Recent advances in carrageenan-based delivery systems for bioactive ingredients: a review. *Trends Food Sci. Technol.,* 112. pp. 348-361, 2021. https:/doi.org/j.tifs.2021.04.012.
20. Guo Z, Wei Y, Zhang Y, Xu Y, Zheng L, Zhu B, Yao Z. Carrageenan oligosaccharides: a comprehensive review of preparation, isolation, purification, structure, biological activities and applications. *Algal Res.* 61, pp. 102593, 2022. https://doi.org/10.1016/j.algal.2021.102593
21. Gholamipour-Shirazi, A., Norton, I. T., & Mills, T. Dual stimuli-sensitive carrageenan-based formulation for additive manufacturing. *Int. J. Biol. Macromol.*, 189, pp. 370-379, 2021.
22. Batista, A.P.; Gouveia, L.; Bandarra, N.M.; Franco, J.M.; Raymundo, A. Comparison of microalgal biomass profiles as novel functional ingredient for food products. *Algal Res.*, 2, pp. 164–173, 2012.
23. Ryzhik IV, Klindukh MP, Dobychina EO. The B-group vitamins in the red alga *Palmaria palmate* (barents sea): composition, seasonal changes and influence of abiotic factors. *Algal Res.*, 59, pp.102473, 2021. https:/doi.org/10.1016/j.algal.2021.102473
24. Genesan AR, Subramani K, Shanmugam M, Seedevi P, Park S, Alfarhan AH, Rajagopal R, Balasubramanian B. A comparison of nutritional value of underexploited edible seaweeds with recommended dietary allowances. *J. King Saud Univ. Sci.,* 32, pp.1206-1211, 2020.
25. Priyadarshani, I.; Rath, B. Commercial and industrial applications of micro algae—A review. *J. Algal Biomass Util.,* 3, p. 89–100, 2012.
26. Smetana, S.; Sandmann, M.; Rohn, S.; Pleissner, D.; Heinz, V. Autotrophic and heterotrophic microalgae and cyanobacteria cultivation for food and feed: Life cycle assessment. *Bioresour. Technol*, 245, pp.162–170, 2017.
27. Mouritsen, O.G.; Rhatigan, P.; Pérez-Lloréns, J.L. The rise of seaweed gastronomy: Phycogastronomy. *Bot. Mar.,* 62, pp. 195–209, 2019.
28. Jibri, S. M., Jakada, B. H., Umar, H. Y., & Ahmad, T. A. Q. Importance of some algal species as a source of food and supplement. *Int J Curr Microbiol Appl Sci, 5*, pp. 186-193, 2016.
29. Yuasa M, Koe M, Maeda A, Eguchi A, Abe H, Tominaga M. Characterization of flavor components in Japanese instant soup stock 'dashi'. *Int. J.* Gastron. Food Sci., 9, pp. 55-61, 2017.
30. Giffin JC, Richards RC, Craft C, Jahan N, Leggiardro C, Chopin T, Szemerda M, MacKinnon SL, Ewart KV. An extract of marine alga *Alaria esculenta* modulates α-synuclein folding and amyloid formation. Neuroscience letters, 644, pp. 87-93, 2017.
31. Tokarek TW, Brownsey DH, Jordan N, Garner NM, Ye GZ, Osthoff HD. Emission of C9-C16 hydrocarbons from kelps apecies on Vancouver Island: *Alaria marginata* (winged kelp) and *Nereocytis luetkeana* (bull kelp) as an atmospheric source of limonene. Atmospheric Environment: X2:100007, 2019.

32. Ren, Y., Deng, J., Huang, J., Wu, Z., Yi, L., Bi, Y., & Chen, F. Using green alga Haematococcus pluvialis for astaxanthin and lipid co-production: Advances and outlook. *Bioresour. Technol*, 340, pp. 125736, 2021.
33. Permatasari HK, Nurkolis F, Augusta PS, Mayulu N, Kuswari M, Taslim NA, Wewengkang DS, Batubara SC, Gunawan WB. Kombucha tea from sea grapes (*Caulerpa recemosa*) potential as a functional anti-aging foods: *In vitro* and *in vivo* study. Heliyon 7:e07944, 2021.
34. Xia X, Hao H, Zhang X, Wong IN, Chung SK, Chen Z, Xu B, Huang R. Immunomodulatory sulfated polysaccharides from Caulerpa racemosa var. peltata induces metabolic shifts in NF-kB signaling pathway in RAW 264.7 macrophages. *Int. J. Biol. Macromol.*, 182, pp. 321-332, 2021.
35. Xu X, Wei C, Yang Y, Liu Meng, Luo A, Song H, Wang Y, Duan X. New discovery of anti-ulcerative ingredients of *Nostoc commune*: p-Hydroxy benzaldehyde. *J. Funct. Foods*, 77, pp. 104327, 2021.
36. Tseng C-C, Yeh H-Y, Liao Z-H, Hung S-W, Chen B, Lee P-T, Nan F-H, Shih W-L, Chang C-C, Lee M-C. An *in vitro* study shows the potential of *Nostoc commune* (Cyanobacteria) polysaccharides extract for wound-healing and anti-allergic use in the cosmetics industry. *J. Funct. Foods*, 87, pp.104754, 2021.
37. Gaonkar, O., Roy-Basu, A., Bharat, G., & Chakraborty, P. Current Scenario of New Persistent Organic Pollutants in India: Review on Quantification Techniques. In *Persistent Organic Pollutants*, pp. 123-154, 2021.
38. Raja, A., Vipin, C. and Aiyappan, A Biological importance of Marine Algae-An overview. *Int. J. Curr. Microbiol. Appl. Sci.*, 2 no.5, pp. 222-257, 2013.
39. Ahrazem, O.; Gómez-Gómez, L.; Rodrigo, M.J.; Avalos, J.; Limón, M.C. Carotenoid Cleavage Oxygenases from Microbes and Photosynthetic Organisms: Features and Functions. *Int. J. Mol. Sci.*, 17, pp. 1781, 2016.
40. Jäpelt, R.B.; Jakobsen, J. Vitamin D in plants: A review of occurrence, analysis, and biosynthesis. *Front. Plant Sci.*, 4, pp.136, 2013.
41. Galasso, C., Gentile, A., Orefice, I., Ianora, A., Bruno, A., Noonan, D. M., Sansone, C., Adriana Albini, A. and Brunet, C. Microalgal Derivatives as Potential Nutraceutical and Food Supplements for Human Health: A Focus on Cancer Prevention and Interception, *Nutrients*, 11, pp. 1226, 2019.
42. Nunes-Alves, C.; Booty, M.G.; Carpenter, S.M.; Jayaraman, P.; Rothchild, A.C.; Behar, S.M. In search of a new paradigm for protective immunity to TB. *Nat. Rev. Microbiol.*, 12, pp. 289–299, 2014.
43. Smerilli, A.; Orefice, I.; Corato, F.; Ruban, A.; Brunet, C. Photoprotective and antioxidant responses to light spectrum and intensity variations on a coastal diatom. *Environ. Microbiol.*, 19, pp. 611–627, 2017.
44. Gerecht A, Carotenutoa Y, Ianora A, Romano G, Fontana A, d'lppolito G, Jakobsen HH, Nejstgaard. (2013). Oxylipin production during a mesocosm bloom of *Skeletonema marinoi*. *J. Exp. Mar. Biol. Ecol.* 446, pp. 159-165.
45. McCauley, J. I., Labeeuw, L., Jaramillo-Madrid, A. C., Nguyen, L. N., Nghiem, L. D., Chaves, A. V., and Ralph, P. J. Management of enteric

methanogenesis in ruminants by algal-derived feed additives. *Curr. Pollut. Rep.*, *6*(3), 188-205, 2020.
46. Gruber, B.M. B-Group Vitamins: Chemoprevention? *Adv Clin Exp Med.,* 25, pp. 561–568, 2016.
47. Devaki, S. J. and Raveendran, R.L. Vitamin C: sources, functions, sensing and analysis. *Vitamin C.* Amal H. Hamza, IntechOpen, 2017. https://www.intechopen.com/chapters/56440
48. Torres-Tiji, Y.; Fields, F.J.; Mayfield, S.P. Microalgae as a future food source. *Biotechnol. Adv.,* 41, pp. 107536, 2020.
49. Wells, M.L.; Potin, P.; Craigie, J.S.; Raven, J.A.; Merchant, S.S.; Helliwell, K.E.; Smith, A.G.; Camire, M.E.; Brawley, S.H. Algae as nutritional and functional food sources: Revisiting our understanding. *J. Appl. Phycol.*, 29, pp. 949–982, 2020.

10

Seasonal Variation of Phytoplanktonic Communities in Fishery Nurseries in the City of Inhumas (GO) and Its Surroundings

Renato Araújo Teixeira[1*], Gustavo de Paula Sousa[1], Josué Nazário de Lima[1], Thaynara de Morais Maia[1], Marajá João Alves de Mendonça Filho[2], Joy Ruby Violet Stephen[3] and Angel José Vieira Blanco[1]

[1]Federal Institute of Education, Science and Technology of Goiás - Campus Inhumas, Inhumas, Brazil
[2]State University of Goiás (UEG), Anápolis, Brazil
[3]Queen Mary's College, Chennai, Tamil Nadu, India

Abstract

In the city of Inhumas, Goiás, in the central region of Brazil, there are commercial enterprises called *Pesque-Pague* (Fish-Pay) which work by offering leisure and food possibilities to the community. It happens that the fish sold in these places are grown in inadequate sanitary conditions, mainly due to the proliferation of cyanophytic algae whose toxins can be very harmful to fish and humans. The objective of the work presented in this chapter was to evaluate the diversity presented by the phytoplankton communities in some fishes of Inhumas and its surroundings. The results revealed the presence of cyanophytic algae in all the establishments analyzed, thus indicating the inadequacy of fish commercialization.

Keywords: Fish and pay, water, inhumas (GO), microalgae

**Corresponding author*: renatoaraujoifg@gmail.com

Charles Oluwaseun Adetunji, Julius Kola Oloke, Naveen Dwivedi, Sabeela Beevi Ummalyma, Shubha Dwivedi, Daniel Ingo Hefft and Juliana Bunmi Adetunji (eds.) Next-Generation Algae: Volume I: Applications in Agriculture, Food and Environment, (241–262) © 2023 Scrivener Publishing LLC

10.1 Introduction

The city of Inhumas, a municipality in the central Goiás state of Brazil, was projected to have a population of 53,501 inhabitants and a territorial area of 615.278 km^2 by 2020, thus the current population density is projected at 86.95419 inhab/km^2. The city of Ihumas is located, according to the IBGE classification, in the Mesoregion of Centro Goiano and in the Microregion of Anápolis and according to SEGPLAN in the Metropolitan Planning Region of Goiânia, as shown in Figure 10.1.

Goiás is the ninth Brazilian economy with a GDP of R$189 billion (estimate for 2017), representing 2.8% of the national GDP. Its per capita income amounts to R$27,457.63. Between 2010 and 2017, the GDP of Goiás grew at an average rate of 1.4% per year, a performance above the national figure, which was 0.48%. This good performance kept Goiás in the select group of the 10 largest economies among the states of the Federation.

The expressive result is due to the evolution of Goiás agribusiness, trade and also the growth and diversification of the industrial sector. This sector counts food and beverage, automobile, medicine manufacturing, mineral processing and, more recently, sugarcane production its main highlights.

The municipality of Inhumas is highlighted in the metropolitan region of Goiânia in the production of sugarcane, oranges, bananas, tomatoes, cassava; as well as its cattle, poultry, swine, milk and eggs. These data confirm the thesis that this municipality has a consolidated dairy basin in the region with several installed slaughterhouses and a sugarcane hub, not to mention being a stronghold of the food industries and farms. However, despite having several companies in the farm sector, corn production in the municipality is low. The most obvious explanation is the competition for land uses in which sugarcane crops advance in areas where they could produce other crops.

According to Schulter and Vieira Filho [29], Brazilian agriculture has undergone important transformations since the 1960s. Public policies were responsible for building an institutional environment favorable to innovation and the adaptation of knowledge and technology [30], and the result was a vertiginous increase in agricultural production. An international comparison [12] showed that several Brazilian regions showed robust growth in productivity; however, this growth trajectory was not observed in aquaculture, the cultivation activity of organisms whose life cycle develops in the aquatic environment.

Schulter and Vieira Filho [29] state that compared to agricultural activity, fish production is divided between extractive fishing and aquaculture.

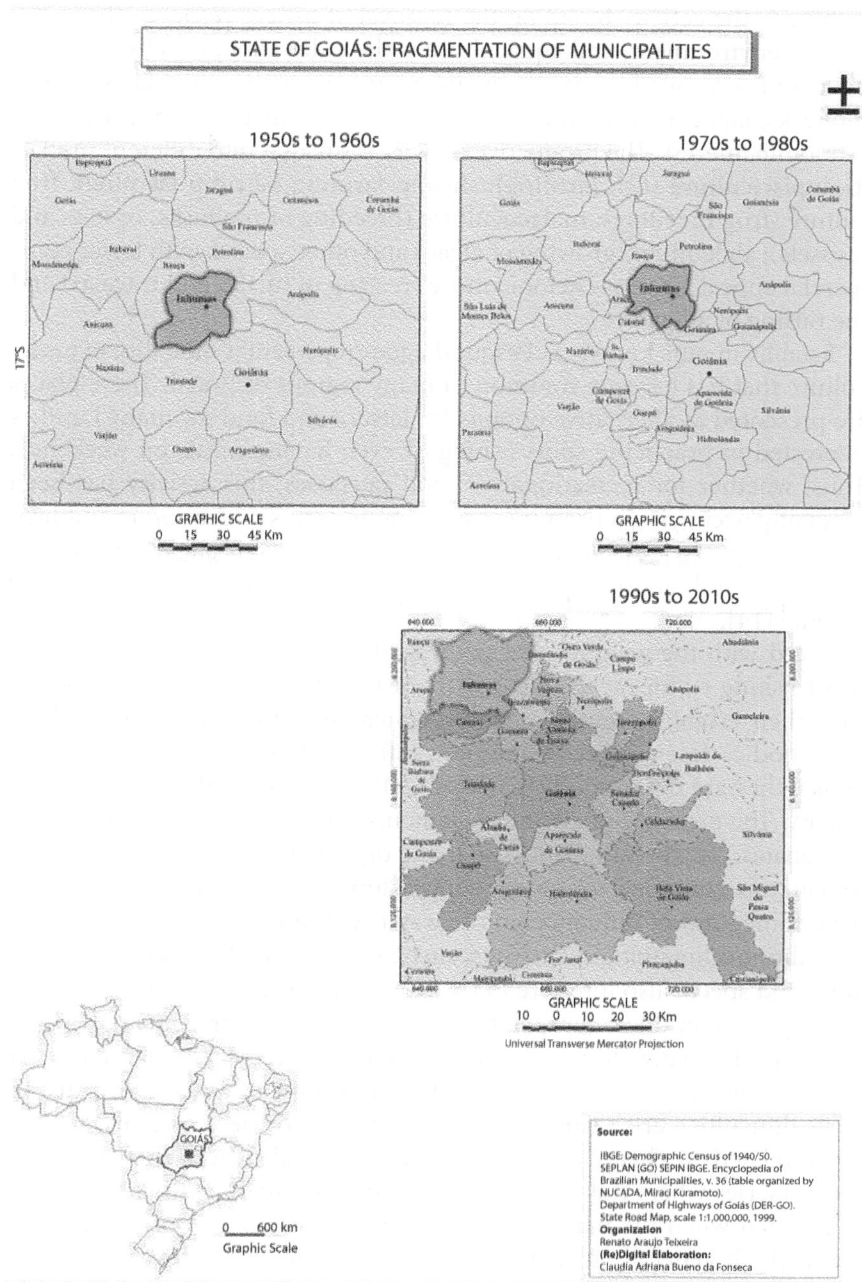

Figure 10.1 Location of Inhumas/GO – Brazil.

Fishing is the activity based on the removal of fishery resources from the natural environment, and aquaculture is the cultivation, usually in a confined and controlled space, of aquatic organisms such as fish, crustaceans, mollusks, algae, reptiles and any other form of aquatic life of productive economic interest. The productive activity is divided into different modalities: pisciculture (fish farming); shrimp farming (shrimp farming); frog culture (frog breeding); malacoculture (breeding of mollusks, oysters and mussels); algiculture (growing of algae) and other species with lesser commercial appeal, such as cheloniculture (rearing of turtles and tracajás) and the raising of alligators.

Establishments known as Pesque-Pagues represent a modality of aquaculture that has become common in many regions of Brazil. These places are presented as alternative options for leisure and food for many families interested in practicing sport fishing or even having meals on weekends. Thus, whether for recreational or food purposes, this activity has been gradually occupying a prominent position in terms of economic and production aspects [18], mainly because it represents an alternative source for many rural landowners to diversify their activities and increase their income [14].

Faced with the need to offer good quality fish—its main product—for sport fishing or for meals, the owners of Pesque-Pagues need to properly treat their establishments' nurseries, which almost never happens. These water bodies deserve special attention, as they can cause public health problems if they are not properly managed [20], especially with regard to water quality. The presence of chemical elements such as nitrogen, phosphorus, carbonates, bicarbonates, and also living organisms, such as bacteria, phytoplankton (microalgae), and zooplankton (microscopic animals), can have some kind of influence on the water within these cropping systems. In addition, climatological aspects, such as air temperature, solar radiation and wind speed, also interfere in the balance between these elements and, consequently, in the water quality of these artificial environments [15].

There are many difficulties in maintaining acceptable standards for water quality considered ideal for the cultivation of aquatic organisms. Such difficulties arise from processes intrinsic to these systems, mainly those related to chemical, physical and biological factors that permeate the nurseries where the crops are carried out. These difficulties tend to be aggravated when the management of these nurseries is done inappropriately, which generally increases the level of minerals and organic matter inside them, a process known as eutrophication.

The eutrophication process in artificial nursery systems is further accelerated by the accumulation of fecal waste, leftover feed and the excretion of ammonia through the gills and urine of farmed fish [13, 16, 21]. As in aquatic ecosystems, the availability of nutrients is one of the factors that acts directly on the dynamics of phytoplankton, and eutrophic nurseries usually present microalgae blooms as a result of the assimilation of nitrogen compounds by these organisms [25].

Microalgal blooms can be characterized as overpopulations that developed in an uncontrolled manner, and where, according to Paerl and Tucker [22], toxic species predominate, which end up refelected in the low quality of the water [28]. These algae belong predominantly to the group of *Cyanophyceae*, also called cyanobacteria, and produce harmful toxins to humans and animals [20]. The numerous blooms of toxic algae in lakes, rivers and oceans are often responsible for the poisoning and death of wild animals, domestic animals and livestock in several countries [2, 26].

Toxins produced by cyanobacteria are, according to Pearson [24], grouped into three categories: neurotoxins, hepatoxins and lipopolysaccharides. Toxic effects and risks to the population due to the presence of cyanobacteria in water sources are very important and critical examples of these effects include diarrhea, nausea, muscle weakness, skin paleness and liver cancer [5, 6, 11]. Gastrointestinal disorders and allergic and respiratory reactions are also diseases related to cyanobacterial toxicity. In fish, these toxins can cause the appearance of diseases and even the death of these animals [20]. Thus, it is possible that there is a process of bioaccumulation and transfer of cyanotoxins through the ingestion of animals such as fish, which, when fed cyanobacteria, continuously accumulate toxins in their tissues, as evidenced by Falconer [10] and Falconer *et al.* [11], thus presenting serious risks to human health.

Knowing that the drop in the quality of the water used in Pesque-Pagues can bring economic losses to the owners of these projects (fish kill), as well as damage to the health of those who frequent these spaces and consume their products, studies of the water bodies from an ecological and also sanitary perspective become relevant. In the municipality of Inhumas, Goiás, and its surroundings there are many Fish-Pay establishments, both in rural properties and in the urban perimeter of the city. In such places, the absence of adequate management practices is noted, at least with regard to the constant renewal of water in the nurseries. In this sense, the purpose of this work was to analyze the temporal and spatial variation composition of the phytoplankton community, using its potential as a bioindicator. Secondarily, it was intended to obtain physicochemical parameters that would be associated with the data to substantiate the results, which, as will be discussed, was not possible in its entirety.

10.2 Material and Methods

10.2.1 Materials

- Portable PH meter – Tecnopon mPA 210
- Digital thermometer – Instrutherm, model TH-075
- Plankton net – 25 μm mesh and 20 cm mouth diameter
- Optical microscope from the manufacturer Physis, model EPX 100
- Portable digital dissolved oxygen meter – Lutron model DO-5519
- Illustrated boards, specialized bibliography and visits to thematic websites on the internet

10.2.2 Methods

Two collections were carried out in periods corresponding to those of drought and also of rain in order to contemplate the seasonality of the seasons. The first collection took place in May and the second in November 2012. A nursery was chosen from each pesque-pague establishment from which samples were collected in at least two points. The physical-chemical parameters, water temperature, pH and dissolved oxygen were evaluated *in situ* using a digital thermometer, distance meter and portable oxygen meter, respectively. Phytoplankton were collected using a plankton net, 25 μm mesh and 20 cm mouth diameter, where enough water was filtered to form sediment. The filtered sediment was concentrated and resuspended in water. Approximately 0.2 mL of each concentrated sample were used as subsamples for analysis under an optical microscope with phase contrast, using objectives of 10, 20, 40 and 100. The typological identification of the phytoplankton was performed based on the morphological characteristics of the cells and their forms of organization in chains or colonies, also considering information on ecology and distribution of the identified taxa. For this purpose, specialized bibliography, illustrated boards [3, 4] were used, in addition to visits to thematic websites on the internet. The identification refinement criteria and the scientific names of species were checked against the international database AlgaeBase [1].

10.3 Results

The reading of physicochemical parameters was limited due to the delay in delivering the equipment requested by the institution. Still, it was possible to rate three parameters (pH, dissolved oxygen, temperature) in the second batch of collected samples. The results of the physicochemical parameters did not show any kind of association with the biological data, although in some cases it is not in agreement with the chemical characteristics of the water suitable for fish farming. It is necessary to ensure the anonymity of the owners of Fish-Pay following Resolution CNS (National Health Council) No. 196, of October 10, 1996, which defines research involving human beings, therefore, we will use as a form of citation the research objects such as: Pay-fish A, Pay-fish B, Pay-fish C, Pay-fish D. In Brazil, it is important to make the research subjects impersonal, avoiding any embarrassment. According to Ceccarelli *et al.* [7], the optimal pH for tropical fish crops should remain between 7 and 8. In general, the pH evaluated in the environments studied in Inhumas and surrounding areas varied between 7 and 8, but in some cases, such as in Pay-fish A, reached 8.3. However, Resende [27] and Craef *et al.* [8] cultivated fish whose pH ranged from 4.9 to 8.3 with success. Thus, the pH variation may not necessarily reflect problems related to the cultivation system.

The water temperature evaluated in the analyzed fish-pays was slightly higher than the maximum limit indicated by Kubitza [17] for the cultivation of tropical fish, whose ideal values vary between 28 and 32 degrees. In the environments studied, the temperature ranged between 34 and 36 °C, which can be interpreted as values that can interfere with pH dynamics, changes in organic matter decomposition rates, increase in concentrations of toxic ammonia and increase in algae metabolism [19].

Dissolved oxygen showed no abnormalities, despite the environments being characteristically eutrophic when visually assessed. Values for dissolved oxygen were between 6.0 and 7.0 mg L-1. The National Environmental Council establishes the value > 5.0 mg L-1 as the acceptable limit of dissolved oxygen for aquaculture. According to Kubitza [17], animal health conditions are better when oxygen levels are close to saturation. Thus, the dissolved oxygen values found in Pesque-Pagues in Inhumas and its surroundings do not represent a limiting factor for the health of fish cultivated in these locations. However, values found for dissolved oxygen, as well as for pH and temperature in this study, must be carefully evaluated, since their representativeness was extremely reduced due to limitations found during the execution of the project.

Microscopic analyses revealed a great diversity of phytoplankton in the sampled environments, although the number of collections was not carried out as initially planned. Taking into account the taxonomic aspect, the phytoplankton communities of the five areas sampled in the two seasons were constituted of 41 genera, 31 families, 19 orders, 8 classes and 5 divisions.

It was not possible to carry out the identification at the species level, as this requires high taxonomic knowledge. Table 10.1 shows the systematic identification of the phytoplanktonic organisms found, according to the consultations carried out in the international database AlgaeBase [1]. As will be discussed later, the largest representation of identified taxa belonged to the Cyanobacterium division. Organisms belonging to the Charophyta, Chlorophyta, Euglenozoa and Ochrophyta divisions were also found in the Fish-Pays evaluated. Of the total number of genera found, almost 50% were exclusive, that is, found in only one of the five sampled areas; and when compared to each other, although many of these genera were present in more than one of these environments, none of them was common to all of the Fish-Pays. These numbers may indicate different bodies of water supplying the nurseries studied in the different locations.

Through an individualized evaluation, it is observed that the establishment that presented the greatest phytoplankton diversity was the Pesquepague called "Chaparral," located within the urban perimeter of inhumas. Of the 41 genera found in the five environments studied in this work, 16 were found in Fish-pay B, of which eight were exclusive (Table 10.2). However, the environmental and sanitary interpretation that is made of this place is that of an inappropriate environment for the practice of sport fishing activity and, above all, for human consumption, since among the organisms found in this establishment there is a genus classified by Palmer [23] as characteristic of polluted water (*Tetraedron*) and five others gathered in the group of cyanophytic algae (*Borzia, Oscillatoria, Spirulina, Cyanothece, Gomphosphaeria*), whose representatives are classified as producers of harmful toxins to humans and animals, in addition to also indicating, according to Palmer [23], environments with polluting elements (Tables 10.1 and 10.2). It is important to point out that of the 15 representatives of cyanophytic algae found, 4 were observed in Fish-Pay B, representing therefore 1/3 of all algae in this group unidentified in this work.

The establishment that presented the lowest diversity of microalgae genera was that of Fish-pay A. Despite being the establishment with the best sanitary and environmental aspect, Fish-pay A presented characteristic bioindicators of unfit water for cultivation and also for consumption of fish by humans. Of the nine genera found in this location, three belong to the group of cyanobacteria (*Oscillatoria, Snowella, Gloeocapsa*). The genus *Oscillatoria* is also classified as an indicator of polluted water according to the classification of Palmer [23], in addition to the fact that it is potentially toxic to humans (Table 10.2).

Fish-Pay C, located in the surroundings of Inhumas, had the smallest number of genera considered characteristic of unsuitable water to

Table 10.1 Taxonomic classification of the phytoplankton genera found in the five points, collected in Fish-Pays in the municipality of Inhumas-GO and its surroundings. Source; www.google.com. Blanco 2022.

Genus	**Family**	**Order**	**Class**	**Division**
Spirogyra	Zygnemataceae	Zygnematales	Conjugatophyceae	Charophyta
Closterium	Closteriaceae	Desmidiales		
Cosmarium	Desmidiaceae			
Micrasterias				
Tetmemorus				
Acanthophaera	Chlorellaceae	Chlorellales	Trebouxiophyceae	
Closteriopsis				
Pandorina	Volvocaceae	Volvocales	Chlorophyceae	

(*Continued*)

Table 10.1 Taxonomic classification of the phytoplankton genera found in the five points, collected in Fish-Pays in the municipality of Inhumas-GO and its surroundings. Source; www.google.com. Blanco 2022. (*Continued*)

Genus	Family	Order	Class	Division
Scenedesmus	Scenedesmaceae	Sphaeropleales		Chlorophyta
Pediastrum	Hydrodictyaceae			
Tetraedron				
Rhizoclonium	Cladophoraceae	Cladophorales	Ulvophyceae	
Arthrospira	Phormidiaceae	Oscillatoriales		
Borzia	Borziaceae			
Oscilatoria	Oscillatoriaceae		Cyanophyceae	Cyanobacteria
Spirulina	Pseudanabaenaceae	Pseudanabaenales		

(*Continued*)

Table 10.1 Taxonomic classification of the phytoplankton genera found in the five points, collected in Fish-Pays in the municipality of Inhumas-GO and its surroundings. Source; www.google.com. Blanco 2022. (*Continued*)

Genus	**Family**	**Order**	**Class**	**Division**
Coelomorum	Merismodediaceae	Synechococcales		
Coelosphaerium				
Eucapsis				
Snowella				
Xenococus	Xenococcaceae			
Cyanothece	Cyanobacteriaceae			
Microcrocis		Chroococcales		

(*Continued*)

Table 10.1 Taxonomic classification of the phytoplankton genera found in the five points, collected in Fish-Pays in the municipality of Inhumas-GO and its surroundings. Source; www.google.com. Blanco 2022. (*Continued*)

Genus	Family	Order	Class	Division
Gomphosphaeria	Gomphosphaeriaceae			
Woronichia				
Gloecapsa	Microcystaceae			
Sphaerocavum				
Lepocinclis	Phacaceae	Euglenales	Euglenophyceae	Euglenozoa
Amphora	Catenulaceae	Thalassiophysales		
Cyclotella	Stephanodiscaceae	Thalassiosirales		
Cymbella	Cymbellaceae	Cymbellales	Bacillariophyceae	Ochrophyta
Anomoeoneis	Anomoeoneis			

(*Continued*)

Table 10.1 Taxonomic classification of the phytoplankton genera found in the five points, collected in Fish-Pays in the municipality of Inhumas-GO and its surroundings. Source; www.google.com. Blanco 2022. (*Continued*)

Genus	Family	Order	Class	Division
Frustulia	Amphipleuraceae	Naviculales		
Navicula	Naviculaceae			
Neidium	Neidiaceae			
Stauroneis	Stauroneidaceae	Naviculales	Bacillariophyceae	Ochrophyta
Pinnularia	Pinnulariaceae			
Surirella	Surirellaceae	Surirellales		
Synedra	Fragilariaceae	Fragilariales		
Tabellaria	Tabellariaceae	Tabellariales		
Tribonena	Tribonem ataceae	Tribonematales	Xanthophyceae	

Table 10.2 Occurrence of phytoplankton genera found in different Fish-Pays located in Inhumas and surrounding areas. Source: www.google.com. Blanco 2022.

Collect	**Dry season**				**Rainy season**				
Place	Inhumas			Surroundings	Inhumas			Surroundings	
Collect point / **Genus**	Fish-Pays C	Fish-Pays D	Fish-Pays B	Fish-Pays A	Fish-Pays C	Fish-Pays	Fish-Pays B	Fish-Pays A	Fish-Pays C
Acanthophaera			X						
Amphora							X		
Anomoeoneis	X								
Arthrospira					X				
Borzia			X						
Closteriopsis					X				
Closterium					X		X	X	X
Coelomorum	X					X			
Coelosphaerium		X							
Cosmarium			X			X			X

(Continued)

Table 10.2 Occurrence of phytoplankton genera found in different Fish-Pays located in Inhumas and surrounding areas. Source: www.google.com. Blanco 2022. (*Continued*)

Collect	Dry season				Rainy season				
Place	Inhumas			Surroundings	Inhumas			Surroundings	
Collect point	Fish-Pays C	Fish-Pays D	Fish-Pays B	Fish-Pays A	Fish-Pays C	Fish-Pays	Fish-Pays B	Fish-Pays A	Fish-Pays C
Genus									
Cyanothece			X						
Cyclotella						X			
Cymbella						X			
Eucapsis									X
Frustulia			X					X	
Gloecapsa				X					
Gomphosphaeria	X		X		X	X	X		
Lepocinclis					X	X			

(*Continued*)

Table 10.2 Occurrence of phytoplankton genera found in different Fish-Pays located in Inhumas and surrounding areas. Source: www.google.com. Blanco 2022. (*Continued*)

Collect	**Dry season**				**Dry season**				
Place	Inhumas			Inhumas	Surroundings				Inhumas
Ponto de coleta / **Gêneros**	Pesque pague C	Pesque pague D	Pesque pague B	Pesque pague A	Pesque pague C	Pesque pague C	Pesque pague D	Pesque pague B	Pesque pague A
Micrasterias			X				X		X
Microcrocis	X								
Navicula						X		X	X
Neidium									X
Oscilatoria							X	X	
Pandorina	X				X				
Pediastrum		X				X			
Pinnularia									X
Rhizoclonium						X		X	
Scenedesmus	X				X	X			

(*Continued*)

Table 10.2 Occurrence of phytoplankton genera found in different Fish-Pays located in Inhumas and surrounding areas. Source: www.google.com. Blanco 2022. (*Continued*)

Collect	**Dry season**				**Dry season**				
Place	Inhumas			Inhumas	Surroundings				Inhumas
Ponto de coleta / **Gêneros**	Pesque pague C	Pesque pague D	Pesque pague B	Pesque pague A	Pesque pague C	Pesque pague C	Pesque pague D	Pesque pague B	Pesque pague A
Snowella				X					
Sphaerocavum						X			
Spirogyra		X							X
Spirulina							X		
Stauroneis			X	X			X	X	
Surirella	X				X		X		X
Synedra	X							X	
Tabellaria							X		
Tetmemorus			X						
Tetraedron					X				
Tribonema							X		
Woronichinia	X				X				
Xenococcus									X

house recreation activities or even food directed at humans. Only two Cyanophyceae (*Eucapsis, Xenococcus*) and a characteristic polluted water alga (*Spirogyra*) were found. Perhaps this result is a reflection of the smaller sampling in this establishment, since the collection procedure was hampered by unsuccessful contact with the owner of this establishment during the dry season collection.

Moreover, research on Fish-Pay B and Fish-Pay C (located within Inhumas) showed inadequate sanitary conditions which were easily observed by the proximity of domestic animals, such as cattle, pigs and chickens, to the nursery where fishing for recreation or eating is performed. In this place, four genera of cyanophyceous algae and three bioindicators of polluted water were found. The diversity of phytoplankton found in this environment was also high, representing almost 35% of the observed genera. Moreover, the research on Fish-Pay B indicated that the amount of cyanophyceous algae observed in this fish-pay was higher than in the other establishments, five in total (*Arthrospira, Coelomorom, Microcrocis, Gomphosphaeria, Woronichinia*). In addition, two other algae considered specific to polluted environments were identified (*Tetraedron, Lepocinclis*).

The survey on Fish-Pay D does not necessarily represent a fish-pay, but rather a fish reservoir used only for human consumption. Even so, the presence of phytoplanktonic organisms found there also suggests that it is a potential source of contamination for those who consume their fish. In addition to four types of Cyanophyceae (*Coelomorom, Gomphosphaeria, Sphaerocavum, Coelosphaerium*), two microalgae characteristic of polluted environments were identified in the samples from this fish reservoir. Regarding the influence of seasonality on the composition of phytoplankton communities, a more detailed analysis in Table 10.2 allows us to observe that only approximately 20% of the taxa found in the five environments studied were identified in the two collections. That is, only 8 of the 41 genera identified were observed in both the dry and the rainy season, suggesting that the phytoplankton communities in fish-pays undergo seasonal variation, most likely as a result of the climatological influence suffered by rivers and streams which supply the cultivation nurseries along their length. However, according to Esteves [9], there is little evidence of the occurrence of fluctuations associated with the seasons of the year, and thus the seasonal variation found in the studied phytoplankton communities is probably associated with local factors.

10.4 Conclusion

In general, it can be said that all the environments analyzed presented unsanitary environmental conditions that were inappropriate for the cultivation and consumption of fish, given that organisms that produce harmful toxins to humans and animals, typical of richly eutrophic environments, were found in all of them. Such conditions, as foreseen in advance of the subject proposed for this study, were mainly due to inadequate management practices, with almost no water renewal, low oxygenation within the nurseries and a constant increase in the amount of decomposing organic matter arising mainly from the nitrogenous excreta of the farmed animals and the feed used to feed them.

The results attest to the initial hypothesis of the project that in fish-pay establishments, fish are generally cultivated in an inadequate manner and in poor unsanitary conditions, proving to be, therefore, an activity that is not compatible with the norms established to ensure the health of people who attend such places. Thus, the object of this work is a case related to public health, requiring special attention from the authorities responsible for regulating the activity of Fish-Pay establishments, so that inadequate management practices, often motivated by financial aspects, do not become a potential risk to those who use such places for leisure or food.

Ar an opportune time, we envision redoing this analyses of Fish-Pays in order to compare which of the establishments reduced unsatifactory sanitary and environmental conditions for the cultivation and consumption of fish. It is known that Brazil is a major producer of beef, pork and chicken, and has all the conditions to enter the fish market; but the promotion of any additional activity depends on public policies that adjust the development and competitiveness of the production chain. This debunks the notion that countries in the world with large hydrographic basins are necessarily large fish producers.

In the case of Brazil, government incentives are needed in order to encourage the public and private sectors to invest in research, as well as in fish production, whether in extractive fishing or aquaculture, which should focus not only on the growth of the domestic market, but fundamentally on expansion into the international market, which presents excellent growth potential.

References

1. Algaebase http://www.algaebase.org/2013 (last accessed on 20/01/2013).
2. Beyruth, Z., Sant'anna, C. L., Azevedo, M. T. D. P., *et al.* Cordeiro-Marino, M. *et al. Algae and Environment: A General Approach.* São Paulo, Brazilian Society of Physiology/CETESB, 1992.
3. Bicudo C.E.M. & Bicudo R.M.T. (1970) *Algae from Brazilian Continental Waters: Illustrated Key to Gender Identification.* Foundation for the Development of Science Teaching, SP.
4. Bicudo C.E.M. & Menezes M. *Algae Genera from Continental Waters in Brazil: Key to Identification and Descriptions.* RIMA, São Carlos, 2005
5. Carmichael, W.W. 'Cyanobacteria secondary metabolites: The cyanotoxins', *Journal of Applied Microbiology,* 72(6), PP. 445-459, 1992.
6. Carmichael, W. 'The Toxins of Cyanobacteria', *Scientific American,* 270(1), PP. 78-86, 1994.
7. Ceccarelli, P.S., Lordini, J.A., Volpato, G. *Tips on Fish Farming.* Botucatu: Santana, 2000.
8. Craef, E., Resende, E., Petry, P; *et al.* 'Polyculture of Matrinchã (Brycon sp.) and Jaraqui (Semaprochilodus sp.) in small dams', *Acta Amazonica,* 17, pp. 33-42, 1987.
9. Esteves, F.A. *Fundamentals of Limnology.* Rio de Janeiro: Interciencia, 1998.
10. Falconer, I.R. 'Tumor promotion and liver injury caused by oral consumption of Cyanobacteria', *Environmental Toxicology and Water Quality,* 6(2), pp. 177-184, 1991. https://doi.org/10.1002/tox.2530060207
11. Falconer I. R., Michael D. B., Dennis A. S *et al.* 'Toxicity of the blue-green algal (Cyanobacterium) Microcystis aeruginosa in drinking water to growing pigs, as an animal model for human injury and risk assessment', *Environmental Toxicology and Water Quality,* 9(2), pp. 131-139, 1994. https://doi.org/10.1002/tox.2530090209
12. Fuglie K. O., Ling S. W., V. Eldon B. *Productivity Growth in Agriculture: An International Perspective.* CAB International, Wallingford (UK) and Cambridge, MA (USA), 2012.
13. Ferragut, C. *Responses of Periphytic and Planktonic Algae to Nutrient Manipulation (N and P) in Urban Reservoir (Lago do IGA, São Paulo).* Thesis (Doctorate) - State University of São Paulo Julio de Mesquita Filho, São Paulo, 2004.
14. Gomes, A. 'Essential care in the implementation of a fish-pay tourism project', *Scientific Journal of Administration,* n.14, VII, 2008.
15. Henry, R., Curi, P. R. 'Influence of climatological parameters on some physicochemical factors of the water of the Parto river dam (Botucatu, SP)', *Brazilian Journal of Biology, Rio de Janeiro,* 42(2), pp. 299-206, 1981.
16. Honda, R.Y. *et al.* Cyanotoxins in Fisheries in the Metropolitan Region of São Paulo. In: Esteves, K.E.; Sant'anna, C.L. *Fisheries under an Integrated Vision of Environment, Public Health and Management.* São Carlos: Rima, 2006.

17. Kubitza, F. *Water Quality in Fish Production.* 3. ed. Jundiaí: Degasparil, 1999.
18. Leitão, M. F. F. 'Chemical and microbiological changes in PACU (Piaractus mesopotamicus) stored under refrigeration at 5ºC', *Food Science and Technology*, 17(2), pp. 160-166, 1997. Available from: https://doi.org/10.1590/S0101-20611997000200018.
19. Martins, Y. K. *Water Quality in a Nile Tilapia (Oreochromis Niloticus) Nursery: Diurnal Characterization of Physical, Chemical and Biological Variables.* Dissertation, 2007.
20. Matsuzaki, M., Honkura, N., Ellis-Davies, G.C.R. *et al.* 'Structural basis of long-term potentiation in single dendritic spines', *Nature*, 429, pp. 761-766, 2004. Available from: https://doi.org/10.1038/nature02617.
21. Ono, E.A., Kubitza, F. *Fish Farming in Net Tanks.* 3rd ed. Jundiaí: F. Kubitza, 2003.
22. Paerl, H.W., Tucker, C.S. 'Ecology of bluegreen algae in aquaculture ponds', *Journal of the Aquaculture Society*, 26(2), pp. 109-131, 1995. Available from: https://doi.org/10.1111/j.1749-7345.1995.tb00235.x.
23. Palmer, C.M. *Algae in Water Supplies: An Illustrated Manual on the Identification, Significance, and Control of Algae in Water Supplies.* Cincinnati, Ohio. UNT Digital Library, 1959.
24. Pearson, M.J. *Toxic blue-green algae. Report of the National Rivers Authority. Water Quality.* Series 2. London, UK. National Rivers Authority, 1990.
25. Pereira, LP., Mercante, C. T. J. 'Ammonia in fish farming systems and its effects on water quality'. *A Review. B. Inst. Fishing*, 31(1), pp. 81-88, 2005.
26. Porfirio, Z., Micheline P. R., Cicero S. E. *et al.* 'Hepatosplenomegaly caused by an extract of cyanobacterium Microcystis aeruginosa bloom collected in the Manguaba Lagoon, Alagoas – Brazil'. *Journal of Microbiolog*, 30(3), pp. 278-285, 1999. Available from: 10.1590/S0001-37141999000300016.
27. Resende, M. 'Applications of pedological knowledge to soil conservation'. *Information Agropec*, 11, pp. 3-18, 1985.
28. Sipaúba-Tavares, L.H., Barros, A.F., Braga, F.M.S. 'Effect of floating macrophyte cover on the water quality in fishpond'. *Acta Scientiarum - Biological Sciences*, 25(1), pp. 101-106, 2003.
29. Schulter, E. P and Vieira Filho, J. E. R. *Evolution of fish farming in Brazil: diagnosis and development of the tilapia production chain. In: Institute of Applied Economic Research.* Brasília, Rio de Janeiro. Ipea, 2017.
30. Vieira Filho, J.E.R. and Fishlow, A. *Agriculture and industry in Brazil: innovation and competitiveness.* Brasilia: Ipea, 2017.

11

Role of Genetical Conservation for the Production of Important Biological Molecules Derived from Beneficial Algae

Charles Oluwasun Adetunji[1]*, Muhammad Akram[2], Babatunde Oluwafemi Adetuyi[3], Umme Laila[2], Muhammad Muddasar Saeed[3], Olugbemi T. Olaniyan[4], Inobeme Abel[5], Ruth Ebunoluwa Bodunrinde[6], Nyejirime Young Wike[7], Phebean Ononsen Ozolua[1], Wadzani Dauda Palnam[8], Olorunsola Adeyomoye[11], Arshad Farid[9] and Shakira Ghazanfar[10]

[1]Applied Microbiology, Biotechnology and Nanotechnology Laboratory, Department of Microbiology, Edo State University Uzairue, Iyamho, Edo State, Nigeria
[2]Department of Eastern Medicine, Government College University Faisalabad, Faisalabad, Punjaba, Pakistan
[3]Department of Natural Sciences, Faculty of Pure and Applied Sciences, Precious Cornerstone University, Ibadan, Nigeria
[4]Laboratory for Reproductive Biology and Developmental Programming, Department of Physiology, Rhema University, Aba, Nigeria
[5]Department of Chemistry, Edo University Iyamho, Auchi, Edo State, Nigeria
[6]Department of Microbiology, Federal University of Technology Akure, Gaga, Nigeria
[7]Department of Human Physiology, Faculty of Basic Medical Science, Rhema University, Aba, Nigeria
[8]Department of Agronomy, Federal University, Gashua, Yobe, Nigeria
[9]Gomal Center of Biochemistry and Biotechnology, Gomal University, Khyber Pakhtunkhwa, Pakistan
[10]National Institute for Genomics & Advanced Biotechnology (NIGAB), National Agricultural Research Centre (NARC), Islamabad, Pakistan
[11]Department of Physiology, University of Medical Sciences, Ondo City, Nigeria

**Corresponding author*: adetunjicharles@gmail.com; adetunji.charles@edouniversity.edu.ng

Charles Oluwaseun Adetunji, Julius Kola Oloke, Naveen Dwivedi, Sabeela Beevi Ummalyma, Shubha Dwivedi, Daniel Ingo Hefft and Juliana Bunmi Adetunji (eds.) Next-Generation Algae: Volume I: Applications in Agriculture, Food and Environment, (263–280) © 2023 Scrivener Publishing LLC

Abstract
Algae has been recognized as one of the important photosynthetic organisms which are found in both freshwater and marine environment. Numerous algae have been reported by several scientists all over the world because of their significant applications in diverse fields. It has been reported that these microalgae possess a lot of secondary and primary metabolites which could be applied for the effective management of different diseases, the production of different food products, as well as application in the manufacturing of different products that have greater industrial relevance. Therefore, this chapter intends to provide comprehensive information on genetical conservation of these beneficial algae, as well as information on their wide applications.

Keywords: Algae, genetical conservation, diseases, food, secondary and primary metabolites

11.1 Introduction

It has been documented since ancient times that the usage of algae is very effective because of the presence of good production of biomass at a rate that is higher compared to other crops which are cereal based. The advantages of algae have been documented most especially in agriculturally based crops and their wide application as 3rd generation biofuels.

The use of algae for the synthesis of energy has been documented for several years, but the production cost of algae biofuels is high because of their limited system associated with cultivation. Tremendous effort has been exerted on how to produce more advancements in numerous techniques that could be applied for effective synthesis and cultivation of algae that could grow in various climates that vary from moderate to tropical regions. Algae have demonstrated significant properties which enable them to reduce carbon-rich gases in the atmosphere via algae acting as CO_2 and NO quenchers which are excreted into various sources [1–3]. Also, most algae use contaminated H_2O containing the maximum phosphorous and nitrogen which have beneficial effects related to biofuels and also decease the level of these elements [4–9].

Interestingly, because algae have a lot of important bioactive component they are a good source for synthesizing significant metabolites which are very effective in different fields. A typical example of this includes *Haematococcus pluvialis,* which is mostly present in freshwater and commercially very effective in the production of astaxanthin pigment. Other important algae which are effective as a good supplement for food are *Chlorella vulgaris* and a species of Dunaliella which possess beta carotene.

The biomass of these marine algae could also be applied for effective production of bioelectricity through bio-oil, biodiesel, bioethanol and co-firing. Almost fifty percent of starch present in algae can be converted into ethanol. Typical examples of these beneficial algae include marimo, diatom, wakame, golden algae, *Arthrospira platensis*, Irish moss, sea lettuce, *Caulerpa lentillifera*, mozuku, giant kelp, coral strands, hijiki, green and yellow algae, fat choy and many more which have effective use in the food sector and other fields [10–17].

11.2 Application of Algae in Various Fuels

Nowadays, global warming and depletion of fossil fuels have played a significant role in investigating alternative sources for the synthesis of bioenergy. The synthesis of biofuels from algal biomass may be a viable alternative to replace fuel fossils. By using various raw materials, different fuels like gaseous, liquid and solid biofuels can be obtained. In this biomass conversion process, various types and amounts of energy are obtained, with a good economic return on the investment [18–20]. Thermochemical process is one of the good conversion processes without the presence of oxygen in which oil is obtained from the conversion of biomass at high temperature. Bio-oils are just like petroleum oils so they can be used as a substitute [21]. Thermochemical liquefication and pyrolysis are two significant methods for the formation of oil [22]. Several microalgae have been investigated to produce bio-oil [23–27] such as species of *Dunaliella, spirulina* [28, 29], *Desmodesmus* and *Scenedesmus* [30–34]. The synthesis of fatty free acids from *E. coli* increases after performing genetic engineering. Production of alkanes also occurs from cyanobacteria, and by using some effective biosynthetic pathway significant production of alkene occurs. So, by using different algal a number of important chemical components are derived which include ethanol, butanol and isobutanol, isoprene, hydrogen, acetone, ethylene, beta-hydroxybutyrate, sugars, lectic acids and many more [35]. The production of biodiesel is very effective and is mostly obtained from the oil of rapeseed, oil of palm and oil of soybean [36], with their cultivation mostly based on the uses of raw material [37]. Two important species which contain a high amount of oil include *Chlorella protothecoides* and *Chlorella vulgaris* [38, 39]. Biohydrogen products are obtained from different cyanobacteria using biological process [40, 41]. Moreover, it has been documented that biogas is gaining more importance by using algae to turn inorganic carbon compounds, such as CO_2, into organic carbon compounds, such as methane, which are effectively used as a chemical and

Table 11.1 Natural compounds available in algae and their relevance.

Presence of natural substances in algae	Significant component obtained from algae
Antioxidant	Polyphenols, catalases, tocopherols and dis-mutase superoxides are obtained
Vitamins	Important vitamins obtained from algae include nicotinic acid, riboflavin, biotin, folic acid and pantothenate
Macromolecules	Toxic products, proteins, amino acids and anti-microbial component
Poly-unsaturated fatty acids	GAL, ARA, EPA, DHA
Carotenoids	Beta-carotene
Pigments	Fucoxanthin, phycoerythrin, phycocyanin, chlorophyll, canthaxanthin, zeaxanthin, lutein and astaxanthin

fuel [42]. Usually, production of biomethane occurs by using various feedstocks which include solid waste, grass, wood and algae. The synthesis of biomethane is enhanced by using the species of chlorella. The synthesis of bioethanol from algae has gained more significance because of the presence of important chemical components, high productivity of biomass and diversity. Polysaccharides and carbohydrates were also derived from algae [43]. The production of ethanol from *Chlorococcum littorale* is very effective. The most commonly used algae used in the production of bioethanol includes *Chlorella, Arthrospira, Dunaliella, Chlamydomonas, Spirulina, Sargassum, Gracilaria, Prymnesium parvum, Scenedesmus* and *Gracilis euglena* sp. [44–47]. In all types of algae, except for the ethanol purposes, the best algae are brown algae [48–51]. A number of important components are obtained from algae, which presented in Table 11.1.

11.3 Algae and Their Pharmaceutical Application

Most of the important secondary and primary chemical metabolites obtained from the algae are considered one of the richest sources for these metabolites and show greater potential for pharmaceutical application [52–54]. The metabolites found in algae show inhibitory effects against various diseases. Significant bioactive metabolite derivatives of microalgae have applications related to commercial fields. Various nutrients, vaccines, proteins and pharmaceutical products are usually obtained from microalgae with varying costs [55–57]. Microalgae are the richest source of pharmaceutical products showing potential in several fields. The components of these microalgae derivatives can be applied to effectively cure various illness [58]. Some important vitamins are also obtained from algae, including tocopherol, riboflavin and ascorbic acid, which are also good sources of nutraceutical and pharmaceutical products [59]. The large amount of microalgae present possesses good metabolites. Different important algae include *Vulgaris chlorella,* which show antibacterial activities. The extract and derivative of algae are very effective. The extract of green algae possesses antifungal characteristics, some of the important microalgae, like species of *Ochromonas* and *Prymnesium parvum,* show toxic components which are effective in pharmaceuticals [60–62]. Important metabolites derived from the species of cyanobacteria show antiviral, antifungal and antibacterial characteristics [63, 64]. Almost 600 species of cyanobacteria were discovered in which the extract derived from 7 different microalgae was found to be very effective for curing viral infections like human immunodeficiency syndrome I & II [65]. One of the best components in cyanovirin is a powerful viricidal which effective against human immune deficiency virus [66]. *Porphyridium* and *Spirulina platensis* are effective

Table 11.2 Different species of algae and their various applications in the production of several macro- and micromolecules.

Species of algae	Component	Uses
Dunaliella salina	Carotenoids	Good supplements for feed and nutraceutical
Porphyridium cruentum	Poly-saccharides	Used in cosmetics and pharmaceutical industry
Odontella aurita	Fatty acids	Effective for baby food, used in cosmetics and pharmaceuticals
Haematococcus pluvialis	Astaxanthin & carotenoids	Additives and nutraceuticals
Chlorella vulgaris	Ascorbic acid	Good surrogate for food
Spirulina platensis	Phycocyanins	Nutraceutical

Table 11.3 Some algae that play an important role against ultraviolet rays.

Ultraviolet screening component	Algae
Mycosporine	*Alexandrium catenella, Stellarima microtrias,* species of *Isochrysis, Dunaliella tertiolecta, Chlorella sorokiniana* and *Ankistrodesmus spiralis*
Scytonemin	*Nostoc punctiforme* & commune, species of *Scytonema* & *Calothrix* and species of *Chlorogloeopsis*
Sporopollenin	*Chlorella fusca, Dunaliella salina, Scotiellopsis rubescens,* species of *Scenedesmus*

antiviral algae in which the sulfur in the polysaccharide component is effective against viral infections [67–70]. Most of the important metabolites which possess anticancer activity are obtained from algae. Almost thousands of the secondary metabolites obtained from the extract of the algae show antineoplastic characteristics. All of these components and extracts are effective against lung carcinoma and leukaemia lymphocytic [71–73]. Cryptophycin is one of the important anticancer metabolites [74]. Scytonemin, which isolated from *Stigonema*, has good anti-inflammatory and antiproliferative characteristics [75]. Table 11.2 shows different species of algae and their various applications while Table 11.3 shows some algae that play an important role against ultraviolet rays.

11.4 Relevance of Some Algae Derivative Components as Well as Their Effects on Human Health

In China, almost 2500 years ago the use of algae was seen to be very effective in the form of food, [76]. Most seaweed or algae in several parts of Asia is used directly in the form of food; and is also eaten in some other countries, including Mexico, America and Africa, due to the presence of an effective number of vitamins and also good nutritional values [77]. The use of these algae is seen in several products, including cookies, noodles, beverages, gums, snacks and pasta, which make these products more delicious [78].

Spirulina platensis is one of the important green and blue algae effective in food due to its being a rich source of nutritional values and good source of phenolic acid, vitamins, pigments, protein and polyunsaturated fatty acids [79–81]. So, all these types of microalgae are very effective and play a beneficial role in the health of humans because of the presence of important components, including lipids, carbohydrates and proteins [82, 83]. Table 11.4 shows some products derived from algae and their applications in the health sector.

Both types of micro- and macroalgae are effective for flowering, germination, and stem or leaf growth and play an effective and protectant role against several diseases. With the passage of time the use of algae has become vast and more effective, and is now used in different forms like tablets, powders, beverages and many other products which show great potential. Some important crop-enhancing components mostly derived from cyanobacteria are discussed below. Also, Table 11.5 shows some examples of algae and their application in the stimulation of crops.

Table 11.4 Products derived from algae and their applications in the health sector.

Products group	Application	Example
Vitamins	Nutrition	Alpha tocopherol, vitamin C and biotin
Polyunsaturated fatty acids	Food supplement & nutraceutical	DHA & EPA
Phycobiliproteins	Vitamins, pigments and cosmetics	Astaxanthin, beta-carotene and phycocyanin

Table 11.5 Crop stimulatory effectiveness of algae.

Cyanobacteria	Growth enhancer
Hapalosiphon & *Nostoc*	Methyl indole, propionic acid indole 3 & acetic acid [84–86]
Nostoc muscorum	Vitamin B12 [87–90]
Tolypothrix tenuis	Vitamin B12 [91–95]
Species of *Cylindrospermum*	Vitamin B12 [96–99]

11.5 Genetic Resources and Algae

Genetic resources are very effective for security and agricultural purposes, and also for sustainable preservation of useful algae. This might be linked to the fact that they could serve as sources of energy, shelter, nutrition and food, which could provide the various ecosystem services necessary for production of a sustainable agriculture system [97, 98]. They also perform functions such as maintaining soil fertility, pest control as well as having a wide application in management of different illnesses. A proper understanding of the genetic diversity of algae has helped numerous scientists in their search for diverse applications for the effective production of food for the survival of life and for providing several effective health benefits [91, 96, 99].

11.6 Conclusions

This chapter has provided detailed information on the role of genetical conservation for the production of important biological molecules derived from beneficial algae. A number of advancements were introduced that could further improve the efficacy of algae in several fields and their application for the effective production of diverse macro- and micromolecules based on their diverse roles in numerous fields.

References

1. Abedin, R. M., Taha, H. M. J. G. J. o. B., & Biochemistry. Antibacterial and antifungal activity of cyanobacteria and green microalgae. Evaluation of medium components by Plackett-Burman design for antimicrobial activity of Spirulina platensis. *3*(1), 22-31, 2008.
2. Adetunji, C.O., Roli, O.I., Adetunji, J.B. Exopolysaccharides Derived from Beneficial Microorganisms: Antimicrobial, Food, and Health Benefits. In: Mishra, P., Mishra, R.R., Adetunji, C.O. (eds) *Innovations in Food Technology*. Springer, Singapore, 2020a. https://doi.org/10.1007/978-981-15-6121-4_10
3. Adetunji, C O. Anani, Bio-fertilizer from Trichoderma: Boom for Agriculture Production and Management of Soil- and Root-Borne Plant Pathogens. Pages 245-256, 2020.
4. Adetunji C.O., Varma A. Biotechnological Application of *Trichoderma*: A Powerful Fungal Isolate with Diverse Potentials for the Attainment of Food

Safety, Management of Pest and Diseases, Healthy Planet, and Sustainable Agriculture. In: Manoharachary C., Singh H.B., Varma A. (eds) Trichoderma: Agricultural Applications and Beyond. *Soil Biology*, vol 61. Springer, Cham., 2020. https://doi.org/10.1007/978-3-030-54758-5_12

5. Olaniyan O,T., and Adetunji, C.O. Biochemical Role of Beneficial Microorganisms: An Overview on Recent Development in Environmental and Agro-Science. Microbial Rejuvenation of Polluted Environment, 2021.
6. Adetunji, C.O., Akram, M., Imtiaz, A., Bertha, E.C., Sohail, A., Olaniyan, O,P., Zahid, R., J Adetunji, J.B., Enoyoze, G.E., Sarin, N.B. Genetically modified cassava; the last hope that could help to feed the world: recent advances. Edition title: Genetically Modified Crops - Current Status, Prospects and Challenges. Editors: Kishor, P. B. Kavi, Rajam, M. V., Pullaiah, T. (Eds.) 2020b. https://www.springer.com/gp/book/9789811559310.
7. Inobeme, A., Jeevanandam, J., Adetunji, C.O., Anani, O.A., Thangadurai, D., Islam, S., Oyawoye, O.M., Oloke, J.K., Yerima, M.B., Olaniyan, O.T. Ecorestoration of soil treated with biosurfactant during greenhouse and field trials. Green Sustainable Process for Chemical and Environmental Engineering and Science. Biosurfactants for the Bioremediation of Polluted Environments, Pages 89-105, 2021. https://doi.org/10.1016/B978-0-12-822696-4.00010-3
8. Anani, O.A., Jeevanandam, J., Adetunji, C.O., Inobeme, A., Oloke, J.K., Yerima, M.B., Thangadurai, D., Islam, S., Oyawoye, O.M., Olaniyan, O.T. Application of biosurfactant as a noninvasive stimulant to enhance the degradation activities of indigenous hydrocarbon degraders in the soil. Green Sustainable Process for Chemical and Environmental Engineering and Science. Biosurfactants for the Bioremediation of Polluted Environments, 2021, Pages 69-87, 2021. https://doi.org/10.1016/B978-0-12-822696-4.00019-X
9. Adetunji, C.O., Inobeme, A., Osikemekha, A.A., Jeevanandam, J., Yerima, M.B., Thangadurai, D., Islam, S., Oyawoye, O.M., Oloke, J.K., Olaniyan, O.T. Isolation, screening, and characterization of biosurfactant-producing microorganism that can biodegrade heavily polluted soil using molecular techniques. Green Sustainable Process for Chemical and Environmental Engineering and Science. Biosurfactants for the Bioremediation of Polluted Environments, 2021, Pages 53-68, 2021a. https://doi.org/10.1016/B978-0-12-822696-4.00016-4
10. Jeevanandam, J., Adetunji, C.O., Selvam, J.D., Anani, O.A., Inobeme, A., Islam, S., Thangadurai, D., Olaniyan, O.T. High industrial beneficial microorganisms for effective production of a high quantity of biosurfactant. In book: *Green Sustainable Process for Chemical and Environmental Engineering and Science*, 2021.
11. Adetunji, C.O., Jeevanandam, J., Anani, O.A., Inobeme, A., Thangadurai, D., Islam, S., Olaniyan, O.T. Strain improvement methodology and genetic

engineering that could lead to an increase in the production of biosurfactants, 2021b.
12. Adetunji, C.O., Jeevanandam, J., Inobeme, A., Olaniyan, O.T., Anani, O.A., Thangadurai, D., Islam, S. Application of biosurfactant for the production of adjuvant and their synergetic effects when combined with different agro-pesticides, 2021c.
13. Adetunji, C.O., Olaniyan, O.T., Anani, O.A., Inobeme, A., Ukhurebor, K.E., Bodunrinde, K.E., Adetunji, J.B., Singh, K.R., Nayak, V., Palnam, W.D., and Singh, R.P. Bionanomaterials for green bionanotechnology. In book: *Bionanomaterials: Fundamentals and Biomedical Applications*. IOP Publishing, 2021d.
14. Adetunji C.O., Inobeme A., Olaniyan O.T., Olisaka F.N., Bodunrinde R.E., Ahamed M.I. Microbial Desalination. In: Inamuddin, Khan A. (eds) *Sustainable Materials and Systems for Water Desalination. Advances in Science, Technology & Innovation (IEREK Interdisciplinary Series for Sustainable Development)*. Springer, Cham, 2021e. https://doi.org/10.1007/978-3-030-72873-1_13
15. Adetunji J.B., Adetunji C.O., Olaniyan O.T. African Walnuts: A Natural Depository of Nutritional and Bioactive Compounds Essential for Food and Nutritional Security in Africa. In: Babalola O.O. (eds) *Food Security and Safety*. Springer, Cham, 2021f. https://doi.org/10.1007/978-3-030-50672-8_19
16. Adetunji, C.O., Olaniyan, O.T., Anani, O.A., Olisaka, F.N., Inobeme, A., Bodunrinde, R.E., Juliana Adetunji, J.B., Singh, K.R., Palnam, W.B., Singh. R.P. Current Scenario of Nanomaterials in the Environmental, Agricultural, and Biomedical Fields. Nanomaterials in Bionanotechnology.. In book: *Nanomaterials in Bionanotechnology: Fundamentals and Applications*. 1. 6. CRC Press, 2021g.
17. Adetunji C.O., Kremer, R.J., Makanjuola, R., BhallaSarin, N. Application of molecular biotechnology to manage biotic stress affecting crop enhancement and sustainable agriculture. *Advances in Agronomy*, 2021h. https://www.sciencedirect.com/science/article/pii/S0065211321000304?dgcid=author
18. Adetunji, C.O., Olaniyan, O.T., Adetunji, J.B., Osemwegie, O., Ubi, B.E. African Mushrooms as Functional Foods and Nutraceuticals. In book: *Fermentation and Algal Biotechnologies for the Food, Beverage and Other Bioproduct Industries..Edition 1st Edition*. CRC Press. Pages 19, 2022.
19. Adetunji, C.O., Ukhurebor, K.E., Olaniyan, O.T., Ubi, B.E., Oloke, J.K., Dauda, W.T., Hefft, D.I. Recent Advances in Molecular Techniques for the Enhancement of Crop Production. In book: *Agricultural Biotechnology, Biodiversity and Bioresources Conservation and Utilization. Edition 1st Edition*. 2022. CRC Press. Pages 20, 2022b.
20. Adetunji C.O., Inobeme A., Olaniyan O.T., Olisaka F.N., Bodunrinde R.E., Ahamed M.I. Microbial Desalination. In: Inamuddin, Khan A. (eds) *Sustainable Materials and Systems for Water Desalination.*

Advances in Science, Technology & Innovation (IEREK Interdisciplinary Series for Sustainable Development). Springer, Cham, 2021e. https://doi.org/10.1007/978-3-030-72873-1_13

21. Adetunji J.B., Adetunji C.O., Olaniyan O.T. African Walnuts: A Natural Depository of Nutritional and Bioactive Compounds Essential for Food and Nutritional Security in Africa. In: Babalola O.O. (eds) *Food Security and Safety*. Springer, Cham, 2021f, https://doi.org/10.1007/978-3-030-50672-8_19
22. Ackman, R. G., Tocher, C., & McLachlan, J. J. J. o. t. F. B. o. C. Marine phyto-plankter fatty acids. *25*(8), 1603-1620, 1968.
23. Adams, J., Ross, A., Anastasakis, K., Hodgson, E., Gallagher, J., Jones, J., & Donnison, I. J. B. t. Seasonal variation in the chemical composition of the bioenergy feedstock Laminaria digitata for thermochemical conversion. *102*(1), 226-234, 2011.
24. Adey, W. H., & Loveland, K. *Dynamic aquaria: building living ecosystems*: Elsevier, 2011.
25. Ayehunie, S., Belay, A., Baba, T. W., Ruprecht, R. M. J. J. o. a. i. d. s., & Association, h. r. o. p. o. t. I. R. Inhibition of HIV-1 replication by an aqueous extract of Spirulina platensis (Arthrospira platensis). *18*(1), 7-12, 1998.
26. Becker, W. 18 microalgae in human and animal nutrition. In *Handbook of microalgal culture: biotechnology and applied phycology* (Vol. 312): Wiley Online Library Hoboken, NJ, USA, 2004.
27. Benemann, J. R., & Oswald, W. J. Systems and economic analysis of micro-algae ponds for conversion of CO {sub 2} to biomass. Final report, 1996.
28. Bhattacharjee, M. J. A. J. P. C. R. Pharmaceutically valuable bioactive com-pounds of algae. *9*, 43-47, 2016.
29. Bigogno, C., Khozin-Goldberg, I., Boussiba, S., Vonshak, A., & Cohen, Z. J. P. Lipid and fatty acid composition of the green oleaginous alga Parietochloris incisa, the richest plant source of arachidonic acid. *60*(5), 497-503, 2002.
30. Boopathy, N. S., & Kathiresan, K. J. J. O. Anticancer drugs from marine flora: an overview. *2010*, 214186, 2010.
31. Borowitzka, M. A. Energy from microalgae: a short history. In *Algae for bio-fuels and energy* (pp. 1-15): Springer, 2013.
32. Borowitzka, M. A., & Borowitzka, L. J. *Micro-algal biotechnology*: Cambridge University Press, 1988.
33. Ogunlana, O. O., Adetuyi, B. O., Rotimi, M., Esalomi, I., Adeyemi A., Akinyemi, J., Ogunlana, O., Adetuyi, O., Adebisi, O., Okpata, E., Baty, R., & Batiha, G. Hypoglycemic Activities of Ethanol Seed Extract of *Hunteria umbellate* (Hallier F.) on Streptozotocin-induced Diabetic Rats. *Clinical Phytoscience*, (1), 1-9, 2021.
34. Ogunlana, O. O., Ogunlana, O. E., Adekunbi, T. S., Adetuyi, B. O., Adegboye, B. E., & Iheagwam, F. N. Anti-inflammatory Mechanism of Ruzu Bitters on Diet-Induced Nonalcoholic Fatty Liver Disease in Male Wistar

Rats. *Evidence-Based Complementary and Alternative Medicine, 2020*, 2020. https://doi.org/10.1155/2020/5246725

35. Ogunlana, O. O., Ogunlana, O. E., Popoola, J. O., Adetuyi, B. O., Adekunbi, T. S., David, O. L., Adeleye, O. J., Udeogu, S. A., & Adeyemi, A. O. Twigs of Andrographis paniculata (Burn. F) Nees attenuates Carbon Tetrachloride (CCl4) Induced Liver Damage in Wistar Albino Rats. *RASAYAN Journal of Chemistry* 14(4): 2598-2603, 2021.
36. Ogunlana, O. O., Adetuyi, B. O., Esalomi, E. F., Rotimi, M. I., Popoola, J. O., Ogunlana, O. E., & Adetuyi, O. A. Antidiabetic and Antioxidant Activities of the Twigs of Andrograhis paniculata on Streptozotocin-Induced Diabetic Male Rats. *BioChem* 1(3):238-249, 2021.
37. Chynoweth, D., Turick, C., Owens, J., Jerger, D. E., Peck, M. J. B., Biochemical methane potential of biomass and waste feedstocks. *Biomass & Bioenergy*. 5(1), 95-111, 1993.
38. Colla, L. M., Reinehr, C. O., Reichert, C., & Costa, J. A. V. J. B. t. Production of biomass and nutraceutical compounds by Spirulina platensis under different temperature and nitrogen regimes. *98*(7), 1489-1493, 2007.
39. Demirbaş, A. J. E. c., & management. Mechanisms of liquefaction and pyrolysis reactions of biomass. *41*(6), 633-646, 2000.
40. Dey, B., Lerner, D. L., Lusso, P., Boyd, M. R., Elder, J. H., & Berger, E. A. J. J. o. v. Multiple antiviral activities of cyanovirin-N: blocking of human immunodeficiency virus type 1 gp120 interaction with CD4 and coreceptor and inhibition of diverse enveloped viruses. *74*(10), 4562-4569, 2000.
41. Edwards, M. R. *Green algae strategy: end biowar I and engineer sustainable food and biofuels*: Lulu.com, 2008.
42. El-Sayed, W., Ibrahim, H., Abdul-Raouf, U., & El-Nagar, M. J. J. B. B. Evaluation of bioethanol production from Ulva lactuca by Saccharomyces cerevisiae. *6*(226), 2, 2016.
43. Furusawa, E., Moore, R., Mynderse, J., Norton, T., & Patterson, G. J. U. P. New purified culture of Scytonema pseudohofmanni ATCC 53141 is used to produce scytophycins A, B, C, D and E, which are potent cytotoxins and antineoplastic agents. (5281533), 1994.
44. Awoyelu, E. H., Oladipo, E. K., Adetuyi, B. O., Senbadejo, T. Y., Oyawoye, O. M., Oloke, J. K. Phyloevolutionary analysis of SARS-CoV-2 in Nigeria. *New Microbes and New Infections* Vol 36. 100717, 2020. https://doi.org/10.1016/j.nmni.2020.100717
45. Batiha, G. B., Awad, D. A., Algamma, A. M., Nyamota, R., Wahed, M. I., Shah, M. A., Amin, M. N., Adetuyi, B. O., Hetta, H. F., Cruz-Marins, N., Koirala, N., Ghosh, A., & Sabatier, J. Diary-derived and Egg White Proteins in Enhancing Immune System against COVID-19 *Frontiers in Nutritionr. (Nutritional Immunology)* 8:629440, 2021.
46. Didunyemi, M. O., Adetuyi, B. O., & Oyewale, I. A. Inhibition of lipid peroxidation and *in-vitro* antioxidant capacity of aqueous, acetone and methanol

leaf extracts of green and red Acalypha wilkesiana Muell Arg. *Int J Biol Med Res.* 11(3):7089- 7094, 2020.

47. Didunyemi, M. O., Adetuyi, B. O., Oyebanjo, O. O. Morinda lucida attenuates acetaminophen-induced oxidative damage and hepatotoxicity in rats, *Journal of Biomedical sciences Vol 8; No,* 2019. https://www.jbiomeds.com/biomedical-sciences/morinda-lucida-attenuates-acetaminopheninduced-oxidative-damage-and-hepatotoxicity-in-rats.php?aid=24482
48. Garcia Alba, L., Torri, C., Samorì, C., Van Der Spek, J., Fabbri, D., Kersten, S. R., . . . fuels. Hydrothermal treatment (HTT) of microalgae: evaluation of the process as conversion method in an algae biorefinery concept. *26*(1), 642-657, 2012.
49. Gerwick, W. H., Roberts, M. A., Proteau, P. J., & Chen, J.-L. J. J. o. A. P. Screening cultured marine microalgae for anticancer-type activity. *6*(2), 143-149, 1994.
50. Guan, Q., Savage, P. E., & Wei, C. J. T. J. o. S. F. Gasification of alga Nannochloropsis sp. in supercritical water. *61*, 139-145, 2012.
51. Gülyurt, M. Ö., Özçimen, D., & İnan, B. J. I. j. o. m. s. Biodiesel production from Chlorella protothecoides oil by microwave-assisted transesterification. *17*(4), 579, 2016.
52. Huleihel, M., Ishanu, V., Tal, J., & Arad, S. M. J. J. o. a. p. Antiviral effect of red microalgal polysaccharides on Herpes simplex and Varicella zoster viruses. *13*(2), 127-134, 2001.
53. Jena, U., Das, K. J. E., & fuels. Comparative evaluation of thermochemical liquefaction and pyrolysis for bio-oil production from microalgae. *25*(11), 5472-5482, 2011.
54. Jung, K. A., Lim, S.-R., Kim, Y., & Park, J. M. J. B. t. Potentials of macroalgae as feedstocks for biorefinery. *135*, 182-190, 2013.
55. Katircioglu, H., Beyatli, Y., Aslim, B., Yüksekdag, Z., & Atici, T. J. M. Screening for antimicrobial agent production of some freshwater. *2*(2), 2006.
56. Kishimoto, M., Okakura, T., Nagashima, H., Minowa, T., Yokoyama, S.-Y., Yamaberi, K., CO2 fixation and oil production using micro-algae. *Journal of Fermentation and Bioengineering, 78*(6), 479-482, 1994.
57. Adetuyi B.O, Oluwole E.O, Dairo J.O Biochemical effects of shea butter and groundnut oils on white albino rats, *International Journal of Chemistry and Chemical Processes* (IJCC). Vol 1(8).:1-17, 2015.
58. Adetuyi BO, Dairo JO, Didunyemi MO Anti-Hyperglycemic Potency of *Jatropha Gossypiifolia* in Alloxan Induced Diabetes. *Biochem Pharmacol (Los Angel)* 4(5):193, 2015.
59. Adetuyi, B. O., Omolabi, F. K., Olajide, P. A., Oloke, J. K. Pharmacological, Biochemical and Therapeutic Potential of Milk Thistle (Silymarin): A Review *World News of Natural Sciences* 37:75-91, 2021.
60. Adetuyi, B. O., & Farombi, E. O. 6-Gingerol, an active constituent of ginger, attenuates lipopolysaccharide-induced oxidation, inflammation,

cognitive deficits, neuroplasticity, and amyloidogenesis in rat. *Journal of Food Biochemistry*, *45*(4), e13660, 2021.

61. Adetuyi, B. O., Adebayo, P. F., Olajide, P. A., Atanda, O. O., Oloke, J. K. Involvement of Free Radicals in the Ageing of Cutaneous Membrane. World News of Natural Sciences 43, 11-37, 2022.
62. Adetuyi, B. O., Adebisi, O. A., Adetuyi, O. A., Ogunlana, O. O., Toloyai, P-B., Egbuna, C., Uche, C. Z., Khan, J., Adumanya, O. C., Patrick-Iwuanyanwu, K.C. (2022), Ficus exasperata Attenuates Acetaminophen-Induced Hepatic Damage via NF-κB Signaling Mechanism in Experimental Rat Model BioMed Research International, 2022.
63. Metting, B., Pyne, J. W. J. E., & Technology, M. Biologically active compounds from microalgae. *8*(7), 386-394, 1986.
64. Minowa, T., & Sawayama, S. J. F. A novel microalgal system for energy production with nitrogen cycling. *78*(10), 1213-1215, 1999.
65. Nguyen, T. L. T., & Gheewala, S. H. J. T. I. J. o. L. C. A. Life cycle assessment of fuel ethanol from cane molasses in Thailand. *13*(4), 301-311, 2008.
66. Noaman, N. H., Fattah, A., Khaleafa, M., & Zaky, S. H. J. M. R. Factors affecting antimicrobial activity of Synechococcus leopoliensis. *159*(4), 395-402, 2004.
67. Olaizola, M. J. B. e. Commercial development of microalgal biotechnology: from the test tube to the marketplace. *20*(4-6), 459-466, 2003.
68. Patterson, G. M., Baker, K. K., Baldwin, C. L., Bolis, C. M., Caplan, F. R., Larsen, L. K., . . . Nelson, C. S. J. J. o. p. Antiviral activity cultured blue-green algae (Cyanophtya) 1. *29*(1), 125-130, 1993.
69. Peng, W., Wu, Q., Tu, P., & Zhao, N. J. B. T. Pyrolytic characteristics of microalgae as renewable energy source determined by thermogravimetric analysis. *80*(1), 1-7, 2001.
70. Peng, W., Wu, Q., & Tu, P. J. J. o. A. P. Pyrolytic characteristics of heterotrophic Chlorella protothecoides for renewable bio-fuel production. *13*(1), 5-12, 2001.
71. Sajilata, M., Singhal, R., & Kamat, M. J. F. C. Fractionation of lipids and purification of γ-linolenic acid (GLA) from Spirulina platensis. *109*(3), 580-586, 2008.
72. Schwartz, R. E., Hirsch, C. F., Sesin, D. F., Flor, J. E., Chartrain, M., Fromtling, R. E., . . . biotechnology. Pharmaceuticals from cultured algae. *5*(2-3), 113-123, 1990.
73. Sharma, P., & Sharma, N. J. J. o. A. i. p. b. Industrial and biotechnological applications of algae: a review. *1*(1), 01, 2017.
74. Spolaore, P., Joannis-Cassan, C., Duran, E., Isambert, A., Commercial applications of microalgae. *Journal of Bioscience and Bioengineering*, *101*(2), 87-96, 2006.
75. Stevenson, C. S., Capper, E. A., Roshak, A. K., Marquez, B., Eichman, C., Jackson, J. R., . . . Therapeutics, E. The identification and characterization of

the marine natural product scytonemin as a novel antiproliferative pharmacophore. *303*(2), 858-866, 2002.
76. Tsai, C.-C., Emau, P., Jiang, Y., Agy, M. B., Shattock, R. J., Schmidt, A., . . . retroviruses, h. Cyanovirin-N inhibits AIDS virus infections in vaginal transmission models. *20*(1), 11-18, 2004.
77. Tseng, C. The past, present and future of phycology in China. In *Asian Pacific Phycology in the 21st Century: Prospects and Challenges* (pp. 11-20): Springer, 2004.
78. Um, B.-H., Kim, Y.-S. J. J. o. I., & Chemistry, E. A chance for Korea to advance algal-biodiesel technology. *15*(1), 1-7, 2009.
79. Vardon, D. R., Sharma, B. K., Blazina, G. V., Rajagopalan, K., & Strathmann, T. J. J. B. t. Thermochemical conversion of raw and defatted algal biomass via hydrothermal liquefaction and slow pyrolysis. *109*, 178-187, 2012.
80. Wang, B., Wang, J., Zhang, W., & Meldrum, D. R. J. F. i. m. Application of synthetic biology in cyanobacteria and algae. *3*, 344, 2012.
81. Wout, R., Greenwell, H., Davies, D., & Theodorou, M. J. G. P. W. Methods of ensiling algae, ensiled algae and uses of ensiled algae, 2013.
82. Adeleke, D. A., Olajide, P. A., Omowumi, O. S., Okunlola, D. D., Taiwo, A. M., & Adetuyi, B. O. Effect of Monosodium Glutamate on the Body System. *World News of Natural Sciences*, *44*, 1-23, 2022.
83. Adetuyi B.O, Oluwole E.O Dairo J.O Chemoprotective Potential of Ethanol Extract of *Ganoderma Lucidum* on Liver and Kidney Parameters in *Plasmodium Beghei*-Induced Mice, *International Journal of Chemistry and Chemical Processes (IJCC)*. Vol 1(8).:29-36, 2015.
84. Adetuyi, B. O., Adebisi, O. A., Awoyelu, E. H., Adetuyi, O. A., Ogunlana, O. O. Phytochemical and Toxicological effect of Ethanol extract of *Heliotropium indicum* on Liver of Male Albino Rats. *Letters in Applied NanoBioscience (LIANB)* 10(2):2085-2095, 2020. https://doi.org/10.33263/LIANBS102.20852095
85. Adetuyi, B. O., Odine, G. O., Olajide, P. A., Adetuyi, O. A., Atanda, O. O., Oloke J. K. Nutraceuticals: role in metabolic disease, prevention and treatment, World News of Natural Sciences 42, 1-27, 2022.
86. Adetuyi, B. O., Ogundipe, A. E., Ogunlana, O. O., Egbuna, C., Estella, O. U., Mishra, A. P., Akram, M., & Achar, R. R. Banana Peel as a Source of Nutraceuticals. Food and Agricultural Byproducts as Important Source of Valuable Nutraceuticals.1st ed. 2022. Springer, Berlin. xxx, 400 p. Gebunden, 2022.
87. Adetuyi, B. O., Okeowo, T. O., Adetuyi, O. A., Adebisi, O. A., Ogunlana, O., Oretade, O. J., Marraiki, N., Beshbishy, A. M., Welson, N. N., & Batiha, G. E Ganoderma lucidum from red mushroom attenuates formaldehyde-induced liver damage in experimental male rat model. *Biology* 2020, *9*(10), 313, 2020. https://doi.org/10.3390/biology9100313

88. Adetuyi, B. O., Olajide, P. A., Awoyelu, E. H., Adetuyi, O. A., Adebisi, O. A., Oloke, J. K. , Epidemiology and Therapeutic measure for COVID-19; A review. *African Journal of Reproductive Health June 2020* (Special Edition on COVID-19); 24 (2):142, 2020. https://www.ajrh.info/index.php/ajrh/article/view/2300
89. Adetuyi, B. O., Olajide, P. A., Omowumi, O. S., Odine, G. O., Okunlola, D. D., Taiwo, A. M., & Opayinka, O. D. Blockage of Alzheimer's gene: Breakthrough effect of Apolipoprotein E4. *African Journal of Advanced Pure and Applied Sciences (AJAPAS)*, 26-33, 2022.
90. Adetuyi, B. O., Toloyai, P. Y., Ojugbeli, E. T., Oyebanjo, O. T., Adetuyi, O. A., Uche, C. Z., Olisah, M. C., Adumanya, O. C., Jude, C., Chikwendu, J. K., Akram, M., Awuchi, C.G., Egbuna, C. Neurorestorative Roles of Microgliosis and Astrogliosis in Neuroinflammation and Neurodegeneration. *Scicom Journal of Medical and Applied Medical Sciences* 1(1):1-5, 2021.
91. Adetuyi, B.O., Olajide, P. A., Oluwatosin, A., Oloke, J. K. Preventive Phytochemicals of Cancer as Speed Breakers in Inflammatory Signaling. *Research Journal of Life Sciences, Bioinformatics, Pharmaceutical and Chemical Sciences* 8 (1) 30-61, 2022.
92. Adewale, G. G., Olajide, P. A., Omowumi, O. S., Okunlola, D. D., Taiwo, A. M., & Adetuyi, B. O. Toxicological Significance of the Occurrence of Selenium in Foods. *World News of Natural Sciences, 44*, 63-88, 2022.
93. Farombi, E. O., Abolaji, A. O., Adetuyi, B. O., Awosanya, O., & Fabusoro, M. Neuroprotective role of 6-Gingerol-rich fraction of Zingiber officinale (Ginger) against acrylonitrile-induced neurotoxicity in male Wistar rats. *Journal of Basic and Clinical Physiology and Pharmacology, 30*(3), 2019.
94. James-Okoro, P. O., Iheagwam, F. N., Sholeye, M. I., Umoren, I. A., Adetuyi, B. O., Ogundipe, A. E., Braimah, A. A., Adekunbi, T. S., Ogunlana, O. E., & Ogunlana, O. O. Phytochemical and *in vitro* antioxidant assessment of Yoyo bitters *World News of Natural Sciences* 37:1-17, 2021.
95. Nazir, A., Itrat, N., Shahid, A., Mushtaq, Z., Abdulrahman, S. A., Egbuna, C., Adetuyi, B. O., Khan, J., Uche, C. Z, Toloyai, P. Y. Orange Peel as a Source of Nutraceuticals. *Food and Agricultural Byproducts as Important Source of Valuable Nutraceuticals.1st ed.* 2022 2022. Springer, Berlin. 400 pp. 2022.
96. Ogunlana, O. O., Adetuyi, B. O., Adekunbi, T. S., Adegboye, B. E., Iheagwam, F. N., Ogunlana, O. E. Ruzu bitters ameliorates high–fat diet induced non-alcoholic fatty liver disease in male Wistar rats. *Journal of Pharmacy and Pharmacognosy Research* 9(3), 251-26, 2021.
97. Olajide, P. A., Adetuyi, O. A., Omowumi, O. S. & Adetuyi, B. O. Anticancer and Antioxidant Phytochemicals as Speed Breakers in Inflammatory Signaling. *World News of Natural Sciences* 44, 231-259, 2022.

98. Olajide, P. A., Omowumi, O. S., Okunlola, D. D., & Adetuyi, B. O. Deadly Pandemia: Monkeypox Disease, a Case Study. *African Journal of Advanced Pure and Applied Sciences (AJAPAS)*, 34-37, 2022.
99. Olajide, P. A., Omowumi, S. O., & Odine, G. O. Pathogenesis of Reactive Oxygen Species: A Review. *World News of Natural Sciences* 44:150-164, 2022.

12

Relevance of Biostimulant Derived from Cyanobacteria and Its Role in Sustainable Agriculture

Charles Oluwaseun Adetunji[1]*, Muhammad Akram[2], Fahad Said[2], Olugbemi T. Olaniyan[3], Inobeme Abel[4], Ruth Ebunoluwa Bodunrinde[5], Nyejirime Young Wike[6], Phebean Ononsen Ozolua[1], Wadzani Dauda Palnam[7], Arshad Farid[8], Shakira Ghazanfar[9], Olorunsola Adeyomoye[10], Chibuzor Victory Chukwu[11] and Mohammed Bello Yerima[12]

[1]Applied Microbiology, Biotechnology and Nanotechnology Laboratory, Department of Microbiology, Edo State University Uzairue, Iyamho, Edo State, Nigeria
[2]Department of Eastern Medicine, Government College University Faisalabad, Punjab, Pakistan
[3]Laboratory for Reproductive Biology and Developmental Programming, Department of Physiology, Rhema University, Aba, Nigeria
[4]Department of Chemistry, Edo University Iyamho, Auchi, Edo State, Nigeria
[5]Department of Microbiology, Federal University of Technology Akure, Gaga, Nigeria
[6]Department of Human Physiology, Faculty of Basic Medical Science, Rhema University, Aba, Nigeria
[7]Department of Agronomy, Federal University, Gashua, Yobe, Nigeria
[8]Gomal Center of Biochemistry and Biotechnology, Gomal University, Khyber Pakhtunkhwa, Pakistan
[9]National Institute for Genomics & Advanced Biotechnology (NIGAB), National Agricultural Research Centre (NARC), Islamabad, Pakistan
[10]Department of Physiology, University of Medical Sciences, Ondo City, Nigeria
[11]Department of Microbiology, Edo State University Uzairue, Iyamho, Edo State, Nigeria
[12]Department of Microbiology, Sokoto State University, Sokoto, Nigeria

**Corresponding author*: adetunjicharles@gmail.com; adetunji.charles@edouniversity.edu.ng

Charles Oluwaseun Adetunji, Julius Kola Oloke, Naveen Dwivedi, Sabeela Beevi Ummalyma, Shubha Dwivedi, Daniel Ingo Hefft and Juliana Bunmi Adetunji (eds.) Next-Generation Algae: Volume I: Applications in Agriculture, Food and Environment, (281–294) © 2023 Scrivener Publishing LLC

Abstract
The demand for food is increasing due to the growing human population. Therefore, in order to address this demand, there is a need to search for a sustainable technology that could boost agricultural production and improve food security while promoting a world-class agricultural system and lifestyle. There is a need to search for an eco-friendly system that will be a permanent replacement to synthetic pesticides that have been reported as having adverse effects. Therefore, this chapter intends to provide detailed information on the application of beneficial cyanobacteria for effective production of several agricultural crops while also providing detailed information on the modes of action through which they perform their activities.

***Keywords*:** Biostimulants, cyanobacteria, agriculture, food, human population

12.1 Introduction

It has been demonstrated that the application of algae could be applied as a biostimulant for adequate support of numerous agricultural crops for continuous supply of nutritious food and sustainability of more secure food [1–19]. Typical examples of these algae include Cyanobacteria, *Nostoc*, and *Oscillatoria angustissima*, which portend the capacity to promote plant growth. Moreover, it has been demonstrated that the macronutrients available in microalgae *Scenedesmus* sp. [72] could influence plant growth while *Anabaena* could also improve the defense of plants [54] as well as improve plant yields [53] by increasing the phytohormones such as auxins, gibberellins, cytokinins, abscisic acid, chlorocysts and cyanobacterial cells.

Interestingly, it has been documented that *Arthrospira platensis* are rich in L-amino acids [61], which accounts for 58% of their total protein content [63] and promotes plant growth [50, 51]. The content of polyamine is one of the biosynthesis of *Spirulina platensis* because biosynthesis is dependent on L-amino acids and promotes cell proliferation as well as cell growth [64]. It has been shown that cyanobacteria can increase digestion by bioactive compounds by promoting plant development [73].

Also, it has been established that these algae portend the potential to secret hormones called polyamines [64] and amino acids [61, 63] as well as having influence on the development of several crops through an increase in crop yield and growth [60, 55].

Cyanobacterial secretions include gibberellins, linoleum, n-acetyl-D-glucosamine, niacinamide which can lead to cell proliferation, cell division, as well as root formation and antioxidants. According to Riahi *et al.* [77], Cyanobacteria provide several benefits to plants by providing amino acids, polypeptides, antioxidants, nutrients, antibacterials and antifungals that increase the growth and performance of plants as well as their significant role in the production of relevant biomolecules, such as chlorophylls, carotenes, protein, glutamine-oxaloacetate, glutamine-pyruvic-transaminase [56], and their role in the development of bio-active compounds present in the cyanobacterial biomass which might contain phytohormone, exopolysaccharide, nitrogen, phosphorus, and potassium [56].

Also, the usage of algae could be applied for adequate improvement of plant growth through the development of chlorophyll and chlorophyll B, which affect photosynthetic levels to form after anabaena treatment. In addition, the application of cyanobacteria can affect the natural properties of soil and its nitrogen content. Therefore, this chapter intends to provide detailed information on the application of algae in the production of biostimulant.

12.2 Biostimulants Derived from Cyanobacteria for Boosting Agriculture

Biostimulants are efficient in the reduction of abiotic stress on plant production [30]. They serve as correctional agents for the various diseased conditions of plants caused by environmental interferences, which result in a poor plant growth and product yield [40, 41]. Metabolites produced from microalgae, especially Cyanobacteria, have found wide application as biostimulants in protecting plants against abiotic stresses [26, 27]. Cyanobacteria contain some materials which are able to enhance the growth of plants, some of which include vitamins, betaines, amino acids, auxins, polyamines, cytokinins, brassinosteroids and gibberellins [24, 25, 36–39].

Much has not been documented on the use of microalgal-based phytohormones in the production of crops [34]. The few studies reported have centered on cyanobacteria like *Nostoc* sp., *Scenedesmus* sp., *Hapalosiphon* sp., *Aulosira fertilissima, Scenedesmus* spp. and *Arthrospira* or *Spirulina* spp. The phytohormones present in both *Scenedesmus* sp. and *Arthrospira* sp. were compared, and it was found that a higher concentration of auxins, cytokinins, salicylic acid, gibberellins and abscisic acid were obtained in the extract of *Scenedesmus* sp. than *Arthrospira* sp. When used as biostimulator, the extract from *Scenedesmus* sp. reportedly showed a greater yield in terms of flowers, leaves and shoots [33]. *Aulosira fertilissima* and *Anabaena doliolum* were used singly and in combination with N-fertilizer which was supplied in the form of urea to improve the nutritive value of rice. An appreciable increase was observed on the root length, chlorophyll content, grain yield, protein and nitrogen content of grain, seed number, plant height, length of leaves and number of panicles. The effect of the cyanobacterium biostimulator was more pronounced when a lower dose of N-fertilizer was accompanied with a quick and more effective response [22].

Since cyanobacteria are known to contain both micro- and macronutrients, apart from phytohormones, a consortium of cyanobacteria containing *Nostoc* sp., *Westiellopsis* sp., *Anabaena* sp., *Scytonema* sp. and *Aulosira* sp. was used against a pure culture of *Anabaena* sp. to inoculate the soil for the purpose of cultivating rice. A higher productivity in terms of grain and straw production was obtained with a percentage increase of grain to be 20.9% while that of straw was 18.1%. The growth performance from the soil inoculated with the consortium was higher than the soil with single culture of *Anabaena* sp. [32]. The synergistic effect from the interaction of the different hormones and nutrients present in each cyanobacterium combined could account for the improved yield with the consortium rather than the single culture.

An improvement in germination of seed, increase in grain weight, protein content, growth of shoot and root of rice was reported when *Nostoc* sp., *Hapalosiphon* sp., and *Aulosira fertilissima* were applied to the soil used for rice plantation [29]. The treatment of rice seedlings (IR-8) with extract from *A. fertilissima* and gibberellic acid for comparison was carried out. There was an increase in the growth of rice seedlings treated with the extract of the cyanobacterium while the seedlings treated with the acid did not show appreciable increase in growth [35].

Apart from plant growth hormones, some cyanobacteria are also known to be rich sources of both micro- and macronutrients with special reference to nitrogen, phosphorus and potassium [21]. Aly and Esawy [20] reported the concentration of nitrogen, phosphorus and potassium in the dried biomass of *Arthrospira* spp. to be 6.70, 2.47 and 1.14 % respectively. Another factor that makes *Arthrospira* spp. a good source of biostimulator is the fact that lead is not present in it and this makes it fit to be used as growth enhancer for plants.

The nutrient content of cyanobacteria has also been presented as a biostimulator which can be employed for foliar application instead of the synthetic compounds used. Extract of *Arthrospira* sp. and a synthetic fertilizer were compared for foliar application on sandy soil on which pepper plant was planted; the applied dosage of 80 gL^{-1} was administered every 15 days. A higher yield was reported from soil treated with a biostimulator from *Arthrospira* compared with the one treated with the organic fertilizer [20]. Similarly, the impact of *Arthrospira* sp.-based biostimulator on winter wheat was compared with *Ascophyllum nodosum*, which is a commercially available biostimulant. Although both products did not exhibit any phytotoxicity on the wheat plant, an increased number of ear grain was obtained with the plant treated with extract of *Arthrospira* sp. at application dose of 1.5 Lha^{-1} [28].

Oliveira *et al.* [31] similarly carried out an investigation on the effect of biostimulant obtained from *Arthrospira* sp. on the production of red beet. The biostimulant alongside the control were employed for foliar spraying at concentrations of 1.5 gL^{-1} and 3.0 gL^{-1} every two weeks. The dry weight, fresh weight and the size of the hypocotyl were compared. An increase in dry and fresh weight of the plant was reported on the *Arthrospira* sp.-based biostimulator.

Foliar spraying of an ornamental plant (Petunia × hybrida) was carried out comparing biostimulator from the extract of *Scenedesmus almeriensis* and *Arthrospira platensis* at a concentration of 10 gL^{-1} for 0, 14, 28, 35 and 42 days post transplanting. An enhanced plant growth in respect to flowering time, leaf, shoot and root growth was obtained with *S. almeriensis* while *Arthrospira* sp.-based biostimulator improved the plant water, quantity of flowers produced and dry matter of the root. The performance of *S. almeriensis*-based biostimulator proved to be more effective in the flowering of Petunia × hybrida [33]. A biostimulator which is able to enhance the flowering of an ornamental plant will be preferred.

Foliar spraying of both tomato and pepper plants was carried out using polysaccharides extracted from *Arthrospira* sp. as a biostimulator. A concentration of 3 gL^{-1} was used in treating both plants. The improvement in terms of root weight was higher in tomato than pepper plants with a percentage increase of 230 and 67%, respectively, while the increase in the size of tomato and pepper plants was 20 and 30% respectively. Improvement in terms of size and number of nodes for tomato plant was 57 and 100% while that of pepper plant was 33 and 50% respectively [23].

12.3 Modes of Action Involved in the Application Microorganism as Biostimulant

Till date, numerous scientists have applied the application of microorganisms (PGPB) as a sustainable tool for effective boosting of several agricultural processes. Studies have shown that cyanobacteria have received increasing attention due to the lack of biodiversity of conservation experts who use their ability to provide biomass for bioenergy, feed (the best feed), and safe carcass cultivation. They perform a beneficial role in increasing the fertility and agriculture productivities because of their ability to regulate the environment and produce phosphates and nutrients necessary for plant growth as well as their role in production of phytohormones (auxins, gibberellins, cytokinins), polypeptides, amino acids [65], polysaccharides [62] and ferrocite monocytes [74], alkalis and micropeptides necessary for cell growth, including their metabolites. Also, some of them also perform an effective role in promoting cyanobacteria that could generate important plant extract [66], which reduces several toxic compounds in the environmental [73, 76].

Numerous scientists have documented the role of microorganisms during their interaction with plants under all possible conditions [42, 43]: (1) Relationship between growth reactions and parasitism; (2) The characteristics of bacteria are that they grow inside soil cells, which are semi-layers called roots; (3) The organism may be temporary or permanent and some microbes of the organism may be vertically transmitted by seed; Increases resistance to abiotic stress and regulates formation as a regulator of plant growth. Roots and related species distribute biological fertilizers, such as microbial vaccines, that stimulate plant supplements. Numerous scientists have established the function of rhizobium-based symbiosis in the literature, most especially their application as biostimulant. PGPR is multifaceted and affects all parts of the plant: stability, evolution, morphology and refinement, responses to biotic and abiotic stresses, interactions with organ tissues in the agricultural system [44–49, 68, 75]. The PGPR trigger depends on the complexity and the variable reactivity of the plant and the conditions being taken. However, the global market for bacterial biostimulants has grown and PGPR vaccines are now considered to be a "plant probiotic," for example, enhancing plant nutrition and immunity [46].

Finally, the mode of action involved in the biological activity of biostimulants includes their capability to increases their resistance to aerodynamic stress (which causes oxidative stress) or increased utility of N (which depends on the limitations of root removal, leading to thickening of additional roots). Agricultural capacity can be converted into economic and environmental benefits, which includes increase in the yields, fertilizer application, product quality and profitability, and ecosystem management. Experiments on all biostimulants with one or more ancillary agricultural activities portend the capacity to improve nutrient capacity, tolerance to abiotic stress, and crop quality characteristics. Indicators and plant quality monitors through the action of some specific biostimulants, such as chitosan, laminarin, some PGPR, etc., could also prevent detrimental actions against some pathogenic microorganisms as well as improve their biocontrol effectiveness [52, 57–59, 67, 69, 70, 71].

12.4 Conclusion and Future Recommendations

This chapter provided detailed information on the application of cyanobacteria as biostimulants in boosting agriculture as well as detailed information on the modes of action involved in the application of cyanobacteria as biostimulant. Detailed information on the application of some other advanced techniques, such as metabolomics, proteomics, genomics, bioinformatics, and the integration of nanotechnology will go a long way in boosting agricultural production.

References

1. Abd El-Baky, H.H.; El-Baz, F.K.; El Baroty, G.S. Enhancing antioxidant availability in wheat grains from plants grown under seawater stress in response to microalgae extract treatments. *J. Sci. Food Agric.*, 90: 299–303, 2010.
2. Adetunji, C.O., Roli, O.I., Adetunji, J.B. Exopolysaccharides Derived from Beneficial Microorganisms: Antimicrobial, Food, and Health Benefits. In: Mishra, P., Mishra, R.R., Adetunji, C.O. (eds) Innovations in Food Technology. Springer, Singapore, 2020a. https://doi.org/10.1007/978-981-15-6121-4_10
3. C. Adetunji, O. Anani, Bio-fertilizer from Trichoderma: Boom for Agriculture Production and Management of Soil- and Root-Borne Plant Pathogens. Pages 245-256. Innovations in Food Technology: Current Perspectives and Future Goals. Editors: Pragya Mishra, Raghvendra Raman Mishra, Charles Oluwaseun Adetunji, 2020.
4. Adetunji C.O., Varma A. Biotechnological Application of *Trichoderma*: A Powerful Fungal Isolate with Diverse Potentials for the Attainment of Food Safety, Management of Pest and Diseases, Healthy Planet, and Sustainable Agriculture. In: Manoharachary C., Singh H.B., Varma A. (eds) Trichoderma: Agricultural Applications and Beyond. Soil Biology, vol 61. Springer, Cham, 2020. https://doi.org/10.1007/978-3-030-54758-5_12
5. Olaniyan Olugbemi Tope and Charles Oluwaseun Adetunji. Biochemical Role of Beneficial Microorganisms: An Overview on Recent Development in Environmental and Agro-Science. *Microbial rejuvenation of Polluted Environment*, 2021.
6. Charles Oluwaseun Adetunji, Muhammad Akram, Areeba Imtiaz, Ehis-Eriakha Chioma Bertha, Adrish Sohail, Oluwaseyi Paul Olaniyan, Rabia Zahid, Juliana Bunmi Adetunji, Goddidit Esiro Enoyoze, Neera Bhalla Sarin Genetically modified cassava; the last hope that could help to feed the world: recent advances. Edition title: Genetically Modified Crops - Current Status, Prospects and Challenges. Editors: Kishor, P. B. Kavi, Rajam, M. V., Pullaiah, T. (Eds.), 2020b. https://www.springer.com/gp/book/9789811559310.

7. Abel Inobeme, Jaison Jeevanandam, Charles Oluwaseun Adetunji, Osikemekha Anthony Anani, Devarajan Thangadurai, Saher Islam, Olubukola Monisola Oyawoye, Julius Kola Oloke, Mohammed Bello Yerima, Olugbemi T. Olaniyan. Ecorestoration of soil treated with biosurfactant during greenhouse and field trials. Green Sustainable Process for Chemical and Environmental Engineering and Science. Biosurfactants for the Bioremediation of Polluted Environments, 2021, Pages 89-105, 2021. https://doi.org/10.1016/B978-0-12-822696-4.00010-3
8. Osikemekha Anthony Anani, Jaison Jeevanandam, Charles Oluwaseun Adetunji, Abel Inobeme, Julius Kola Oloke, Mohammed Bello Yerima, Devarajan Thangadurai, Saher Islam, Olubukola Monisola Oyawoye, Olugbemi T.Olaniyan. Application of biosurfactant as a noninvasive stimulant to enhance the degradation activities of indigenous hydrocarbon degraders in the soil. Green Sustainable Process for Chemical and Environmental Engineering and Science. Biosurfactants for the Bioremediation of Polluted Environments, 2021, Pages 69-87, 2021. https://doi.org/10.1016/B978-0-12-822696-4.00019-X
9. Charles Oluwaseun Adetunji, Abel Inobeme, Osikemekha, Anthony Anani, Jaison Jeevanandam, Mohammed Bello Yerima, Devarajan Thangadurai, Saher Islam, Olubukola Monisola Oyawoye, Julius Kola Oloke, Olugbemi T.Olaniyan Isolation, screening, and characterization of biosurfactant-producing microorganism that can biodegrade heavily polluted soil using molecular techniques. Green Sustainable Process for Chemical and Environmental Engineering and Science. Biosurfactants for the Bioremediation of Polluted Environments, 2021, Pages 53-68, 2021a. https://doi.org/10.1016/B978-0-12-822696-4.00016-4
10. Jaison Jeevanandam, Charles Oluwaseun Adetunji, Jaya DivyaSelvam Osikemekha Anthony Anani Abel Inobeme, Saher Islam, Devarajan Thangadurai, Olugbemi T.Olaniyan. High industrial beneficial microorganisms for effective production of a high quantity of biosurfactant. In book: Green Sustainable Process for Chemical and Environmental Engineering and Science, 2021.
11. Charles Oluwaseun Adetunji, Jaison Jeevanandam, Osikemekha Anthony Anani, Abel Inobeme, Devarajan Thangadurai, Saher Islam, Olugbemi T.Olaniyan. Strain improvement methodology and genetic engineering that could lead to an increase in the production of biosurfactants, 2021b.
12. Charles Oluwaseun Adetunji, Jaison Jeevanandam, Abel Inobeme, Olugbemi T.Olaniyan, Osikemekha Anthony Anani, Devarajan Thangadurai, Saher Islam. Application of biosurfactant for the production of adjuvant and their synergetic effects when combined with different agro-pesticides, 2021c.
13. Charles Oluwaseun Adetunji, Olugbemi T Olaniyan, Osikemekha Anthony Anani, Abel Inobeme, Kingsley Eghonghon Ukhurebor, Ruth Ebunoluwa Bodunrinde, Juliana Bunmi Adetunji, Kshitij RB Singh, Vanya Nayak, Wadzani Dauda Palnam and Ravindra Pratap Singh. Bionanomaterials for

green bionanotechnology. In book: Bionanomaterials: Fundamentals and Biomedical Applications. Publisher: IOP Publishing, 2021d.

14. Adetunji C.O., Inobeme A., Olaniyan O.T., Olisaka F.N., Bodunrinde R.E., Ahamed M.I. Microbial Desalination. In: Inamuddin, Khan A. (eds) Sustainable Materials and Systems for Water Desalination. Advances in Science, Technology & Innovation (IEREK Interdisciplinary Series for Sustainable Development). Springer, Cham, 2021e. https://doi.org/10.1007/978-3-030-72873-1_13
15. Adetunji J.B., Adetunji C.O., Olaniyan O.T., African Walnuts: A Natural Depository of Nutritional and Bioactive Compounds Essential for Food and Nutritional Security in Africa. In: Babalola O.O. (eds) Food Security and Safety. Springer, Cham, 2021f. https://doi.org/10.1007/978-3-030-50672-8_19
16. Charles Oluwaseun Adetunji, Olugbemi T. Olaniyan, Osikemekha Anthony Anani, Frances. N. Olisaka, Abel Inobeme, Ruth Ebunoluwa Bodunrinde, Juliana Bunmi Adetunji, Kshitij RB Singh, Wadzani Dauda Palnam, Ravindra Pratap Singh. Current Scenario of Nanomaterials in the Environmental, Agricultural, and Biomedical Fields. Nanomaterials in Bionanotechnology.. In book: Nanomaterials in Bionanotechnology: Fundamentals and Applications. 1. 6. CRC Press, 2021g.
17. Charles Oluwaseun Adetunji, Robert J.Kremer, RasheedMakanjuola, Neera BhallaSarin. Application of molecular biotechnology to manage biotic stress affecting crop enhancement and sustainable agriculture. Advances in Agronomy, 2021h. https://www.sciencedirect.com/science/article/pii/S0065211321000304?dgcid=author
18. Charles Oluwaseun Adetunji, Olugbemi Tope Olaniyan, Juliana Bunmi Adetunji, Osarenkhoe O. Osemwegie, Benjamin Ewa Ubi. African Mushrooms as Functional Foods and Nutraceuticals. In book: Fermentation and Algal Biotechnologies for the Food, Beverage and Other Bioproduct Industries. 1st Edition. CRC Press. pp. 19, 2022.
19. Charles Oluwaseun Adetunji, Kingsley Eghonghon Ukhurebor, Olugbemi Tope Olaniyan, Benjamin Ewa Ubi, Julius Kola Oloke, Wadzani Palnam Dauda, Daniel Ingo Hefft. Recent Advances in Molecular Techniques for the Enhancement of Crop Production. In book: Agricultural Biotechnology, Biodiversity and Bioresources Conservation and Utilization. 1st Edition. CRC Press, Pages 20, 2022.
20. Aly, M.S., and Esawy, M.A. Evaluation of Spirulina platensis as bio stimulator for organic farming systems. J. Gen. Eng. Biotechnol., 6: 1–7, 2008.
21. Coppens, J.; Grunert, O.; Van Den Hende, S.; Vanhoutte, I.; Boon, N.; Haesaert, G.; De Gelder, L. The use of microalgae as a high-value organic slow-release fertiliser results in tomatoes with increased carotenoid and sugar levels. *J. Appl. Phycol.*, 28: 2367–2377, 2016.
22. Dubey A.K. and Rai, A.K. Application of algal biofertilizers (Aulosira fertilissima tenuis and Anabaena doliolum bhardawaja for sustained paddy cultivation in northern India. *Israel Journal of Plant Sciences*, 43(1): 41-51, 2013.

23. Elarroussia, H.; Elmernissia, N.; Benhimaa, R.; El Kadmiria, I.M.; Bendaou, N.; Smouni, A.; Wahbya, I. Microalgae polysaccharides a promising plant growth biostimulant. *J. Algal Biomass Utln.*, 7:55–63, 2016.
24. Garcia-Gonzalez, J. Sommerfeld, M. Biofertiliser and biostimulant properties of the microalga *Acutodesmus dimorphus*. *J. Appl. Phycol.*, 28: 1051–1061, 2016.
25. Gebser, B.; Pohnert, G. Synchronised Regulation of Di_erent Zwitterionic Metabolites in the Osmoadaption of Phytoplankton. *Mar. Drugs* 2013, 11, 2168–2182, 2013.
26. Karthikeyan, S.; Balasubramanian, R.; Iyer, C.S.P. Evaluation of the marine algae Ulva fasciata and Sargassum sp. for the biosorption of Cu (II) from aqueous solutions. *Bioresour. Technol.*, 98: 452–455, 2007.
27. Kowalczyk, K.; Zielony, T.; Gajewski, M. Effect of aminoplant and asahi on yield and quality of lettuce grown on rockwool. In Biostimulators in Modern Agriculture. Vegetable Crop; Da̒browski, Z.T., Ed.; Wies´ Jutra: Warsaw, Poland, pp. 35–43, 2008.
28. Michalak, I.; Chojnacka, K.; Dmytryk, A.; Wilk, R.; Gramza, M.; Rój, E. Evaluation of supercritical extracts of algae as biostimulants of plant growth in field trials. *Front. Plant Sci.*, 7: 1591, 2016.
29. Misra, S.; Kaushik, B.D. Growth promoting substances of cyanobacteria. In vitamins and their influence on rice plant. *Proc. Indian Natl. Sci. Acad. Part B Biol. Sci.*, 55: 295–300, 1989.
30. Muday, G.K.; Murphy, A.S. An emerging model of auxin transport regulation. *Plant Cell*, 14: 293–299, 2002.
31. Oliveira, J.; Mógor, G.; Mógor, A. Produtividade de beterraba em função da aplicação foliar de biofertilisante.Cadernos de Agroecologia, 8: 1–4, 2013.
32. Paudel, Y.P.; Pradhan, S.; Pant, B.; Prasad, B.N. Role of blue green algae in rice productivity. *Agric. Biol. J. N. Am.*3: 332–335, 2012.
33. Plaza, B.M., Gómez-Serrano, C., Acién-Fernández, F.G., Jimenez-Becker, S. Effect of microalgae hydrolysate foliar application (*Arthrospira platensis* and *Scenedesmus* sp.) on *Petunia x hybrida* growth. *J Appl Phycol* 30, 2359–2365, 2018.
34. Ronga D., Biazzi E., Parati, K., Carminati D., and Tava A., Microalgal Biostimulants and Biofertilizers in Crop Productions. *Agronomy 9: 192-214*, 2019.
35. Singh, V.P.; Trehan, K. Effect of extracellular products of Aulosira fertilissima on the growth of rice seedlings. *Plant Soil*, 38, 457–464, 1973.
36. Spolaore, P.; Joannis-Cassan, C.; Duran, E.; Isambert. A. Commercial applications of microalgae. *J. Biosci. Bioeng.*, 101, 87–96, 2006.
37. Stirk,W.A.; Ördög, V.; Novák, O.; Rolèík, J.; Strnad, M.; Bálint, P.; Staden, J. Auxin and cytokinin relationships in 24 microalgal strains. *J. Phycol.*, 49:459–467, 2013.

38. Stirk,W.A.; Bálint, P.; Tarkowská, D.; Novák, O.; Strnad, M.; Ördög, V.; van Staden, J. Hormone profiles in microalgae: Gibberellins and brassinosteroids. *Plant Physiol. Biochem.*, 70: 348–353, 2013.
39. Tate, J.J.; Gutierrez-Wing, M.T.; Rusch, K.A.; Benton, M.G. The effects of plant growth substances and mixed cultures on growth and metabolite production of green algae *Chlorella sp.*: A review. J. Plant Growth Regul., 32: 417–428, 2013.
40. Tsavkelova, E.A.; Klimova, S.Y.; Cherdyntseva, T.A.; Netrusov, A.I. Hormones and hormone-like substances of microorganisms: A review. *Appl. Biochem. Microbiol*, 42: 229–235, 2006.
41. Van Oosten, M.J.; Pepe, O.; De Pascale, S.; Silletti, S.; Maggio, A. The role of biostimulants and bioeffectors as alleviators of abiotic stress in crop plants. *Chem. Biol. Technol. Agric.*, 4(5): 1-12, 2017.
42. A. Vaishampayan, R.P. Sinha, D.P. Häder, T. Dey, A.K. Gupta, U. Bhan, A.L. Rao, Cyanobacterial biofertilizers in rice agriculture, *Bot. Rev.* 67, 453–516, 2001.
43. Ahmad, I., Pichtel, J., Hayat, S., Plant-Bacteria Interactions. Strategies and-Techniques to Promote Plant Growth. WILEY-VCH Verlag GmbH and Co., KGaA,Weinheim, 2008.
44. Arora, N.K., Khare, E., Maheshwari, D.K., Plant growth promotingrhizobacteria: constraints in bioformulation, commercialization, and futurestrategies. In: Maheshwari, D.K. (Ed.), Plant Growth and Health PromotingBacteria. Springer, Berlin/Heidelberg, pp. 97–116, 2011.
45. Babalola, O.O., Beneficial bacteria of agricultural importance. *Biotechnol. Lett.* 32, 1559–1570, 2010.
46. Berendsen, R.L., Pieterse, C.M., Bakker, P.A., The rhizosphere microbiome andplant health. *Trends Plant Sci.* 17, 1360–1385, 2012.
47. Berendsen, R.L., Pieterse, C.M., Bakker, P.A., The rhizosphere microbiome andplant health. Trends Plant Sci. 17, 1360–1385, 2012.
48. Berg, G., Grube, M., Schloter, M., Smalla, K., Unraveling the plant microbiome:looking back and future perspectives. *Front. Microbiol.* 5, 1–7, Article 148, 2014.
49. Bhattacharyya, P.N., Jha, D.K., Plant growth-promoting rhizobacteria (PGPR):emergence in agriculture. *World J. Microbiol. Biotechnol.* 28, 1327–1350, 2012.
50. Brahmaprakash, G.P., Sahu, P.K., Biofertilizers for Sustainability. J. *Indian Inst.Sci.* 92, 37–62, 2012.
51. Calvo, P., Nelson L. and Kloepper, J.W. Agricultural Uses of Plant Biostimulants. *Plant Soil*, 383, 3-41, 2014. https://doi.org/10.1007/s1110
52. D. Wang, S. Yang, F. Tang, H. Zhu, Symbiosis specificity in the legume – rhizobial mutualism, *Cell. Microbiol.* 14, 334–342, 2012.
53. Garcia-Gonzalez, J. and Sommerfeld, M. Biofertilizer and Biostimulant Properties of the Microalga Acutodesmus dimorphus. *Journal of Applied Phycology*, 28, 1051-1061, 2016. https://doi.org/10.1007/s10811-015-0625-2

54. Hussain, A. and Hasnain, S. Phytostimulation and Biofertilization in Wheat by Cyanobacteria. *Journal of Industrial Microbiology and Biotechnology*, 38, 85-92, 2011. https://doi.org/10.1007/s10295-010-0833-3
55. Irisarri P, Gonnet S, Monza J, Cyanobacteria in Uruguayan rice fields: diversity, nitrogen fixing ability and tolerance to herbicides and combined nitrogen. *J Biotechnol* 91:95–103, 2001.
56. Ismail, G., and Abo-Hamad, S., Effect of Different Anabaena variabilis (Kütz) Treatments on Some Growth Parameters and Physiological Aspects of Hordeum vulgare L. and Trigonella foenum-graecum L. Egyptian Journal of Botany 57(3): 507-516, 2017.
57. J.S. Singh, A. Kumar, A.N. Rai, D.P. Singh, Cyanobacteria: a precious bioresource in agriculture, ecosystem, and environmental sustainability, *Front. Microbiol.* 7, 1–19, 2016.
58. J.S. Singh, Cyanobacteria: a vital bio-agent in eco-restoration of degraded lands and sustainable agriculture, *Climate Change Environ. Sustain.* 2, 133–137, 2014.
59. J.S. Singh, V.C. Pandey, D.P. Singh, Efficient soil microorganisms: a new dimension for sustainable agriculture and environmental development, *Agric. Ecosyst. Environ.* 140, 339–353, 2011.
60. Layer, G., Reichelt, J., Jahn, D. and Heinz, D.W. Structure and Function of Enzymes in Heme Biosynthesis. Protein Science, 19, 1137-1161, 2010. https://doi.org/10.1002/pro.405
61. Lisboa, C.R., Pereira, A.M. and Costa, J.A.V. Biopeptides with Antioxidant Activity Extracted from the Biomass of Spirulina sp. LEB 18. African *Journal of Microbiology Research*, 10, 79-86, 2016. https://doi.org/10.5897/AJMR2015.7760
62. M.P. Maqubela, P.N.S. Mnkeni, O. Malam Issa, M.T. Pardo, L.P. D'Acqui, Nostoc cyanobacterial inoculation in South African soils enhances soil structure, fertility, and maize growth, *Plant Soil* 315, 79–92, 2009.
63. Machado, A.R., Graça, C.S., Assis, L.M. and Souza-Soares, L.A. An Approach on Characteristics and Potential Assessment of Antioxidant Extracts from Phenolic Microalgae Spirulina sp. LEB-18 and Chlorella pyrenoidosa. Revista de Ciências Agrárias, 40, 264-278, 2017. https://doi.org/10.19084/RCA16011
64. Mógor, A.F., Ördög, V., Lima, G.P.P., Molnár, A. and Mógor, G. Biostimulant Properties of Cyanobacterial Hydrolysate Related to Polyamines. *Journal of Applied Phycology*, 29, 453-46, 2017. https://doi.org/10.1007/s10811-017-1242-z
65. N. Karthikeyan, R. Prasanna, A. Sood, P. Jaiswal, S. Nayak, B.D. Kaushik, Physiological characterization and electron microscopic investigations of cyanobacteria associated with wheat rhizosphere, *Folia Microbiol.* 54, 43–51, 2009.

66. N. Renuka, A. Guldhe, R. Prasanna, P. Singh, F. Bux, Microalgae as multifunctional options in modern agriculture: current trends, prospects and challenges, *Biotechnol. Adv.* 36, 1255–1273, 2018.
67. N. Thajuddin, G. Subramanian, Cyanobacterial biodiversity and potential applications in biotechnology, *Curr. Sci.* 89, 50–57, 2005.
68. Philippot, L., Raaijmakers, J.M., Lemanceau, P., Putten, W.H.V.D., Going back to the roots: the microbial ecology of the rhizosphere. *Nat. Rev. Microbiol.* 11, 789–799, 2013.
69. R. Ortiz-Castro, H.A. Contreras-Cornejo, L. Macías-Rodríguez, J. López-Bucio, The role of microbial signals in plant growth and development, *Plant Signal. Behav.* 4, 1–12, 2009.
70. Riahi, H., Shariatmadari, Z., Khanjir, M., and Azizi, A., Effect of Anabaena vaginicola inoculum on growth of pot plants. In International Symposium on Growing Media, Composting and Substrate Analysis 1013 (pp. 507-513), 2013.
71. S. Mazhar, S. Hasnain, Screening of native plant growth promoting cyanobacteria and their impact on Triticum aestivum var. Uqab 2000 growth, *Afr. J. Agric. Res.* 6, 3988–3993, 2011.
72. Shaaban, M.M., El-Saady, A.M. and El-Sayed, A.E.B., Green Microalgae Water Extract and Micronutrients Foliar Application as Promoters to Nutrient Balance and Growth of Wheat Plants. *Journal of American Science*, 6, 631-636, 2010.
73. Singh, J.S., Kumar, A., Rai, A.N. and Singh, D.P. Cyanobacteria: A Precious Bio-Resource in Agriculture, Ecosystem, and Environmental Sustainability. *Frontiers in Microbiology*, 7, 1-19, 2016. https://doi.org/10.3389/fmicb.2016.00529
74. T. Rezanka, A. Palyzová, K. Sigler, Isolation and identification of siderophores produced by cyanobacteria, *Folia Microbiol.* 63, 569–579, 2018.
75. Vacheron, J., Desbrosse, G., Bouffaud, M.-L., Touraine, B., Moënne-Loccoz, Y.,Muller, D., Legendre, L., Wisniewski-Dye, P., rigent-combaret, C., Plantgrowth-promoting rhizobacteria and root system functioning. *Front. Plant Sci.* 4, 1–19, Article 356, 2013.
76. Z. Shariatmadari, H. Riahi, M. Seyed Hashtroudi, A. Ghassempour, Z. Aghashariatmadary, Plant growth promoting cyanobacteria and their distribution in terrestrial habitats of Iran, *Soil Sci. Plant Nutr.* 59, 535–547, 2013.
77. Cyanobacteria from paddy fields in Iran as a biofertilizer in rice plants. H Saadatnia, *H Riahi Plant Soil Environ.* 55 (5), 207-212, 2009.

13

Biofertilizer Derived from Cyanobacterial: Recent Advances

Charles Oluwaseun Adetunji[1]*, Muhammad Akram[2], Babatunde Oluwafemi Adetuyi[3], Fahad Said Khan[2], Abid Rashid[3], Hina Anwar[2], Rida Zainab[2], Mehwish Iqbal[4], Victoria Olaide Adenigba[5], Olugbemi T. Olaniyan[6], Inobeme Abel[7], Ruth Ebunoluwa Bodunrinde[8], Nyejirime Young Wike[9], Olorunsola Adeyomoye[5], Wadzani Dauda Palnam[10], Phebean Ononsen Ozolua[1], Arshad Farid[11], Shakira Ghazanfar[12], Chibuzor Victory Chukwu[13] and Mohammed Bello Yerima[14,15]

[1]Applied Microbiology, Biotechnology and Nanotechnology Laboratory, Department of Microbiology, Edo State University Uzairue, Iyamho, Edo State, Nigeria
[2]Department of Eastern Medicine, Government College University Faisalabad, Punjab, Pakistan
[3] Department of Natural Sciences, Faculty of Pure and Appled Sciences, Precious Cornerstone University, Ibadan, Nigeria
[4]Faculty of Medical Sciences, Government College University Faisalabad, Punjab, Pakistan
[5]Department of Science Laboratory Technology, Ladoke Akintola University of Technology, Ogbomosho, Oyo State, Nigeria
[6]Institute of Management Sciences, Dow University of Health Sciences, Karachi, Pakistan
[7]Laboratory for Reproductive Biology and Developmental Programming, Department of Physiology, Rhema University, Aba, Nigeria
[8]Department of Chemistry, Edo University Iyamho, Auchi, Edo State, Nigeria
[9]Department of Human Physiology, Faculty of Basic Medical Science, Rhema University, Aba, Nigeria
[10]Department of Agronomy, Federal University, Nigeria
[11]Gomal Center of Biochemistry and Biotechnology, Gomal University, Khyber Pakhtunkhwa, Pakistan
[12]National Institute for Genomics & Advanced Biotechnology (NIGAB), National Agricultural Research Centre (NARC), Islamabad, Pakistan

**Corresponding author*: adetunjicharles@gmail.com; adetunji.charles@edouniversity.edu.ng

Charles Oluwaseun Adetunji, Julius Kola Oloke, Naveen Dwivedi, Sabeela Beevi Ummalyma, Shubha Dwivedi, Daniel Ingo Hefft and Juliana Bunmi Adetunji (eds.) Next-Generation Algae: Volume I: Applications in Agriculture, Food and Environment, (295–320) © 2023 Scrivener Publishing LLC

[13]*Department of Physiology, University of Medical Sciences, Ondo City, Nigeria*
[14]*Department of Microbiology, Edo State University Uzairue, Iyamho, Edo State, Nigeria*
[15]*Department of Microbiology, Sokoto State University, Sokoto-Nigeria*

Abstract
The ongoing rise in human population and the depletion of energy resources pose a threat to the environment's demands, as well as to the sustainable production of food and energy. The most environmentally friendly method, or "green technology", has been used to prepare biofertilizer. The prokaryotic creature that has evolved and persisted the longest is the cyanobacterium. They are regarded as one of the earliest living organisms on Earth. New candidates for effectively converting radiative energy into chemical energy include cyanobacteria. As a byproduct, this biological system creates oxygen. The large-scale production of food, energy, biofertilizers, secondary metabolites, cosmetics, and pharmaceuticals can all be accomplished with cyanobacterial biomass. As a result, cyanobacteria are utilized in environmentally friendly, sustainable farming practices to produce biomass with a very high value and reduce CO_2 levels. This chapter cover the methods for cyanobacterial biofertilizers' mass production as well as their uses in industrial and agricultural settings.

***Keywords*:** Bioferterlizer, cyanobacterial, soil, environment, industry

13.1 Introduction

Apart from water, nitrogen is an important nutrient required for plant growth, and there is a need for adequate provision of this nutrient for plants to grow optimally. It is impossible for plants to utilize the elemental nitrogen available, which is why chemical-based or industrial fertilizer has been used to make nitrogen available to plants, although in a limited quantity. The chemical-based fertilizer which has been in constant use for plant yield improvement has limitations in terms of its adverse effect on soil texture and environmental pollution [90]. The need for other sources of fertilizer which can alleviate the challenges posed by chemical fertilizer is the reason why biological-based fertilizer has been sourced [1–18].

Biological-based fertilizer, also known as biofertilizer, refers to materials which contain microorganisms which are able to attach to the rhizosphere or the interior of plants with a consequential increase in plant yield by adequate provision of essential nutrients to the plants when applied to the seed, soil or even the surface of the plant [20]. The mode of operation of these microorganisms used as biofertilizer involves the release of nutrients to plants through nitrogen fixation, enhancement of plant growth and phosphorus solubilization [79].

Some of the microorganisms that have been employed in the formulation of biofertilizer are *Azospirillum*, *Azolla*, Mycorrhizae, *Azotobacter*, and Phosphorus-solubilizing microorganisms, *Sinorhizobium* and Blue-green algae [35]. These microorganisms, which can be used in the formulation of biofertilizer, enhance the physical, chemical and biological characteristics of the soil [30, 79]. Cyanobacteria are blue-green algae with wide distribution at global level [24]. Their metabolic activity is close to that of bacteria and they also possess chlorophyll with which they carry out photosynthesis like plants. They are autotrophic organisms and can be found in both freshwater and marine environment [81]. Cyanobacteria are known for their role in atmospheric nitrogen fixation [90]. They can exist as either free-living or in a symbiotic relationship with ferns, lichens and cycads, their ability for nitrogen fixation is close to that of the symbiotic relationship which existed between *Rhizobium* and legume [86].

Cyanobacteria contains three unique layers in their outermost section, these are the cell wall, plasma membrane and mucilaginous layer [77]. Different pigments are present in the cytoplasm of the blue-green algae, which are: carotenes, c-phycocyanin, c-phycoerythrin, chlorophyllys and xanthophylls [87]. Cyanobacteria have been found useful as microorganisms in the formulation of biofertilizer. Song *et al.* [64] described the role of cyanobacteria-based biofertilizer in maintaining and enhancing the fertility of the soil; the application of biofertilizer resulted in increased plant yield.

Cyanobacteria operates as a biofertilizer by: its ability to excrete some plant hormones, amino acids and vitamins [56], enhancing soil porosity and producing substances which are adhesives [60], resulting in increase in soil biomass after cell death and decay [60], reduction in salt concentration of the soil and inhibiting growth of weeds [60], enhancing water retention capacity via jelly-like structures [57], excretion of organic acids leading to increase in phosphate level [76], ability to absorb heavy metals [36], improvement in electrical conductivity and pH of soil [84] and enhancement of physicochemical properties of the soil [84]. Some cyanobacteria species that have been identified for their high potential in the fixing of atmospheric nitrogen are: *Anabaena variabilis, Nostoc linckia, Scytonema* sp., *Calothrix* sp., *Aulosira fertilisima* and *Tolypothrix* sp. [49].

Biofertilizer (also bio-compost) is a substance that contains to a lesser extent the life forms that, when applied to seeds, plant surfaces, or soil, colonize the root zone or inner part of the plant and promote growth, with the expansion of development or the availability of additives necessary for the plant. Biofertilizers include additives through the usual processes of obsessive nitrogen, dissolving phosphorous and stimulating plant growth through a group of growth-promoting substances. The microorganisms in biofertilizers restore the characteristic integration cycle of dirt, and create natural soil. By using biofertilizers you can develop tough plants, and increase repair and resistance to dirt. Biofertilizers are expected to reduce the use of compost and pesticides, but they are not yet ready to replace their use.

Because they fulfill multiple functions, the preferred logical term for these beneficial microscopic microorganisms is "plant root bacteria" (PGPR). Blue-green growth occurs with a common family of cyanobacteria, *Nostoc*, *Anabaena*, *Tolipotrix*, or *Aulosira*, which stabilizes nitrogen parameters and is used as an inoculation of rice crops growing in mountain and swamp conditions. For aquatic plants, azole brings nitrogen up to 60 kg/ha/season and also increases land with natural substances.

Cyanobacteria are known to be the first microorganisms to carry out photosynthesis on Earth with oxygen, and have added to the formation of oxygen in Earth's air over the past three billion years [64]. Other than that, they are known as blue-green growths and are considered important photosynthetic microorganisms found in a variety of conditions, including freshwater, sea, open soil and rocks. These microorganisms exist as single cells, colonies or fibers. Although cyanobacteria are small in nature, they can appear when present in conditions like in warm, nutrient rich (high phosphorous and nitrogen) environments.

Cyanobacteria have cell systems that make them versatile for natural change and effectively thrive in a fleshy variety at an impressive rate. However, their rapid dynamics are based on a variety of additive levels, biotic variables, environmental changes or abnormal climate change. There are several unique traits of cyanobacteria that have led experts and researchers to consider their options on a recent scale [46, 47].

13.2 Biological Fertilizers

Human practices can result in soil waste, degradation and salinization, which in turn reduce the amount of arable land and harm food security [40]. Cyanobacteria can be used as a common biological fertilizer and to increase efficiencies in a range of horticultural and environmental conditions [64]. In the horticulture department, phosphorous is the second most important component necessary for plant development and improvement. Cyanobacteria can expand soil phosphates by improving aggressive natural formation [75]. In total, there are about 1 billion soils, which, under the influence of salinity, can be restored using a mixture of physical and synthetic properties [26, 27].

Cyanobacterial compounds, such as *Nostoc* and *Anabaena*, have an impressive ability to treat dust pollution by optimizing barometric nitrogen and creating good networks and solutions [40]. Cyanobacteria can help reduce soil salinity, increase weeds and decay, as well as increase soil fertility [60].

Cyanobacteria have the capability to improve nitrogen and other nutrients in the soil [52]. For example, *Anabaena, Aulosira, Nostoc, Calothrix, Plectonema* and others are nitrogen-fixing animals. Where compost is used, it helps in maintaining long-term nutrition and nourishing the soil by improving air permeability, absorption of ammonium exchange, eliminating volatiles, modifying or converting insoluble phosphorous and clay into plant-containing structures [90]. *Spirulina platensis* has been found to increase the chlorophyll content in natural ingredients and carrot seeds found in ongoing reports [53, 54]. This discovery demonstrates the potential for large quantities of algae and cyanobacteria and their application in horticulture and tillage. Products from the *Spirulina* spp. family can be used to prepare crops, thus creating organic foods for natural growth and protection needs [91].

Biomass of cyanobacteria and soil microalgae has been documented to improve drainage and soil regeneration. In general, organic fertilizers, such as cyanobacteria, are a natural choice and have the potential to be used in agriculture due to the negligible contribution of capital and pollution, unlike fertilizers, which make water like urea [55]. Regardless of what may be expected, we should consider introducing cyanobacteria into the groundwater due to the expulsion of the large biomass of cyanobacteria (around one million tons) and the lack of therapeutic interventions in plants [19]. This may be the source of cyanotoxin contamination in the community. This type of use of cyanobacteria biomass has been tested as a common food in China [19].

Herbicides, particularly fixative nitrogen, cyanobacteria and weeds, are used for their ability to complete photosynthesis and nitrogen, in contrast to their natural environment as a natural nitrogen treasure [25]. Scientists have clearly shown that one of the best ways to fertilize nitrogen in paddy fields is to grow blueberries, which provide about 25–30 kg of nitrogen/ha/oge [73]. High-density microalgae correlation and coordination of photobioreactors (PBR) are designed, manufactured and applied to achieve high photosynthetic rates [58]. This device is designed to produce large quantities of oxygen and biomass, and it can also be used in waste production [50].

Azotobacter chroococcum, a dipotrophic, forest-dwelling bacterium, leads to beneficial effects on plants through a number of factors including biosynthesis of organic alteration, degradation of microorganisms in the roots, changes in nutritional supplements and increased fertility [39]. Cyanobacteria can be initiated through a photosynthesis process (where the growth of green cells awaits the light and the nature of the new biomass) [61, 62]. It is difficult to implement living organisms without the support of microorganisms and requires special developers or bioreactors due to the ability of cyanobacteria to deliver sugars that help the microscopic life to obtain food [63]. To ensure the presence of small amounts of microorganisms in the area surrounding the plant material, this system will allow them to grow green in the growth phase. This will reduce the number of dead algae cells and other wastes being delivered under unfavorable conditions. Second, better numbers of bacteria can be transported using only good centrifugation. After this treatment, green growth will reduce pathogens, and use of suitable material may be considered for further examination [65].

Moreover, the issue of land ownership is a global concern due to the decline in agricultural production, with the goal of land reclamation, waste disposal, salt collection and more. Plant water from the water can increase the use of weeds, which has a negative effect [67, 68]. There are numerous activities capable of attacking these herbaceous plants [48], food and urea. After all that has happened, the use of herbaceous plants has been shown to improve the quality of the product, so it can encourage development so that it does not die. From now on, depending on the results of this study, cyanobacteria can be incorporated as biofertilizers using or promoting a low-fat diet. In any case, the results obtained should be confirmed by conducting field tests (including soil types and soil systems) before spreading to the affected species [72].

Cyanobacteria also have a new power to contribute to the well-being and state of the earth and its environment. Cyanobacteria are found in many soils, growing in the soil and outdoors. Most paddy soils have characteristic cyanobacteria that allow for nitrogen uptake [74]. Combined cyanobacteria oil can be obtained for free distribution or ammonia ($NH4^+$) in the same process [78]. The amino acids arginine, asparagine and glutamine have been described for liver function as nitrogen derivatives. Organics and nitrites are important sources, which ultimately reduce the salt content [80]. Most cyanobacteria are prepared for the use of dinitrogen (N_2) as a nitrogen fertilizer, which is called nitrogen. Like many other acidic compounds, nitrogen and energetic cyanobacteria are absorbed by the lower body, a substance known as nitrogenase [82]. Nitrogenase lowers subatomic acid and attracts nearby salts to hydrogen [85]. So, it is important to highlight the effects of nitrogen, cyanobacteria and agriculture to improve the environment. Numerous studies have been conducted on the use of dried cyanobacteria as a method to fertilize the soil, and products including cyanobacteria and rice were first considered in the 1950s in Japan. The term "algalization" has been incorporated into the treatment of cyanobacterial animals for vaccination, but compilation research is important. In the quality of all inspected products, the grain size increased by 15–20% in the experiment. It is recommended to produce cyanobacteria as a result of self-regulation by yourself if the administration of the computer system is carried out 3-4 times [88].

The most common form of cyanobacteria in 1985 were four groups of microorganisms known as Chroococcales, Nostocales, Oscillatoriales, Stigonematales; and the phyla were Chroococcales, Gloeobacterales, and Pleurocapsales. Cyanobacteria are associated with the growing season. They are important in determining the course of progress and environmental change around the world. During the Proterozoic period or the early Cambodian period, cyanobacteria arose in some eukaryotic plants, which caused endosymbiosis in early eukaryotes. Finally, they can reduce the natural nitrogen, so a lot of plants, such as rice and beans, grow in summer. The outer layer of cyanobacteria generally consists of three types of layers, such as the adhesive layer, the cell membrane, and the inner plasma film. The cytoplasm in porcine lamellae does not constitute a gram. The varieties contain carbohydrates, carotene, xanthophylls, c-phycoerythrin, and c-phycocyanin, all of which are found in green plants [41].

Cyanobacteria contain a number of natural compounds that can charge a wide variety of energies. The body is thought to include the absorption of light-absorbing compounds such as phycobilisomes, body polyphosphates [69], cyanophosphate granules [34], polyhydroxyalkanoate (PHA) deposits, Carbohydrate/polypropylene body [70], adipose [59], thylakoid tissue [28], dressings containing DNA [71] and ribose [32]. The current population will continue to grow to 9.7 billion in 30 years. The majority of the population is Indian. Home supplements are relevant and do not constitute healthy, non-polluting foods. The World Health Organization will shut down half of the industrialized world by 2029. Evidence from the "green revolution" is also being used to increase agricultural capacity, reduce opportunities for fertilizer application, and respect human life as a whole. In this context, analysts have used "green panels" to generate the natural and destructive properties of microorganisms. In some aspects, green innovation uses cyanobacteria to improve crop quality and soil fertility [65].

Cyanobacteria can break down various toxins and perform various functions in biological systems to support soil stability [31]. Cyanobacteria are microorganisms of agricultural growth using a hypothetical model showing the effects of cyanobacteria on aquaculture and its environment [91]. Diazotroph is a blue bacterium, it is both effective and less effective, due to the lifespan of the natural biofertilizer in it. It can regulate plant nitrogen uptake, increase natural gas distribution, prevent water loss, and even provide nourishment [64].

The most effective nitrogen-adding cyanobacteria available for growing plant parts are *Nostoc linckia, Anabaena variabilis, Aulosira Calothrix* sp., *Tolypothrix* sp. and *Scytonema* sp. [49]. *Anabaena* and *Nostoc* do the work outside the ground. *Anabaena* can repair 60 kg/ha/nitrogen, and increase soil and environmental problems [21]. Cyanobacteria do not need to be a host for the development, growth and creation of normal organisms. The *Azolla-Anabaena* alliance is the cause of the effective nitrogen fixation reactions and the addition of additives in the field. They have demonstrated cellular apoptotic degeneration and mutant mechanisms that produce results that provide a variety of benefits [44]. This biological enrichment has been suggested for use in tortillas, oatmeal, tomatoes, radishes, carrots, sugar, corn, dill, and lettuce [73]. Melody *et al.* [73], observed that cyanobacteria play an important role in the maintenance and development of fertile fruit, which is produced as a specific biofertilizer.

Important findings in the development of functional horticulture practices using cyanobacteria in a "blue growth" strategy include: a) making soil permeable and producing a sticky substance; b) secreting plant hormones (auxin, gibberellin, etc.), nutrients, amino acids [20]; c) increasing groundwater restrictions through the watch structure; d) increasing soil biomass after completion and degradation [60]; e) reducing salt in the soil [60]; f) range development administration [64]; g) using soil phosphates obtained by producing natural acids [75]; h) employing the efficacy of large minerals on microbial surfaces (biological treatment) [36].

Cyanobacteria include the geochemical structures of carbon, nitrogen, and oxygen [75]. They can enter wetlands and can affect health status, core strength and crop yield [51]. During expansion, there are no abnormal changes in atomic levels required for high UV resistance (280–400 nm), percussion, explosive temperatures, and high salinity conditions [19, 66]. Each of these cases is beneficial and provides insurance to a variety of competitors and slow eaters [42], the different metabolites and their classifications. These metabolites with cyanobacteria are important for cultivation and financing [91].

Biofertilizers are characterized as substances that contain microorganisms that irritate plants that promote expansion by developing flexible or necessary attachments, and increase expansion in the context of objective harvesting when applied to seeds and cultivated surfaces of the Earth [20]. The standard treatments for biofertilizers are microorganism-based inoculants for green growth with parasites alone or mixed with boosting plant-based nutritional supplements. The biological composition of microorganisms and parasites is good for improving the composition and quality of organics and phosphates and for cultivation [30, 76]. The use of biofertilizers in composting formulations provides economic and environmental benefits in terms of soil nutrient. Biofertilizers are nitrogen-fixers and phosphorous solubilizers, and have a host of advanced materials for developing motile plants. Biofertilizers can be used to reduce the use of chemical compounds and pesticides.

Biodegradable biofertilizers are used to improve rural practices [24]. Most biofertilizers contain microorganisms that stimulate nitrogen synthesis, positive degradation and plant growth. The beneficial bio-harvesting fertilizers are *Azotobacter*, *Azospirillum*, Celestial algae, *Azolla*, P-digestive microorganisms, Mycorrhizae, and *Sinorhizobium* [35]. Various microorganisms are used as biofertilizers. Some of them are found in nitrogen killers, for example, *Azotobacter*, *Malva*, *Rhizobium*, and *Azospirillum*. *Rhizobium* is used to set limits for nitrogen release in its formula. *Azotobacter* is used as a biofertilizer to boost a variety of vegetables such as mustard, corn, wheat, cotton, and more. *Azospirillum* is used in rice, sorghum, seed, maize and wheat. Blue and green plants, such as *Nostoc, Tolypothrix, Anabaena,* and *Aulosira*, regulate airborne nitrogen and improve pollutant intake (Table 13.1).

Cyanobacteria have an important role to play in maintaining and improving soil fertility after promoting grain growth and acting as a biofertilizer [64]. The visible effects of this green growth include: 1) producing follicles with wider structure and the formation of sticky particles; 2) promoting drug development such as hormones (auxin, gibberellin), nutrients, and amino acids [56]; 3) increasing limited water retention by standard plugs [57]; 4) increasing biomass in the soil after death and the lack of animals; 5) decreased salinity; 6) preventing plant improvement through acidification [89]; 7) increasing soil salinity [75].

The benefits of cyanobacteria vaccine have been carefully studied in many different products, such as wheat, oats, tomatoes, radishes, cotton, sugarcane, corn, corn porridge and lettuce [73]. Microalgae (along with BGA or cyanobacteria) are a group of phototrophic microorganisms of large, clustered, and multilayered species collected from most

Table 13.1 Important microorganisms constituting biofertilizer and their application in a variety of crops.

Microorganisms use as biofertilizer	Nutrient fixed (Kg/ha/year)	Beneficiary crops
Rhizobium	50–300 kg N/ha	Ground vegetables, soybeans, red gram, green gram, black gram, lentils, beans, Bengal gram, vegetables
Azotobacter	0.026–20 kg N/ha	Cucumber, vegetables, roses, crops, rice, corn, maize, ragi, jowar, mustard, safflower, Niger fruits, tobacco, fruits, spices, food, rain flowers
Azospirillum	10–20 kg N/ha	Sugarcane, vegetables, corn, wheat, food, sour seeds, fruits, flowers
Blue-green algae	25 kg N/ha	Rice, banana
Azolla	900 kg N/ha	Rice
Phosphate-solubilizing bacteria and fungi	Solubilizes about 50–60% of them fixed	All phosphorous (unspecified) in the ground

columnar trees. Almost all microalgae are found in water, but when there is no water, seawater, or salt, they can be found in a variety of similar situations, such as in groundwater [21].

Cyanobacteria, or green and blue growths (BGA), are a group of microorganisms that can improve the nitrogen environment. BGA can be adapted to a wide range of soils and conditions that make expansion difficult. Nitrogen absolute nitrogen-free *Nostoc linckia, Anabaena variabilis, Calothrix* sp., *Tolypothrix* sp., and *Scytonema* sp. are adapted to different climatic zones and used in rice production [49]. After irrigation, nitrogen is a constant constraint to plant growth in many areas, and the shortage of these products is compounded by fertilizers [43]. Cyanobacteria play an important role in promoting and regulating soil fertility, as well as in grain growth and as a biofertilizer [64]. Blue-green algae (BGA) are photosynthetic and nitrogen-fixing. They are found in the cargo in India. They are supplemented with nutrient B12 absorption, improved chemical expression, improved soil and water retention in air, and biomass in the absence of post-life.

Azolla is a small aquatic fern found in shallow waters and farm dams. It has good contact with BGA and helps to distinguish cereals or grains from soil by double compaction or green manure. Because they contain chloroplasts, they manufacture their food via photosynthesis, therefore they can be independent. The nitrogenous, green, and blue growths treated with heterocystosis consist of two types of cells: heterocysts, those that respond to alkaline compounds, and tumor cells that exhibit both photosynthesis and reproduction. Cyanobacteria are found to be very diverse and biodegradable [31]. These animals will be guests for creating natural products, as they were directly developed in advance. Malliga *et al.* [44] reported that when *Anabaena azollae* is used as biofertilizer, the phenolic compounds that begin to enrich the structure of life show lignin degradation and delivery. This report uses cyanobacteria as a vehicle for bacterial biofuels by supporting chemical studies of the degradation potential of lignin and the use of lignocellulose disperser as superior and intermediate compounds for bacterial biofuels. There are BGAs, such as *Anabaena* and *Nostoc*, that live on land and in rocks. They have a stable nitrogen barometer volume of up to 20–25 kg/ha. The blue-green growth of the genus *Nostoc*, *Anabaena*, *Tolypothrix* and *Aulosira* regulates the nitrogenous climate and is used as a pollinator for rice plants cultivated in mountain and marshy layers. *Anabaena azollae* provide up to 60 kg/ha/season of nitrogen in aquatic vegetables and naturally increases soil [21].

In addition, animal species with cyanobacteria have cooperative relationships with plants (green growth, that is, diatoms; parasites, that is, lichens; parasites, that is, liver, growth, and annihilation; pteridophytes, namely *Azolla*; gymnosperms, i.e., rotations); Angiosperms (e.g., Gunnera), organisms (marine napkins, achiruoid worms), non-photosynthetic protists (consistent with the Glaucophyta group), microorganisms, and normal white polar hairs. The water plant *Azolla Anabaena azollae*, containing cyanobacteria N2-fixing, is a bacterial biofuel. Dry green growth contains high levels of macronutrients, essential trace elements and amino acids [29]. It can be used in wastewater and hard water and is partially disposable in synthetic fertilizers to prevent environmental pollution. Kulk [37] and Adam [22] showed that the development of *Nostoc muscorum* could add detail to nitrogen, as well as train nitrate reductase on plant-associated cyanobacteria or amino acids and peptides formed. Such filters as cyanobacteria increase the range of rates that increase productive plant growth. In addition to being a source of N2, BGA are associated with a variety of factors; for example, accumulation of alkaline biomass as a normal problem;

production of growth materials that stimulate rice growth; resistance to insecticides and pesticides, and also helps restore salty and soluble soils.

13.3 Biofuel Production Technology

Often the acquisition of any innovation depends on the ability to generate funds technologically. Indian officials announced that several algae scientists created a delicate and sensitive algae plant. This innovation has the potential to drive more than the supply of alkaline biofertilizers. Typically, four alkali-forming strategies are considered: a) the rod or reservoir strategy, b) the pit technique, c) the field technique and d) the nursery leaf formation technique. The previous two strategies relate to individual farm owners, while the last two were created from a business perspective.

Some areas related to biofuels include:

- The first and most important purpose of using biofuels is that they are beneficial and unlike mixed fertilizers do not harm the soil.
- They are cheap.
- The use of soil and dirt repairs can be prolonged.
- Although they do not show quick results, the results after some time are amazing.
- Microorganisms convert complex natural substances into essential compounds, allowing plants to receive supplements without delay. These compounds dissolve the barometric nitrogen and allow direct access to the plant.
- Biofertilizers increase the pulse of the roots due to growth hormone.
- Help to increase yield by 10-25%.

Biofertilizers have different advantages. In addition to obtaining nutritional supplements, which will now be recognized as waste, typical biofertilizers also provide plant-developed ingredients, some of which are effective in treating soil and using mixes with success. By controlling diseases of treated soil and improving the well-being of dirt and soil, this life form aids in their elimination, in addition to the successful use of compounds and the production of better crops. Cyanobacteria play an amazing role in the fields of biofertilizers, bioproduction, human food, animal feed, sugars, biochemistry, medicines and natural alternatives, among others. Cyanobacteria provide economical nitrogen to plants and increase their yields by creating fertile and profitable soil. The BGA biofertilizers in the area known as "algalization" help create an environmentally friendly agricultural environment that ensures the financial viability of field development while minimizing the input intensity. The bacterial compost facilitates soil adaptation, including natural problems, removes advanced materials, and improves the synthetic physical properties of soil and insoluble phosphates. These innovations can easily be adopted by educators in order to improve them.

13.4 Significant of Biofertilizers

In view of all the side effects from extended utilizations of conventional fertilizers, organic agriculture has appeared as a powerful substitutive region in terms of the increasing requirements for a healthy food supply, allowing a fast growth and survival, and fears related to pollution of environment. Though the utilization of conventional fertilizers is inevitable to address the increasing requirements of foodstuff in the world, there is the possibility of choosing some crops and niche regions where crops can be well-grown using natural farming [18, 19]. A biofertilizer contains living microorganisms which, when applied to plants, seeds, or soil, colonize the rhizosphere or the interior of the plant and promotes plant growth by increasing the supply of nutrients to the host plant [20–22]. Biofertilizers are lucrative, eco-friendly, and their extended application makes the fertility of soil considerably better [23, 24]. It has been stated that the utilization of biofertilizers increase crop yield about 10 to 40% by enhancing constituents of vitamins, EAA, proteins and NF. Utilization of microorganisms as biofertilizers is regarded on some level to be a substitute for chemical fertilizers in the agriculture sector because of their widespread possibility in increasing the production of crop and food safety. It has been found that a number of microorganisms, including PGPB, cyanobacteria, fungi, etc., have demonstrated activities just like biofertilizer in the agriculture sector. Comprehensive studies on biofertilizers have shown their ability to supply the necessary nutrients to crops in adequate quantities to result in the improvement in crop yields. Biofertilizer plays a significant role in fixation of nitrogen, sequestration of iron, and solubilization of phosphate; thereby building complex natural molecules which are accessible for use by plants. Nitrogen is amongst the most fundamental nutrients for plant productivity and development. Though 78% of N2 is found in the air, it continues to be inaccessible for use by plants. In turn, in order to utilize the N2 of air, the nitrogen has to be changed to NH_3 which can be simply absorbed by plants all the way through the procedure of BNF. Nitrogen fixation (NF) organisms are classified as non-symbiotic and symbiotic. Symbiotic organisms comprise the components of *Rhizobiaceae* which creates a symbiotic association with the plants of the family *Leguminosae*. Contrary to this, non-symbiotic organisms consist of freely existing and endophytic types of microbes such as *Azospirillum, Cyanobacteria, Azotobacter*, etc.

13.5 Relevance of Cyanobacteria

Cyanobacteria fit into eight diverse families and are photographic in character. They are the most plentiful group of creatures on Earth. Cyanobacteria are autotrophic and present in various environments, particularly in fresh and marine water. Seawater is the richest nutrient source for the development of cyanobacteria [7–10, 81]. Currently, cyanobacteria are accessible as dietary supplements on the market in a range of forms like tablets, capsules, and liquids [91]. At the present time, the most common strain of cyanobacteria utilized for an individual's nutrition value is *Spirulina*, because of its enormous content of protein and considerable nutritive value [11–13, 84]. It consists of more than 60% protein, a wide spectrum of therapeutic and prophylactic nutrients, including minerals, beta-carotene, vitamin B1, vitamin B2, and other B complex vitamins, trace elements and lots of astonishing bioactive compounds [18, 19]. These bacteria are the source of vitamins and minerals which are beneficial for teeth, bones and blood. They contain vitamin A, thiamine, riboflavin, niacin, B6, vitamin E, vitamin H, pantothenic acid, folacin, inositol, and an abundance of cyanocobalamine [14–17].

13.6 Cyanobacteria as Biofertilizer

Cyanobacteria promote plant growth by producing IAA, auxin, gibberellic acid, and fix about 20 to 30 kg of N/ha in immersed fields of rice as they are plentiful in paddy fields [92]. N2 is amongst the chief nutrients needed in huge amounts for rice production. The most important sources for N2 are nitrogen from soil and organic nitrogen fixation by related microorganisms. Youssef and Ali [93] stated that three blue-green algae, *Nostoc calcicola, Anabaena oryzae* and *Spirulina* sp., decreased the quantity of galls and egg collections created by the root-knot nematode *Meloidogyne incognita* contaminating cowpea and enhancing the growth criteria of plant. They are little and commonly unicellular and frequently produced in big colonies. Cyanobacteria are made up of a wide variety of bacteria with diverse forms and sizes. They can include 150 biological genera that have been recognized so far. The taxonomy of cyanobacteria was suggested in 1985, in which the four classes of bacteria are known as Nostocales, Chroococcales, Stigonematales, and Oscillatoriales; and their phyla are Gloeobacterales, Chroococcales, and Pleurocapsales. They have the ability to fix nitrogen in atmosphere so that they can be utilized as a biofertilizer for the significantly cost-effective development of crops, for instance beans and rice. Cyanobacteria might be utilized as prospective food supplements and offer nutritional, healing and helpful values. There are a number of features which make cyanobacteria an appealing alternative for long-term food production: worldwide distribution, nutrient-rich content, require a small amount of water for development and growth (seawater can also be consumed), require smaller quantities of land which may be unproductive and inappropriate for other crops, simply digestible, product reliability over a considerable temperature and pH range, etc. [41]. These bacteria can also generate secondary metabolites which are the resource of bioactive molecules, including antitumor (13%), cytotoxic (41%), antibacterial (12%), antiviral (4%), and other complexes (18%), for instance, antimalarials, antifungals, MDR reversers, insecticides, herbicides, algaecides, and immunosuppressive agents. All of the above essential features make cyanobacteria a promising and attractive substitutive source for long-lasting production of food. Research studies have also accepted the outcome of cyanobacteria as a fractional alternative to conventional fertilizers. De cano [18] disscovered that injection of soil with trichostrongylus tenuis and fertilization with urea enhanced length of stem and growth of rice. It was established that fertilization of soil with urea and inoculation of soil with *Nostoc muscorum* and trichostrongylus tenuis enhanced carbon content and desiccated weight and shoot length of rice in comparison with control. Zaccaro *et al.* [3–6, 20] stated that biofertilization with algae (blue green) and *Azolla*, and fertilization by means of urea, considerably enhanced content of chlorophyll in plant and yield of rice as well. Furthermore, Pereira *et al.* [1, 2, 20] reported that biofertilization with a blend of nitrogen-fixing cyanobacteria (*Nostoc linckia, Nostoc commune, Anabaena iyengarii* var. tenuis and *Nostoc* sp.) minimized the utilization of nitrogen fertilizer by 50% to obtain the same output of grain and rice quality contrasted with the complete dose of conventional fertilizer. The contents of tetraterpenoids, chlorophyll a, overall N, whole protein, exopolysaccharide and the activity of nitrogenase were higher in *Nostoc entophytum* in comparison with *O. angustissima*; the opposite was accomplished for the contents of entire carbohydrate and overall phosphorus (P). Moreover, *N. entophytum* demonstrated advanced levels of IAA, abscisic acid and cytokinin (nanogram/mg desiccated weight) in comparison with *O. angustissima*, while the latter consisted of GA in higher amounts (nanogram/mg desiccated weight) in comparison with the previous one. Filtrates of cyanobacteria can encourage production of gibberellic acid and cytokinin and auxin in sprouted pea. Ördög and Pulz [94] demonstrated that suspension of cyanobacteria contains a particular set of organically active complexes counting regulators of plant growth, which can reduce aging, transpiration and increase the substance of leaf chlorophyll. The entire fertilization treatment enhanced overall P and N contents of pea seedlings throughout the stage of germination. The maximum amplifications were created by suspension of cyanobacterial in +100% chemical fertilizer, followed by suspension of cyanobacterial in +50% chemical fertilizer. The increase in the total nitrogen content could be because of the nitrogen fixation and nitrate reductase activities of cyanobacteria as stated above, or because of the uptake of ammonium and uptake of the peptides and amino acids created by cyanobacteria [22]. Inoculation with the two combinations of the two species of cyanobacteria were more efficient and gave a greater yield of pea than the solitary inoculation of cyanobacteria, as confirmed by Sinha and Kumar [91–94]. The investigated species of cyanobacteria (*O. angustissima* and *N. entophytum*) consists of a number of biologically dynamic compounds like plant hormones, N, and fractions of carbohydrate [95–98]. The inoculation with a blend of the two species of cyanobacteria was quite efficient compared to the distinct application of cyanobacteria. The application of cyanobacterial biofertilizers is advised as it can save almost about 50% of the suggested dose of chemical fertilizers. "Green Revolution" practices are also effective in amplifying the efficiency of agriculture and minimizing the hazards of chemical-based fertilizers on the health of humans in addition to the environment. Green technology provides numerous ways to utilize cyanobacteria to enhance the productivity of crops and fertility of soil. Cyanobacteria can lessen an extensive variety of pollutants and perform diverse roles in the ecosystem of soil to maintain its fertility [31, 99]. Cyanobacteria are promising microbes for long-lasting development in agriculture [91, 100]. Diazotrophs are cyanobacteria that are helpful for the production of environment-friendly biofertilizers which are easily obtainable and less expensive [101–103]. They can manage insufficient nitrogen in plants, the aeration of soil, the holding capacity of water, and add cyanocobalamine [43, 64, 104]. The most competent NF cyanobacteria are *Anabaena variabilis*, *Tolypothrix* sp., *Nostoc linckia*, *Aulosira fertilisima*, *Scytonema* sp. and *Calothrix* sp., which are found in the agriculture area of rice crop [49, 105]. Use for these biofertilizers have been accounted for in oats, barley, lettuce, tomato, cotton, radish, sugarcane, maize and chilies [73]. Song *et al.* [64, 106] reported that cyanobacteria in organic biofertilizer play a principal role in the preservation and rise in the fertility of soil, and therefore the yield.

13.7 Conclusion

Blue-green algae bacteria have the potential to grow as a biofertilizer. They can use carbon dioxide, water, and additives to divert sunlight from biomass. The effective use of cyanobacteria has been considered in agricultural practice to reduce unwanted climate changes by reducing carbon dioxide. Public research shows that biomass based on blue-green algae can be used to improve food properties, physical properties of soil, and control infectious soil diseases, including natural problems, adaptive expansion materials, insoluble phosphates used as nutrients, and other substances used in pharmacies. As a result, biofertilizers made of cyanobacteria are vulnerable to conditions and conditions.

References

1. Adetunji, C.O., Roli, O.I., Adetunji, J.B. Exopolysaccharides Derived from Beneficial Microorganisms: Antimicrobial, Food, and Health Benefits. In: Mishra, P., Mishra, R.R., Adetunji, C.O. (eds) *Innovations in Food Technology*. Springer, Singapore, 2020. https://doi.org/10.1007/978-981-15-6121-4_10
2. Adetunji, C.O., Anani, O.O. Bio-fertilizer from Trichoderma: Boom for Agriculture Production and Management of Soil- and Root-Borne Plant Pathogens. Pages 245-256. *Innovations in Food Technology: Current Perspectives and Future Goals*. Editors: Pragya Mishra, Raghvendra Raman Mishra, Charles Oluwaseun Adetunji, 2020.
3. Adetunji C.O., Varma A. Biotechnological Application of *Trichoderma*: A Powerful Fungal Isolate with Diverse Potentials for the Attainment of Food Safety, Management of Pest and Diseases, Healthy Planet, and Sustainable Agriculture. In: Manoharachary C., Singh H.B., Varma A. (eds) *Trichoderma: Agricultural Applications and Beyond. Soil Biology*, vol 61. Springer, Cham, 2020. https://doi.org/10.1007/978-3-030-54758-5_12
4. Olaniyan O.T., and Adetunji, C.O. Biochemical Role of Beneficial Microorganisms: An Overview on Recent Development in Environmental and Agro-Science. *Microbial Rejuvenation of Polluted Environment*, 2021.
5. Adetunji, C.O., Akram, M., Imtiaz, A., Bertha, B.E., Sohail, A., Olaniyan, O.P., Zahid, R., Adetunji, J.B., Enoyoze, G.E., Sarin, N.B. Genetically modified

cassava; the last hope that could help to feed the world: recent advances. Edition title: Genetically Modified Crops - Current Status, Prospects and Challenges. Editors: Kishor, P. B. Kavi, Rajam, M. V., Pullaiah, T. (Eds.), 2020b. https://www.springer.com/gp/book/9789811559310.

6. Inobeme,A., Jeevanandam, J., Adetunji, C.O., Anani, O.A., Thangadurai, D., Islam, S., Oyawoye, O.A., Oloke, J.K., Yerima, M.B., Olaniyan, O.T. Ecorestoration of soil treated with biosurfactant during greenhouse and field trials. Green Sustainable Process for Chemical and Environmental Engineering and Science. Biosurfactants for the Bioremediation of Polluted Environments, pp. 89-105, 2021. https://doi.org/10.1016/B978-0-12-822696-4.00010-3
7. Anani, O.A., Jeevanandam, J., Adetunji, C.O., Inobeme, A., Oloke, J.K., Yerima, M.B., Thangadurai, D., Islam, S., Oyawoye, O.M., Olaniyan, O.T. Application of biosurfactant as a noninvasive stimulant to enhance the degradation activities of indigenous hydrocarbon degraders in the soil. Green Sustainable Process for Chemical and Environmental Engineering and Science. *Biosurfactants for the Bioremediation of Polluted Environments*, 2021, Pages 69-87, 2021. https://doi.org/10.1016/B978-0-12-822696-4.00019-X
8. Adetunji, C.O., Inobeme, A., Osikemekha, A.A., Jeevanandam, J., Yerima, M.B., Thangadurai, D., Islam, S., Oyawoye, O.M., Oloke, J.K., Olaniyan, O.T. Isolation, screening, and characterization of biosurfactant-producing microorganism that can biodegrade heavily polluted soil using molecular techniques. Green Sustainable Process for Chemical and Environmental Engineering and Science. *Biosurfactants for the Bioremediation of Polluted Environments*, 2021, Pages 53-68, 2021a. https://doi.org/10.1016/B978-0-12-822696-4.00016-4
9. Jeevanandam, J., Adetunji, C.O., Selvam, J.D., Anani, O.A., Inobeme, A., Islam, S., Thangadurai, D., Olaniyan, O.T. High industrial beneficial microorganisms for effective production of a high quantity of biosurfactant. In book: *Green Sustainable Process for Chemical and Environmental Engineering and Science*, 2021.
10. Adetunji, C.O., Jeevanandam, J., Anani, O.A., Inobeme, A., Thangadurai, D., Islam, S., Olaniyan, O.T. Strain improvement methodology and genetic engineering that could lead to an increase in the production of biosurfactants, 2021b.
11. Adetunji, C.O., Jeevanandam, J., Inobeme, A., Olaniyan, O.T., Anani, O.A., Thangadurai, D., Islam, S. Application of biosurfactant for the production of adjuvant and their synergetic effects when combined with different agro-pesticides, 2021c.
12. Adetunji, C.O., Olaniyan, O.T., Anani, O.A., Inobeme, A., Ukhurebor, K.E., Bodunrinde, K.E., Adetunji, J.B., Singh, K.R., Nayak, V., Palnam, W.D., and Singh, R.P. Bionanomaterials for green bionanotechnology. In book: *Bionanomaterials: Fundamentals and Biomedical Applications*. IOP Publishing, 2021.

13. Adetunji C.O., Inobeme A., Olaniyan O.T., Olisaka F.N., Bodunrinde R.E., Ahamed M.I. Microbial Desalination. In: Inamuddin, Khan A. (eds) Sustainable Materials and Systems for Water Desalination. *Advances in Science, Technology & Innovation (IEREK Interdisciplinary Series for Sustainable Development).* Springer, Cham, 2021e. https://doi.org/10.1007/978-3-030-72873-1_13
14. Adetunji J.B., Adetunji C.O., Olaniyan O.T. African Walnuts: A Natural Depository of Nutritional and Bioactive Compounds Essential for Food and Nutritional Security in Africa. In: Babalola O.O. (eds) *Food Security and Safety*. Springer, Cham, 2021f. https://doi.org/10.1007/978-3-030-50672-8_19
15. Adetunji, C.O., Olaniyan, O.T., Anani, O.A., Olisaka, F.N., Inobeme, A., Bodunrinde, R.E., Juliana Adetunji, J.B., Singh, K.R., Palnam, W.B., Singh. R.P. Current Scenario of Nanomaterials in the Environmental, Agricultural, and Biomedical Fields. Nanomaterials in Bionanotechnology. In book: *Nanomaterials in Bionanotechnology: Fundamentals and Applications*. 1. 6. CRC Press, 2021g.
16. Adetunji C.O., Kremer, R.J., Makanjuola, R., BhallaSarin, N. Application of molecular biotechnology to manage biotic stress affecting crop enhancement and sustainable agriculture. *Advances in Agronomy*, 2021h. https://www.sciencedirect.com/science/article/pii/S0065211321000304?dgcid=author
17. Adetunji, C.O., Olaniyan, O.T., Adetunji, J.B., Osemwegie, O., Ubi, B.E. African Mushrooms as Functional Foods and Nutraceuticals. In book: *Fermentation and Algal Biotechnologies for the Food, Beverage and Other Bioproduct Industries*. 1st Edition. 2022. CRC Press. pp. 19, 2022a
18. Adetunji, C.O., Ukhurebor, K.E., Olaniyan, O.T., Ubi, B.E., Oloke, J.K., Dauda, W.T., Hefft, D.I. Recent Advances in Molecular Techniques for the Enhancement of Crop Production. In book: *Agricultural Biotechnology, Biodiversity and Bioresources Conservation and Utilization*.1st Edition. 2022. CRC Press. pp. 20, 2022b.
19. Gröniger, A., Sinha, R.P., Klisch, M., Häder, D.P. Photoprotective compounds in cyanobacteria, phytoplankton and macroalgae—a database, *J. Photochem. Photobiol. B Biol.* 58 (2-3) 115–122, 2000. https://doi.org/10.1016/S1011-1344(00)00112-3.
20. Rodríguez, A.A., Stella, A.M., Storni, M.M., Zulpa, G., Zaccaro, M.C. Effects of cyanobacterial extracellular products and gibberellic acid on salinity tolerance in Oryza sativa L, *Saline Syst.* 2 (1) 7, 2006. https://doi.org/10.1186/1756-1448-2-7.
21. A.W. Moore. biology and agronomic significance, *Bot. Rev.* 35 (1) 17–34, 1969.
22. Adam M.S. The promotive effect of cyanobacterium Nostoc muscorum on the growth of some crop plants. *Acta Microbiologica Poloinca* 48(2) 163-171, 1999.
23. Anderson R.A. Algal Culturing Techniques, Elsevier Academic press. p. IX. ASTM International. ASTM D5373-08 Standard test methods for

instrumental determination of carbon, hydrogen nitrogen in laboratory samples of coal, 2005.

24. Bloemberg G.V., Wijfijes, A.H.M., Lamers, G.E.M., Stuurman, N., and Lugtenberg, B.J.J. Simulataneous imaging of Pesudomonas flourescens WCS 3655 populations expressing three different autofluorescent proteins in rizosphere: new perspective for studying microbial communities. *Molecular Plant Microbe Interaction* 13, 1170–6, 2000.
25. Catherine, Q., Susanna, W., Isidora, E.S., Mark, H., Aurélie, V., Jean-François, H. A. review of current knowledge on toxic benthic freshwater cyanobacteria—Ecology, toxin production and risk management. *Water Res.* 47, 5464–5479, 2013.
26. Chen, W., Jia, Y., Li, E., Zhao, S., Zhou, Q., Liu, L., Song, L. Soil-based treatments of mechanically collected cyanobacterial blooms fromlake taihu: Efficiencies andpotential risks. *Environ. Sci. Technol.* 46, 13370–13376, 2012.
27. Chen, W., Li, L., Gan, N., Song, L. Optimization of an effective extraction procedure for the analysis of microcystins in soils and lake sediments. *Environ. Pollut.* 143, 241–246, 2006.
28. Kunkel, D.D. Thylakoid centers: structures associated with the cyanobacterial photosynthetic membrane system, *Arch. Microbiol.* 133 97–99, 1982. https://doi.org/10.1007/BF00413518.
29. El-Fouly, M.M., Abdalla, F.E., and Shaaban, M.M. Multipurpose large scale production of microalgae biomass in Egypt Proceedings on 1st Egyptian Etalian Symptoms on Biotechnology, Assiut, Egypt (Nov., 21-23) 305–14, 1992.
30. El-Habbasha, S.F., Hozayn, M., Khalafallah, M.A. Integration effect between phosphorus levels and biofertilizers on quality and quantity yield of faba bean (Vicia faba L.) in newly cultivated sandy soils. *Research Journal of Agriculture and Biological Science* 3(6) 966-971, 2007.
31. Subramanian, G., Uma, L. Cyanobacteria in pollution control, *J. Sci. Ind. Res.* 55, 685–692, 1996.
32. Ris, H., Singh, R.N. Electron microscope studies on blue-green algae, *J. Biophys. Biochem. Cytol.* 9, 63–80, 1961.
33. Saadatnia, H. Riahi, H. Cyanobacteria from paddy fields in Iran as a biofertilizer in rice plants, *Plant Soil Environ.* 55 (5) 207–212, 2009.
34. Pankratz, H.S., Bowen, C.C. Cytology of blue-green algae. I. The cells of Symploca muscorum, *Int. J. Bot.* 50, 387–399, 1963.
35. Hegde, D.M., Dwivedi, B.S and Babu, S.N.S. Bio-fertilizers for cereal production in India- A review. Indian *Journal of Agriculture Science* 69, 73–83, 1999.
36. Ibraheem, I.B. Cyanobacteria as alternative biological conditioners for bioremediation of barren soil. Egypt, *J. Phycol.* 8 (100), 99–116, 2007.
37. Kulk M.M. The potential for using cyanobacteria (blue-green algae) and algae in the biological control of plant pathogenic bacteria and fungi. *European Journal of Plant Pathology* 101(6) 85-599, 1995.

38. Wilson, L.T. Cyanobacteria: a potential nitrogen source in rice fields, Texas Rice 6, 9–10, 2006.
39. Lakshmann, M. Azotobacter inoculation and crop productivity. In: Narula N (Eds.) *Azotobacter in Sustainable Agriculture*. CBS Publishers, New Delhi, India, pp. 109-116, 2000.
40. Li, H., Zhao, Q., Huang, H. Current states and challenges of salt-a ected soil remediation by cyanobacteria. *Sci. Total Environ.* 669, 258–272, 2019. [CrossRef]
41. Meena, M., Swapnil, P., Barupal, T., Sharma, K. A review on infectious pathogens and mode of transmission, *J. Plant Pathol. Microbiol.* 10, 472, 2019. https://doi. org/10.4172/2157-7471.1000472.
42. Adeleke, D. A., Olajide, P. A., Omowumi, O. S., Okunlola, D. D., Taiwo, A. M., & Adetuyi, B. O. Effect of Monosodium Glutamate on the Body System. *World News of Natural Sciences*, *44*, 1-23, 2022.
43. Malik, F.R., Ahmed, S., Rizki, Y.M. Utilization of lignocellulosic waste for the preparation of nitrogenous biofertilizer. Pakistan *Journal of Biological Sciences* 4 1217–1220, 2001.
44. Malliga, P., Uma, L., and Subramanian, G. Lignolytic activity of the cyanobacterium Anabena azollae ML2 and the value of coir waste as a carrier for biofertilizer. Microbios 86 pp 175-183, 1996.
45. Moore, A.W. Azolla: biology and agronomic significance. *Botanical Review* 3517–30, 1969.
46. Thajuddin, N., Subramanian, G. Cyanobacterial biodiversity and potential applications in biotechnology, *Curr. Sci.* 89, 47–57, 2005.
47. Malliga, P., Uma, L., Subramanian, G. 1996. Lignolytic activity of the cyanobacterium Anabaena azollae ML2 and the value of coir waste as a carrier for BGA biofertilizer, *Microbios* 86, 175–183, 1996.
48. Paul, S., Verma, O.P. and Rathi, M. S. Potential of homologus and heterologus Azotobacter chroococcum strains as bio-inoculants for cotton. *New Botanist* 29 p: 169-174, 2002.
49. Prasad, R.C., and Prasad, B.N. Cyanobacteria as a source Biofertilizer for sustainable agriculture in Nepal. *Journal in Plant Science Botanica Orientalis* 127-133, 2001.
50. Pulz, O. Photobioreactors: production systems for phototrophic microorganisms. *Applied Appl. Microbiol. Biotechnol.* (57), p: 287-293, 2001.
51. Nisha, R., Kaushik, A., Kaushik, C.P. Effect of indigenous cyanobacterial application on structural stability and productivity of an organically poor semi-arid soil, *Geoderma* 138 (1-2), 49–56, 2007. https://doi.org/10.1016/j. geoderma.2006.10.007.
52. Adetuyi B.O, Oluwole E.O Dairo J.O Chemoprotective Potential of Ethanol Extract of *Ganoderma Lucidum* on Liver and Kidney Parameters in *Plasmodium Beghei*-Induced Mice, *International Journal of Chemistry and Chemical Processes* (IJCC). Vol 1(8).:29-36, 2015.

53. Prasad, R.C., and Prasad, B.N. Cyanobacteria as a source biofertilizer for sustainable agriculture in Nepal, *J. Plant Sci. Bot. Orient.* 1, 127–133, 2001.
54. Singh, R.N. Role of Blue-Green Algae in Nitrogen Economy of Indian Agriculture, Indian Council of Agricultural Research, New Delhi, 1961.
55. Rasmussen, B., Fletcher, I.R., Brocks, J.J., Kilburn, M.R. Reassessing the first appearance of eukaryotes and cyanobacteria. *Nature* 2008, 455, 1101–1104, 2008. [CrossRef] [PubMed]
56. Rodriguez, A.A., Stella, A.A., Storni, M.M., Zulpa, G., Zaccaro, M.C. Effects of cyanobacterial extracelular products and gibberellic acid on salinity tolerance in Oryza sativa L. *Saline System*, 2: 7, 2006.
57. Roger, P.A., Reynaud, P.A. Free-living Blue-green Algae in Tropical Soils. Martinus Nijh off Publisher, La Hague, 1982.
58. El-Sohaimy, S.A. Functional foods and nutraceuticals – modern approach to food science, *World Appl. Sci. J.* 20 (5) (2012) 691–708, 2012.
59. Adetuyi B.O, Oluwole E.O, Dairo J.O Biochemical effects of shea butter and groundnut oils on white albino rats, *International Journal of Chemistry and Chemical Processes* (IJCC). Vol 1(8).:1-17, 2015.
60. Adetuyi, B. O., Omolabi, F. K., Olajide, P. A., Oloke, J. K. Pharmacological, Biochemical and Therapeutic Potential of Milk Thistle (Silymarin): A Review *World News of Natural Sciences* 37:75-91, 2021.
61. Adetuyi, B. O., & Farombi, E. O. 6-Gingerol, an active constituent of ginger, attenuates lipopolysaccharide-induced oxidation, inflammation, cognitive deficits, neuroplasticity, and amyloidogenesis in rat. *Journal of Food Biochemistry, 45*(4), e13660, 2021.
62. Adetuyi, B. O., Adebisi, O. A., Awoyelu, E. H., Adetuyi, O. A., Ogunlana, O. O. Phytochemical and Toxicological effect of Ethanol extract of *Heliotropium indicum* on Liver of Male Albino Rats. *Letters in Applied NanoBioscience (LIANB)* 10(2):2085-2095, 2020. https://doi.org/10.33263/LIANBS102.20852095
63. Saadatnia and Riahi. Cyanobacteria from paddy fields in Iran as a biofertilizer in rice plants, Plant Soil and Environment 55(5) 207-212, 2009.
64. Adetuyi, B. O., Ogundipe, A. E., Ogunlana, O. O., Egbuna, C., Estella, O. U., Mishra, A. P., Akram, M., & Achar, R. R. Banana Peel as a Source of Nutraceuticals. *Food and Agricultural Byproducts as Important Source of Valuable Nutraceuticals.*1st ed. Springer, Berlin. 400 p. Gebunden, 2022.
65. Adetuyi, B. O., Okeowo, T. O., Adetuyi, O. A., Adebisi, O. A., Ogunlana, O., Oretade, O. J., Marraiki, N., Beshbishy, A. M., Welson, N. N., & Batiha, G. E Ganoderma lucidum from red mushroom attenuates formaldehyde-induced liver damage in experimental male rat model. *Biology* 2020, *9*(10), 313, 2020. https://doi.org/10.3390/biology9100313
66. Adetuyi, B. O., Olajide, P. A., Awoyelu, E. H., Adetuyi, O. A., Adebisi, O. A., Oloke, J. K. , Epidemiology and Therapeutic measure for COVID-19; A review. *African Journal of Reproductive Health June 2020* (Special Edition on

COVID-19); 24 (2):142, 2020, https://www.ajrh.info/index.php/ajrh/article/view/2300

67. Awoyelu, E. H., Oladipo, E. K., Adetuyi, B. O., Senbadejo, T. Y., Oyawoye, O. M., Oloke, J. K. Phyloevolutionary analysis of SARS-CoV-2 in Nigeria. *New Microbes and New Infections* Vol 36. 100717, 2020. https://doi.org/10.1016/j.nmni.2020.100717
68. Batiha, G. B., Awad, D. A., Algamma, A. M., Nyamota, R., Wahed, M. I., Shah, M. A., Amin, M. N., Adetuyi, B. O., Hetta, H. F., Cruz-Marins, N., Koirala, N., Ghosh, A., & Sabatier, J. Diary-derived and Egg White Proteins in Enhancing Immune System against COVID-19 *Frontiers in Nutritionr. (Nutritional Immunology)* 8:629440, 2021.
69. Didunyemi, M. O., Adetuyi, B. O., & Oyewale, I. A. Inhibition of lipid peroxidation and *in-vitro* antioxidant capacity of aqueous, acetone and methanol leaf extracts of green and red Acalypha wilkesiana Muell Arg. *Int J Biol Med Res.*11(3):7089- 7094, 2020.
70. Didunyemi, M. O., Adetuyi, B. O., Oyebanjo, O. O. Morinda lucida attenuates acetaminophen-induced oxidative damage and hepatotoxicity in rats, *Journal of Biomedical sciences Vol 8; No,* 2019. https://www.jbiomeds.com/biomedical-sciences/morinda-lucida-attenuates-acetaminopheninduced-oxidative-damage-and- hepatotoxicity-in-rats.php?aid=24482
71. Farombi, E. O., Abolaji, A. O., Adetuyi, B. O., Awosanya, O., & Fabusoro, M. Neuroprotective role of 6-Gingerol-rich fraction of Zingiber officinale (Ginger) against acrylonitrile-induced neurotoxicity in male Wistar rats. *Journal of Basic and Clinical Physiology and Pharmacology, 30*(3), 2019.
72. James-Okoro, P. O., Iheagwam, F. N., Sholeye, M. I., Umoren, I. A., Adetuyi, B. O., Ogundipe, A. E., Braimah, A. A., Adekunbi, T. S., Ogunlana, O. E., & Ogunlana, O. O. Phytochemical and *in vitro* antioxidant assessment of Yoyo bitters *World News of Natural Sciences* 37:1-17, 2021.
73. Nazir, A., Itrat, N., Shahid, A., Mushtaq, Z., Abdulrahman, S. A., Egbuna, C., Adetuyi, B. O., Khan, J., Uche, C. Z, Toloyai, P. Y. Orange Peel as a Source of Nutraceuticals. Food and Agricultural Byproducts as Important Source of Valuable Nutraceuticals. 1st ed. 2022 2022. Springer, Berlin. 400 pp. Gebunden, 2022.
74. Ogunlana, O. O., Adetuyi, B. O., Adekunbi, T. S., Adegboye, B. E., Iheagwam, F. N., Ogunlana, O. E. Ruzu bitters ameliorates high–fat diet induced non-alcoholic fatty liver disease in male Wistar rats. *Journal of Pharmacy and Pharmacognosy Research* 9(3), 251-26, 2021.
75. Ogunlana, O. O., Adetuyi, B. O., Rotimi, M., Esalomi, I., Adeyemi A., Akinyemi, J., Ogunlana, O., Adetuyi, O., Adebisi, O., Okpata, E., Baty, R., & Batiha, G. Hypoglycemic Activities of Ethanol Seed Extract of *Hunteria umbellate* (Hallier F.) on Streptozotocin-induced Diabetic Rats. *Clinical Phytoscience* 7 (1), 1-9, 2021.
76. Ogunlana, O. O., Ogunlana, O. E., Adekunbi, T. S., Adetuyi, B. O., Adegboye, B. E., & Iheagwam, F. N. Anti-inflammatory Mechanism of Ruzu Bitters

on Diet-Induced Nonalcoholic Fatty Liver Disease in Male Wistar Rats. *Evidence-Based Complementary and Alternative Medicine,* 2020. https://doi.org/10.1155/2020/5246725

77. Ogunlana, O. O., Ogunlana, O. E., Popoola, J. O., Adetuyi, B. O., Adekunbi, T. S., David, O. L., Adeleye, O. J., Udeogu, S. A., & Adeyemi, A. O. Twigs of Andrographis paniculata (Burn. F) Nees attenuates Carbon Tetrachloride (CCl4) Induced Liver Damage in Wistar Albino Rats. *RASAYAN Journal of Chemistry* 14(4): 2598-2603, 2021.
78. Ogunlana, O. O., Adetuyi, B. O., Esalomi, E. F., Rotimi, M. I., Popoola, J. O., Ogunlana, O. E., & Adetuyi, O. A. Antidiabetic and Antioxidant Activities of the Twigs of Andrograhis paniculata on Streptozotocin-Induced Diabetic Male Rats. *BioChem* 1(3):238-249, 2021.
79. Olajide, P. A., Adetuyi, O. A., Omowumi, O. S. & Adetuyi, B. O. Anticancer and Antioxidant Phytochemicals as Speed Breakers in Inflammatory Signaling. *World News of Natural Sciences* 44, 231-259, 2022.
80. Olajide, P. A., Omowumi, O. S., Okunlola, D. D., & Adetuyi, B. O. Deadly Pandemia: Monkeypox Disease, a Case Study. *African Journal of Advanced Pure and Applied Sciences (AJAPAS),* 34-37, 2022.
81. Olajide, P. A., Omowumi, S. O., & Odine, G. O. Pathogenesis of Reactive Oxygen Species: A Review. *World News of Natural Sciences* 44:150-164, 2022.
82. Song, T., Mårtensson, L., Eriksson, T., Zheng,W., Rasmussen, U. Biodiversity and seasonal variation of the cyanobacterial assemblage in a rice paddy field in Fujian, China. FEMS *Microbiol. Ecol.* 54, 131–140, 2005.
83. Sukenik, A.; Hadas, O.; Kaplan, A.; Quesada, A. Invasion of Nostocales (cyanobacteria) to subtropical and temperate freshwater lakes - physiological, regional, and global driving forces. *Front. Microbiol.* 3, 86, 2012.
84. T. Barupal, M. Meena, K. Sharma, A study on preventive effects of Lawsonia inermis L. bioformulations against leaf spot disease of maize, *Biocatal. Agric. Biotechnol.* 23 101473, 2020. https://doi.org/10.1016/j.bcab.2019.101473.
85. Adetuyi, B. O., Adebayo, P. F., Olajide, P. A., Atanda, O. O., Oloke, J. K. Involvement of Free Radicals in the Ageing of Cutaneous Membrane. *World News of Natural Sciences* 43, 11-37, 2022.
86. Jensen, T.E., Bowen, C.C. Organization of the centroplasm in Nostoc pruniforme, *Proc. Iowa Acad. Sci.* 68 (1961) 86–89, 1961.
87. Adetuyi BO, Dairo JO, Didunyemi MO Anti-Hyperglycemic Potency of *Jatropha Gossypiifolia* in Alloxan Induced Diabetes. *Biochem Pharmacol (Los Angel)* 4(5):193, 2015.
88. Adetuyi, B. O., Adebisi, O. A., Adetuyi, O. A., Ogunlana, O. O., Toloyai, P-B., Egbuna, C., Uche, C. Z., Khan, J., Adumanya, O. C., Patrick-Iwuanyanwu, K.C., Ficus exasperata Attenuates Acetaminophen-Induced Hepatic Damage via NF-κB Signaling Mechanism in Experimental Rat Model BioMed Research International, 2022, 2022.
89. Thajuddin, N., Subramanian, G. Cyanobacterial biodiversity and potential applications in biotechnology. *Current Science* 89 47–57, 2005.

90. Wilson, L.T. Cyanobacteria: A Potential Nitrogen Source in Rice Fields. *Texas Rice* 6 9–10, 2006.
91. Yosefi, K., Galavi, M., Ramrodi, M., Mousavi, S.R. Effect of bio-phosphate and chemical phosphorus fertilizeraccompanied with micronutrient foliar application on growth, yield and yield components of maize (Single Cross 704). Australian *Journal of Crop Sciences* 5(2) 175-180, 2011.
92. Chittora, D., Meenaa, M., Barupala,T., Swapnil, P., and Sharma, K. Cyanobacteria as a source of biofertilizers for sustainable agriculture. *Biochemistry and Biophysics Reports,* 22 : 1-10, 2020.
93. El-Habbasha, S.F., Hozayn, M., and Khalafallah, M.A. Integration effect between phosphorus levels and biofertilizers on quality and quantity yield of faba bean (*Vicia faba* L.) in newly cultivated sandy soils. *Research Journal of Agriculture and Biological Science,* 3(6) 966-971, 2007.
94. Goel, A.K., Laura, R.D.S, Pathak, G., Anuradha, G. and Goel, A. Use of bio-fertilizers: potential, constraints and future strategies review. *International Journal of.Tropical Agriculture* 17: 1–18, 1999.
95. Hegde, D.M., Dwivedi,, B.S and Babu, S.N.S Bio-fertilizers for cereal production in India- A review. *Indian Journal of Agriculture Science,* 69: 73–83, 1999.
96. Hoekman, S.K., Broch,A., Robbins,C., Ceniceros, E.and Natarajan, M. Review of biodiesel composition, properties, and specifications, *Renew. Sustain. Energy Rev.* 16 (1):143–169, 2012.
97. Kaushik B.D. Developments in Cyanobacterial Biofertilizer. *Proc Indian Natn Sci Acad* 80 (2): 379-388, 2014.
98. Meena, M., Divyanshu, K., Kumar, S., Swapnil, P., Zehra, A., Shukla, V., Yadav, M., and Upadhyay, R.S. Regulation of L-proline biosynthesis, signal transduction, transport, accumulation and its vital role in plants during variable environmental conditions, *Heliyon* 5 (12): e02952, 2019.
99. Meena, M., Swapnil, P., Barupal, T., and Sharma, K.A review on infectious pathogens and mode of transmission, *J. Plant Pathol. Microbiol.* 10: 472, 2019. https://doi.org/10.4172/2157-7471.1000472.
100. Prasad,R.C., and Prasad, B.N. Cyanobacteria as a source biofertilizer for sustainable agriculture in Nepal, *J. Plant Sci. Bot. Orient.* 1:127–133, 2001.
101. Rodriguez, A.A., Stella, A.M., Storni, M.M, Zulpa, G., and Zaccaro, M.C. Effects of cyanobacterial extracellular products and gibberellic acid on salinity tolerance in *Oryza sativa* L, *Saline Syst.* 2 (1):7, 2006. https://doi.org/10.1186/1756-1448-2-7.
102. Roger, P.A., and Reynaud, P.A. Free—living blue—green algae in tropical soils, *Microbiology of Tropical Soils and Plant Productivity,* Springer, Dordrecht, 1982, pp. 147–168, 1982.
103. Saadatnia, H., H. Riahi, Cyanobacteria from paddy fields in Iran as a biofertilizer in rice plants, *Plant Soil Environ.* 55 (5): 207–212.

104. Sahu, D., Priyadarshani, I., and Rath, B. Cyanobacteria - as potential biofertilizer. *Journal of Microbiology,* 1(2-3):20-26, 2012.
105. Singh, A.K., Singh,A.P., Gaurav,N., Srivastava,A., and Kumar, A.A review on blue green algae as biofertilizer (site characteristics, location and boundaries of Vindhyabasin, geology of Rewa, physiography, climatology). *International Journal of Botany Studies,* 1(4): 01-07, 2016.
106. Youssef, M.M.A., Ali, M.S. Management of*Meloidogyne incognita* infecting cowpea by using some native blue green algae. *Anz. Schadlingskde., Pflanzenschutz, Umweltschutz* 71, 15–16, 1998. https://doi.org/10.1007/BF02770565

14

Relevance of Algae in the Agriculture, Food and Environment Sectors

Olotu Titilayo[1]* and Charles Oluwasun Adetunji[2]

[1]*Department of Microbiology, Adeleke University, Ede, Osun State, Nigeria*
[2]*Applied Microbiology, Biotechnology and Nanotechnology Laboratory, Department of Microbiology, Edo State University Uzairue, Iyamho, Edo State, Nigeria*

Abstract
Innovations are needed in today's world to tackle the limitations and environmental issues associated with agricultural and industrial practices. In order to meet the future demands of the growing global population and adjust to the current innovations in agrotechnology, sustainable commodities are needed that can replace the less sustainable ones, which will minimize negative environmental impacts in addition to creating new solutions to meet these demands. Therefore, this chapter intends to provide detailed information on the relevance of algae as a sustainable solution to several challenges in diverse sectors such as agriculture, food and environment.

Keywords: Agriculture, food, environment, agrotechnology, algae, environmental issues

14.1 Introduction

Three generations of biofuels have emerged: the first generation is biofuel generated from corn ethanol, starch, and soy, the second generation is generated from cellulosic biomass, and the third generation is generated from algae, which is known as algae biofuel [1, 2]. Algae are primarily photosynthetic organisms of the Protista kingdom, which are eukaryotic photosynthesized organisms without a true leave, root or stem. Algae are

**Corresponding author*: tmolotu@adelekeuniversity.edu.ng

Charles Oluwaseun Adetunji, Julius Kola Oloke, Naveen Dwivedi, Sabeela Beevi Ummalyma, Shubha Dwivedi, Daniel Ingo Hefft and Juliana Bunmi Adetunji (eds.) Next-Generation Algae: Volume I: Applications in Agriculture, Food and Environment, (321–330) © 2023 Scrivener Publishing LLC

classified based on color (green, red and brown), which are a reflection of their chloroplast pigments, though some algae lack the chloroplast and photosynthesis features distinguishing them from other organisms, and some are evolutionary related with protozoa and fungi rather than with other algae.

Algae are abundant in their distribution around the world and are categorized based on their habitats; planktonic algae (grow in suspended form in water), neustonic algae (grow on water surface), cryophilic algae (grow in ice), thermophilic algae (grow in hot springs), edaphic algae (grow in soil), epizoic algae (grow on animals), epiphytic algae (grow on algae or other plants), endolithic algae (grow on porous rocks), and chasmolithic algae (grow in rock fissures).

Algae capture energy from the sunlight that forms the basis of the food chain, making all sea animals dependent on algae for their survival, and are the producer of about 50% of the oxygen for respiration. Crude oil, including natural gas, are derived from remnants of ancient algae modified by hydrocarbonoclastic bacteria like the coccolithophore algae *Botryococcus*. Algae, especially the red algae (*Porphyra*), serves as food in both processed and unprocessed forms for direct consumption in Asian and Pacific Island countries and have a market value of billions of dollars annually [3]. Microalgae are microorganisms having about 200,000 species with an astonishing biological diversity [4] with high ecological adaptations thriving in low nutrient environment, and have evolved their metabolic adaptations, creating a favorable condition for their growth [5]. Algae possess higher photosynthetic adeptness than plants to generate biomass [6]. Growing algae in a pond or photobioreactor is water-efficient because a minimal amount of freshwater or seawater is needed and they grow on less fertile soil. Microalgae as a next-generation application creates innumerable sustainable products because its biofactories are less expensive and produce plant-derived products [7, 8]. Since algae have a comparatively small industrial application, algal-based bioenergy and model algae-based biorefineries could assist in tapping the untapped algae-based potential, and it is foreseen that an algal-based bioeconomy will create a great many solutions to meet the demands of the growing global society [9].

Algae production industries have contributed to economic growth in many countries, including the EU countries and Chad, where sustainable transition in the area of algal development and production is currently among the top priorities in EU policies [10].

14.2 Fourth Generation Biofuel: Next Generation Algae

Fourth generation biofuels are the result of developments in plant biology and biotechnology (metabolic engineering) in the field of carbon capture and storage technique. This requires an advanced method for biofuel production. In this generation, bio-engineered plants/tree/algae function as a carbon capture machine to lock carbon in their different parts, that is, leaves, branches, etc., for the feedstock generation of biofuel [11].

14.3 Next Generation Algae: Application in Agriculture

For about 70 years, the algae industry has been known as a producer of extracts for processed foods and raw materials in cosmetics and medicine, bringing innovations like the use of biodegradable packaging produced from seaweed to replace the use of plastic packaging. Pure biomass production is associated with algae production as well as a lot of the supporting research innovations, patents and product developments, which is a great step towards a bioeconomy [2].

Algae production is considered to be a relatively new branch in agriculture that could proffer a solution to the challenges in food security and which has the ability to produce oil spatial distribution greater than that of terrestrial plants. Blue-green algae (cyanobacteria) have been used as inoculum in rice farming to increase its yields, as natural symbionts because of its N-fixing potential and for phosphorous recovery in wastewater [12]. A vast amount of agricultural land is not needed for algae production since they are easily grown on degraded land, in seawater and wastewater, etc.

Algal technology is presently the major source of biofuel generation and has the potential to replace diesel from fossil diesel, with a microalgae oil content that exceeds 80% biomass by dry weight. Algae production predominantly has short harvesting period cycles of approximately 1 to 10 days, helps to overcome greenhouse gas by consuming CO_2 in the atmosphere, and grow well on poor or non-fertile soil. They also grow on freshwater, marine water, and contaminated wastewater, and therefore do not compete with agriculture land meant for food production. Area of algae plantation does not require treatment with herbicide or pesticide and the algae is used for producing fertilizer, biopolymers, animal feed and polysaccharides [13]. Globally, there are about 100,000 algae species, 300 species are diatoms of the genera and green algae and are good sources of biodiesel. The microalgae in the likes of diatoms are widely distributed in saltwater, constituting a larger percentage of phytoplankton biomass. Microalgae commonly used for biodiesel generation are *Botryococcus braunii*, *Nitzschia* sp, *Schizochytrium* sp., *Pleurochrysis carterae*, *Chlorella*, *Gracilaria*, *Nannochloropsis* sp, *Dunaliella tertiolecta* and Ulva [14, 15].

Microalgae are sources of biostimulants and biofertilizers in agriculture augmented with synthetic fertilizers for plant protection and growth regulation, higher yields in crops, and root enhancer, and they also enhance plant tolerance to drought and salinity; however, more exploration is needed since very little research has been conducted in these areas globally [16]. A high level of macro- and micronutrient content in microalgae assists it function as an optimal growth enhancer for plants, has attracted the interest of growers and agricultural industries as a means of sustainable crop production [17–19].

Even though microalgae cultivation is high cost, its essential uses tend to overcome the cost issues and make its economically sustainable. By optimization and using low-cost resources for cultivation like wastewater, organic fertilizer tends to assist its cost-effectiveness [9, 20, 21]. Co-cultivation of *Chlorella infusionum* (microalga) and *Solanum lycopersicum* (tomato) were enhanced by microalgae supplying oxygen that assisted in root respiration to the tomato. A symbiotic association was observed between the tomato and microalga and the algae supplies of CO_2 and O_2 for optimum growth and water, and light and nutrients were available for the microalga [22].

In a similar study by Barone *et al.* [23] of *Chlorella vulgaris* (microalgae) and tomato plants, an increased shoot height of the tomato plant and higher biomass of the microalgae were observed [24–26]. It was observed that *Arthrospira* spp. (microalgae) has a great tendency to grow under high salinity concentration and extreme pH condition. Microalgae also has the capacity to bioremediate heavy metals in wastewater; plant growth-promoting compounds such as betaines, polyamines, amino acids cytokinins, brassinosteroids and gibberellins are components [13, 27].

14.4 Next Generation Algae: Application in the Environment

Since algae production requires water, carbon dioxide and other macronutrients, algae production could be produced from nutrient-rich streams owing to the fact that a considerable amount of nutrients from agricultural runoff end up in lakes and seas, and seaweed farming decreases nutrient run-offs in deltas, river mouths, and bay areas [28]. Applications of microalgae for bioremediations is very essential since it reduces the cost of energy and greenhouse gas emissions. This is especially the case when using phototrophic algae to produce biomass processed into biofuel in an aerobic fermentation process and CO_2 generated during the burning of the fuel as feed for the bioreactor to increase algal growth, thereby reducing the environmental impact through wastewater treatments [29–45]. Certain studies reported that algae biofuel has the capacity to yield more of the second and third generation of biofuel, producing a hundred times more oil per acre than soyabean or other oil crops [46].

14.5 Conclusion

Globally, energy from biomass is the fourth largest form of energy. Research on biomass energy has been expedited to meet the demands of the growing population, as algae biofuel will not compromise food production derived from crops. Algal technology has more advantages than disadvantages and deserves more research development than the present development and it could serve as a more reliable source of energy in the future. Therefore, algae biofuel is a good alternative to replace fossil fuel in order to meet future demands.

References

1. Roland, A. L., Jean-Michel Lavoie, From first- to third-generation biofuels: Challenges of producing a commodity from a biomass of increasing complexity, *Animal Frontiers*, Volume 3, Issue 2, April 2013, pp. 6–11, 2013. https://doi.org/10.2527/af.2013-0010
2. Michele, A., Raffaela A., Mathieu, P., Donna L. S., Audrey, C.S., Christopher, C. H., Leen, L., McCauley, J.I., Unnikrishnan, K., Parijat,R., Tim, K. and Peter J.R. Emerging Technologies in Algal Biotechnology: Toward the Establishment of a Sustainable, Algae-Based Bioeconomy, *Frontiers in Plant Science*, 11, 2020.
3. Andersen, R. A. and Lewin, R. A. "Algae". Encyclopedia Britannica, 2022. https://www.britannica.com/science/algae. Accessed 5 May 2022.
4. Guiry, How many species of algae are there? 2012. http://dx.doi.org/10.1111/j.1529-8817.2012.01222.x

5. Litchman, E. Resource competition and the ecological success of phytoplankton," in *Evolution of Primary Producers in the Sea*, P. G. Falkowski and A. H. Knoll, Academic Press, Cambridge, Mass. USA, 351–375, 2007.
6. Benedetti, M., Vecchi, V., Barera, S. *et al.* Biomass from microalgae: the potential of domestication towards sustainable biofactories. *Microb Cell Fact* 17, 173, 2018. https://doi.org/10.1186/s12934-018-1019-3
7. Rasala, B. A., Lee, P. A., Shen, Z., Briggs, S. P., Mendez, M., and Mayfield, S. P. Robust expression and secretion of Xylanase1 in *Chlamydomonas reinhardtii* by fusion to a selection gene and processing with the FMDV 2A peptide. *PLoS One* 7:e43349, 2012.
8. Vavitsas, K., Fabris, M., and Vickers, C. E. Terpenoid metabolic engineering in photosynthetic microorganisms. *Genes* 9:E520, 2018.
9. Michele, A., Raffaela A., Mathieu, P., Donna L. S., Audrey, C.S., Christopher, C. H., Leen, L., McCauley, J.I., Unnikrishnan, K., Parijat,R., Tim, K. and Peter J.R. Emerging Technologies in Algal Biotechnology: Toward the Establishment of a Sustainable, Algae-Based Bioeconomy, *Frontiers in Plant Science*, 11, 2020.
10. European Union, Harvest of hope: spirulina from Lake Chad, 2020. https://gcca.eu/stories/harvest-hope-sprirulina-lake-chad. Accessed 6 Oct 2020.
11. Acheampong, M., Ertem, F.C., Kappler, B. and Neubauer, P. In pursuit of Sustainable Development Goal (SDG) number 7: Will biofuels be reliable?, *Renewable and Sustainable Energy Reviews*, 75: 929-937, 2017.
12. Mishra, U., Pabbi, S. Cyanobacteria: a potential biofertilizer for rice. *Resonance* 9(6):6–10, 2004.
13. Stirk, W.A., Ördög, V., Novák, O., Rolèík, J., Strnad, M., Bálint, P., Staden, J. Auxin and cytokinin relationships in 24 microalgal strains. *J. Phycol.*, 49, 459–467, 2013.
14. Amaro, H.M., C. Ângela, A.C. Macedo and F.X. Malcata. Microalgae: An alternative as sustainable source of biofuels? *Energy*, 44, 158-166, 2012.
15. Miranda, C. T., Pinto, R.F., de Lima, D. V., Viegas, C.V., da Costa, S. M. and Azevedo, S.M. Microalgae Lipid and Biodiesel Production: A Brazilian Challenge. *American Journal of Plant Sciences* 66, 2522-2533, 2015.
16. Ronga, D., Biazzi, E.,Parati, K., Carminati, D., Carminati, E. and Tava, A. Microalgal Biostimulants and Biofertilisers in Crop Productions, *Agronomy*, 9, 192, 2019.
17. Calvo, P., Nelson, L., Kloepper, J.W. Agricultural uses of plant biostimulants. *Plant Soil*, 383, 3–41, 2014.
18. Elarroussia, H., Elmernissia, N., Benhimaa, R.; El Kadmiria, I.M.; Bendaou, N.; Smouni, A.; Wahbya, I. Microalgae polysaccharides a promising plant growth biostimulant. *J. Algal Biomass*, Utln., 7, 55–63, 2016.
19. Mata, T.M., Martins, A.A., Caetano, N.S. Microalgae for biodiesel production and other applications: A review. Renew. *Sustain. Energy Rev.* 2010, 14, 217–232, 2010.

20. Pragya, N., Pandey, K.K., Sahoo, P.K. A review on harvesting, oil extraction and biofuels production technologies from microalgae. *Renew. Sustain. Energy Rev.* 2013, 24, 159–171, 2013.
21. Gong, Y., Jiang, M. Biodiesel production with microalgae as feedstock: From strains to biodiesel. *Biotechnol. Lett.* 2011, 33, 1269–1284, 2011.
22. Zhang, J. Wang, X.; Zhou, Q. Co-cultivation of Chlorella spp. and tomato in a hydroponic system. *Biomass Bioenerg.* 2017, 97, 132–138, 2017.
23. Barone, V., Puglisi, I., Fragalà, F., Piero, A.R.L., Giuffrida, F., Baglieri, A. Novel bioprocess for the cultivation of microalgae in hydroponic growing system of tomato plants. J. *Appl. Phycol.* 2019, 31, 465–470, 2019.
24. Yang, C., Liu, H., Li, M.; Yu, C., Yu, G. Treating urine by Spirulina platensis. *Acta Astronaut.*, 63, 1049–1054, 2008.
25. Çelekli, A., Yavuzatmaca, M. Predictive modeling of biomass production by Spirulina platensis as function of nitrate and NaCl concentrations. *Bioresour. Technol.*, 100, 1847–1851, 2009.
26. Ogbonda, K.H., Aminigo, R.E., Abu, G.O. Influence of temperature and pH on biomass production and protein biosynthesis in a putative Spirulina sp. *Bioresour. Technol.*, 98, 2207–2211, 2007.
27. Ronga, D., Biazzi, E.; Parati, K., Carminati, D., Carminati, E., Tava, A. Microalgal Biostimulants and Biofertilisers in Crop Productions. *Agronomy*, 9, 192, 2019.
28. Michalak A.M., Anderson, E.J., Beletsky, D., Boland, S., Bosch, N.S. and Bridgeman, T.B. *et al.*, Record-setting algal bloom in Lake Erie caused by agricultural and meteorological trends consistent with expected future conditions. *Proc Natl Acad Sci.*, 110(16):6448–6452, 2013.
29. Lundquist, T.J., Woertz, I.C., Quinn, N.W.T., Benemann, J.R. A realistic technology and engineering assessment of algae biofuel production. Energy Biosciences Institute, 1, 2010.
30. Adetunji, C.O., Roli, O.I., Adetunji, J.B. Exopolysaccharides Derived from Beneficial Microorganisms: Antimicrobial, Food, and Health Benefits. In: Mishra, P., Mishra, R.R., Adetunji, C.O. (eds) *Innovations in Food Technology.* Springer, Singapore, 2020a. https://doi.org/10.1007/978-981-15-6121-4_10
31. Adetunji, C. and Anani, O. Bio-fertilizer from Trichoderma: Boom for Agriculture Production and Management of Soil- and Root-Borne Plant Pathogens. Pages 245-256. Innovations in Food Technology: Current Perspectives and Future Goals. Editors: Pragya Mishra, Raghvendra Raman Mishra, Charles Oluwaseun Adetunji, 2020.
32. Adetunji C.O., Varma A. Biotechnological Application of *Trichoderma*: A Powerful Fungal Isolate with Diverse Potentials for the Attainment of Food Safety, Management of Pest and Diseases, Healthy Planet, and Sustainable Agriculture. In: Manoharachary C., Singh H.B., Varma A. (eds) *Trichoderma: Agricultural Applications and Beyond. Soil Biology*, vol 61. Springer, Cham, 2020. https://doi.org/10.1007/978-3-030-54758-5_12

33. Olaniyan O. T. and Adetunji C.O. Biochemical Role of Beneficial Microorganisms: An Overview on Recent Development in Environmental and Agro-Science. *Microbial rejuvenation of Polluted Environment*, 2021.
34. Abel Inobeme, Jaison Jeevanandam, Charles Oluwaseun Adetunji, Osikemekha Anthony Anani, Devarajan Thangadurai, Saher Islam, Olubukola Monisola Oyawoye, Julius Kola Oloke, Mohammed Bello Yerima, Olugbemi T.Olaniyan. Ecorestoration of soil treated with biosurfactant during greenhouse and field trials. Green Sustainable Process for Chemical and Environmental Engineering and Science. *Biosurfactants for the Bioremediation of Polluted Environments*, 2021, Pages 89-105, 2021. https://doi.org/10.1016/B978-0-12-822696-4.00010-3
35. Osikemekha Anthony Anani, Jaison Jeevanandam, Charles Oluwaseun Adetunji, Abel Inobeme, Julius Kola Oloke, Mohammed Bello Yerima, Devarajan Thangadurai, Saher Islam, Olubukola Monisola Oyawoye, Olugbemi T. Olaniyan. Application of biosurfactant as a noninvasive stimulant to enhance the degradation activities of indigenous hydrocarbon degraders in the soil. Green Sustainable Process for Chemical and Environmental Engineering and Science. *Biosurfactants for the Bioremediation of Polluted Environments*, 2021, Pages 69-87, 2021. https://doi.org/10.1016/B978-0-12-822696-4.00019-X
36. Charles Oluwaseun Adetunji, Abel Inobeme, Osikemekha, Anthony Anani, Jaison Jeevanandam, Mohammed Bello Yerima, Devarajan Thangadurai, Saher Islam, Olubukola Monisola Oyawoye, Julius Kola Oloke, Olugbemi T.Olaniyan Isolation, screening, and characterization of biosurfactant-producing microorganism that can biodegrade heavily polluted soil using molecular techniques. Green Sustainable Process for Chemical and Environmental Engineering and Science. *Biosurfactants for the Bioremediation of Polluted Environments*, 2021, Pages 53-68, 2021a. https://doi.org/10.1016/B978-0-12-822696-4.00016-4
37. Charles Oluwaseun Adetunji, Jaison Jeevanandam, Osikemekha Anthony Anani, Abel Inobeme, Devarajan Thangadurai, Saher Islam, Olugbemi T. Olaniyan. Strain improvement methodology and genetic engineering that could lead to an increase in the production of biosurfactants, 2021b.
38. Charles Oluwaseun Adetunji, Jaison Jeevanandam, Abel Inobeme, Olugbemi T. Olaniyan, Osikemekha Anthony Anani, Devarajan Thangadurai, Saher Islam. Charles Oluwaseun Adetunji, Olugbemi T Olaniyan, Osikemekha Anthony Anani, Abel Inobeme, Kingsley Eghonghon Ukhurebor, Ruth Ebunoluwa Bodunrinde, Juliana Bunmi Adetunji, Kshitij RB Singh, Vanya Nayak, Wadzani Dauda Palnam and Ravindra Pratap Singh. 2021d. Bionanomaterials for green bionanotechnology. In book: *Bionanomaterials: Fundamentals and Biomedical Applications*. Publisher: IOP Publishing, 2021c.
39. Adetunji C.O., Inobeme A., Olaniyan O.T., Olisaka F.N., Bodunrinde R.E., Ahamed M.I. Microbial Desalination. In: Inamuddin, Khan

A. (eds) *Sustainable Materials and Systems for Water Desalination. Advances in Science, Technology & Innovation (IEREK Interdisciplinary Series for Sustainable Development)*. Springer, Cham, 2021d. https://doi.org/10.1007/978-3-030-72873-1_13

40. Adetunji J.B., Adetunji C.O., Olaniyan O.T. African Walnuts: A Natural Depository of Nutritional and Bioactive Compounds Essential for Food and Nutritional Security in Africa. In: Babalola O.O. (eds) *Food Security and Safety*. Springer, Cham, 2021e. https://doi.org/10.1007/978-3-030-50672-8_19
41. Charles Oluwaseun Adetunji, Olugbemi T. Olaniyan, Osikemekha Anthony Anani, Frances. N. Olisaka, Abel Inobeme, Ruth Ebunoluwa Bodunrinde, Juliana Bunmi Adetunji, Kshitij RB Singh, Wadzani Dauda Palnam, Ravindra Pratap Singh. Current Scenario of Nanomaterials in the Environmental, Agricultural, and Biomedical Fields. Nanomaterials in Bionanotechnology. In book: *Nanomaterials in Bionanotechnology: Fundamentals and Applications*. 1. 6. CRC Press, 2021f.
42. Charles Oluwaseun Adetunji, Robert J.Kremer, RasheedMakanjuola, Neera BhallaSarin. Application of molecular biotechnology to manage biotic stress affecting crop enhancement and sustainable agriculture. *Advances in Agronomy*, 2021h. https://www.sciencedirect.com/science/article/pii/S0065211321000304?dgcid=author
43. Jaison Jeevanandam, Charles Oluwaseun Adetunji, Jaya DivyaSelvam Osikemekha Anthony Anani Abel Inobeme, Saher Islam, Devarajan Thangadurai, Olugbemi T. Olaniyan. High industrial beneficial microorganisms for effective production of a high quantity of biosurfactant. In book: *Green Sustainable Process for Chemical and Environmental Engineering and Science*, 2021.
44. Charles Oluwaseun Adetunji, Olugbemi Tope Olaniyan, Juliana Bunmi Adetunji, Osarenkhoe O. Osemwegie, Benjamin Ewa Ubi. African Mushrooms as Functional Foods and Nutraceuticals. In book: *Fermentation and Algal Biotechnologies for the Food, Beverage and Other Bioproduct Industries*. 1st Edition. CRC Press. pp. 19, 2022a.
45. Charles Oluwaseun Adetunji, Kingsley Eghonghon Ukhurebor, Olugbemi Tope Olaniyan, Benjamin Ewa Ubi, Julius Kola Oloke, Wadzani Palnam Dauda, Daniel Ingo Hefft. Recent Advances in Molecular Techniques for the Enhancement of Crop Production. In book: *Agricultural Biotechnology, Biodiversity and Bioresources Conservation and Utilization*. 1st Edition. 2022. CRC Press. pp. 20, 2022b.
46. Kim, S., and B. E. Dale,—Global potential bioethanol production from wasted crops and crop residues. *Biomass and Bioenergy*, vol. 26, no. 4, pp. 361–375, 2004.

15

Application of Biofuels for Bioenergy: Recent Advances

Charles Oluwaseun Adetunji[1*], Muhammad Akram[2], Babatunde Oluwafemi Adetuyi[3], Fahad Said[2], Tehreem Riaz[2], Olugbemi T. Olaniyan[4], Inobeme Abel[5], Phebean Ononsen Ozolua[1], Ruth Ebunoluwa Bodunrinde[6], Nyejirime Young Wike[7], Wadzani Dauda Palnam[8], Arshad Farid[9], Shakira Ghazanfar[10], Olorunsola Adeyomoye[11], Chibuzor Victory Chukwu[12] and Mohammed Bello Yerima[13]

[1]Applied Microbiology, Biotechnology and Nanotechnology Laboratory, Department of Microbiology, Edo State University Uzairue, Iyamho, Edo State, Nigeria
[2]Department of Eastern Medicine, Government College University Faisalabad, Punjab, Pakistan
[3]Department of Natural Sciences, Faculty of Pure and Applied Sciences, Precious Cornerstone University, Ibadan, Nigeria
[4]Laboratory for Reproductive Biology and Developmental Programming, Department of Physiology, Rhema University, Aba, Nigeria
[5]Department of Chemistry, Edo University Iyamho, Auchi, Edo State, Nigeria
[6]Department of Microbiology, Federal University of Technology Akure, Gaga, Nigeria
[7]Department of Human Physiology, Faculty of Basic Medical Science, Rhema University, Aba, Nigeria
[8]Department of Agronomy, Federal University, Gashua, Yobe, Nigeria
[9]Gomal Center of Biochemistry and Biotechnology, Gomal University, Khyber Pakhtunkhwa, Pakistan
[10]National Institute for Genomics Advanced Biotechnology (NIGAB), National Agricultural Research Centre (NARC), Islamabad, Pakistan
[11]Department of Physiology, University of Medical Sciences, Ondo City, Nigeria
[12]Department of Microbiology, Edo State University Uzairue, Iyamho, Edo State, Nigeria
[13]Department of Microbiology, Sokoto State University, Sokoto, Nigeria

**Corresponding author*: adetunjicharles@gmail.com; adetunji.charles@edouniversity.edu.ng

Charles Oluwaseun Adetunji, Julius Kola Oloke, Naveen Dwivedi, Sabeela Beevi Ummalyma, Shubha Dwivedi, Daniel Ingo Hefft and Juliana Bunmi Adetunji (eds.) Next-Generation Algae: Volume I: Applications in Agriculture, Food and Environment, (331–360) © 2023 Scrivener Publishing LLC

Abstract
Algae are a very important source of biofuels. Production of biofuels from algae as a renewable source is considered a remarkable sustainable alternative for fuels, which are an important source of economics and environmental sustainability. Microalgae, macroalgae and cyanobacteria are currently being promoted as an ideal source for biofuels because they have significant growing ability, carbon dioxide fixation ability and the capability to produce important substances such as lipids, carbohydrates and proteins by metabolizing waste streams. By metabolizing a waste stream, algae can produce a variety of useful products. Lipids are used for producing biodiesels, proteins are used for consumption by human beings and animals, and carbohydrates are used in the production of ethanol and many other useful products. This chapter is a detailed account of the use of algae for the synthesis of biofuels. The aim of this review is to evaluate whether algae have great potential for synthesizing biofuels and how they are grown and used. Throughout this chapter we want to enhance environmental awareness about algal production and their use in industry, which will alternatively help human beings fulfill their energy needs.

***Keywords*:** Biofuels, microalgae, macroalgae, cyanobacteria, biodiesels, global warming, ethanol, biogases

15.1 Introduction

The potential ecological advantages that can be gotten from replacing oil fuels with biofuels, bioenergy and sustainable biomass have been the primary impetus for advancing their creation as well as utilization. Energy and greenhouse gas (GHG) equalization of bioenergy frameworks is contingent upon the kind of feedstock sources, transformation advancements, end-use advances, and framework limits, including the energy framework with which the bioenergy chain is brought about. Territorial contrasts can be likewise critical, particularly for countryside adoption where organic matter is utilized to create fuel and the reference energy framework, and the life periodicity estimate consequence may be alternative to advance develop. Besides, any fuel that is obtained through biomass (biofuel) manufacturing ordinarily brings about the age of co-items, which can supplant ordinary items, giving further ecological advantages to the biofuel procedure chain.

Over time, numerous choices have been examined and actualized, practically speaking, with various degrees of success and in various periods of study and usage. Models are incorporated sun-oriented energy, either warm or photovoltaic, hydroelectric, geothermal, wind, biofuels, and carbon sequestration, among others [20]. Everyone has their own focus and issues as well, which are contingent upon the territory being used, and various alternatives will be more qualified. One significant objective is to measure the decrease in transportation emanations; for example, the slow substitution of petroleum derivatives by sustainable power sources, where biofuels are viewed as genuine supporters of arriving at those objectives, especially for the time being.

Biofuels creation is relied upon to present new chances to broaden financial gain and biofuels' graceful provenances, to advance work into country zones, to grow long haul substitution of petroleum derivatives, along with diminishing greenhouse gas discharges, boosting the decarbonization of transportation energies as well as expanding the security of energy flexibly. The main well-known biofuels are biodiesel and bioethanol, which may supplant diesel and gas, individually, in today's vehicles along with little or no alteration of vehicle motors. Biofuels are for the most part delivered from biomass or sustainable power sources and add to bring down burning emanations from petroleum products per comparable force yield, which may be delivered utilizing existing innovations and conveyed through the accessible circulation framework. Biodiesel is created from vegetable oils (consumable or non-palatable) or creature fats. Since vegetable oils may likewise be utilized by humans, it can prompt an expansion in cost of food-grade oils, causing the expense of biodiesel to increase and forestalling its use, regardless of whether or not it has points of interest contrasting with diesel fuel.

The potential market for biodiesel far outperforms the accessibility of plant oils not assigned for different markets. For instance, to satisfy a 10% objective of the EU for local creation, the real feedstocks aren't sufficient to fulfill the present need and the land prerequisites for biofuels creation, would be more than the potential accessible arable land for bioenergy crops [46]. The broad range and weight for land use change and increment of developed fields may prompt land rivalry and biodiversity misfortune, because of the cutting of existing backwoods and the usage of biologically significance regions [44]. Biodiesel may likewise be disadvantageous when substituting crops utilized for human utilization or if its feedstocks are developed in timberlands and other basic living spaces with related organic assorted variety.

Behind just coal and oil, biomass remains the third largest energy source on the planet [22]. One of the significant confinements of utilizing biomass as a feedstock for bioenergy items is its low mass thickness (wet premise), which ordinarily extends from 80–100 kg/m^3 for rural straws and grasses and 150–200 kg/m^3 for woody assets like wood chips and sawdust [47].

The low densities of biomass regularly makes the material hard to store, transport, and interface with biorefinery infeed frameworks. For instance, when low-thickness biomass is co-fired with coal, the distinction in thickness causes problems in taking care of the fuel into the heater and decreases consuming efficiencies [50]. The compound organization of biomass, which incorporates cellulose, hemicelluloses, protein, starch, lignin, unrefined fiber, fat, and debris, likewise influences the densification procedure. During pressure at high temperatures, the protein and starch plasticizes and becomes a fastener, which helps with expanding the quality of the pellet [25]. Lignin in the biomass at temperatures above about 140 °C relaxes and improves the original particles [1]. Examining electron magnifying instruments (SEMs) have been utilized to comprehend the strong kind of extensions framed during briquetting and pelleting of corn stover and switchgrass [36]. More investigations at a small-scale level utilizing strategies like SEM and transmission electron magnifying lens (TEM) will be valuable in comprehending intramolecule cavities, material properties, and procedure variable connections on the quality traits of densified biomass.

In addition to investigation, life-cycle assessment (LCA) has also discovered that a noteworthy total decrease in greenhouse gas (GHG) emissions and skeleton deposits are mainly recognized as being due to the use of biofuel in vehicles as well as the biofuel utilized to replace ordinary diesel fuel for engine ignition [37, 42]. A few LCAs have additionally measured the effects of biofuel from more natural angles, taking into consideration neighbourhood ozone contamination, fermentation, excess nutrients in lakes, air exhaustion, farm uses and so forth [30, 41].

15.2 General Overview

The world has been facing an energy crisis due to the deficiency of fossil fuel resources. Therefore, biomass and biofuels are being considered as some of the most important energy sources for renewable energy. Biofuels obtained from biomass are preferred over other fuels because they contain high energy security, have fewer socioeconomic issues, decrease environmental impact and result in foreign exchange savings [2–19, 53, 54, 57]. According to types, biomass have one to four generations. Among the biofuels, according to the United State Department of Energy, biodiesel and bioethanol are the main biofuels that represent first generation biofuels.

Billions of years ago, the Earth was filled with carbon dioxide and there was no life. Life on earth started with cyanobacteria and algae. These photosynthetic organisms started the release of oxygen and sucked up the carbon dioxide present in the environment. Due to decreased carbon dioxide concentration and significant present of oxygen, life began on Earth.

Algae are a diverse and large group of organisms. They are unicellular to multicellular organisms used for the synthesis of organic products from inorganic compounds by using energy obtained from chemical reactions of inorganic molecules, photosynthesis and fermentation [59]. The term algae refers as blue-green algae, including macroalgae, microalgae and cyanobacteria. Algae resides in aqueous and terrestrial environment conditions such as freshwater, marine water, brackish water and hypersaline environment. Algae have two populations; one is filamentous and the other is phytoplankton. Algae are classified into four types; diatoms, blue-green, green and golden algae. Microalgae are unicellular organisms and are divided into three groups; diatoms, golden and green algae. Cyanobacteria also come under the category of microalgae. Macroalgae are further classified into brown seaweed, red seaweed and green seaweed. All these photosynthetic organisms have great potential for energy production. This chapter intends to provide detailed information on algal biofuels production by using different processes; thermochemical processes are used for the synthesis of oil and gas and biochemical process is used for the synthesis of biodiesels, ethanol and biohydrogen.

15.3 Algae Production and Cultivation

The simplest and oldest system for cultivation of algae is open ponds. The pond was designed in such a way to contains a wheel which circulates in the pond and mixes algal cell and other nutrients. Nutrients are added into the pond via the paddle wheel or through run-off water. The pond water is kept in motion and nutrients are mixed due to circulatory motion. The other way of algal cultivation is photobioreactor system. Raceway ponds and photobioreactor open pond systems are considered the best methods for production of algae. Open pond system is used for the cultivation of the majority of algae and photobioreactor system used for production of high algae productivity [25, 62, 63]. The major benefit of algal production through open ponds is that they are easily constructed and are operated at minimum cost. The limitation of the open pond system is contamination containing algae and bacteria, poor light utilization, diffusion of carbon dioxide to atmosphere and evaporative losses. For scientific and commercial production of algae, photobioreactor system is used due to good quality control and pure culture environment. The photobioreactor system provides all growth requirements which are needed due to it being a closed system. Photobioreactor system also requires minimum cost for construction and operation of the system. For designing photobioreactor system the conditions which are optimized are volume size to surface area ratio, pure culture contaminant, contaminant for temperature control, distribution of air and carbon dioxide, transfer rate of CO_2 and automated flow through sensor. Both open pond system and photobioreactor system have some limitations along with advantages but they are the best way to produce algae.

15.3.1 Harvesting

Harvesting is a process used to obtain algal cell from water content present in culture medium. Different methods are used for harvesting algal cell. These methods are filtration, flocculation, flotation, centrifugation, sonication and precipitation. Sometimes, two or more techniques are used for getting the best result with no harm. Dewatering technique is also used for harvesting. In dewatering technique water contents are removed from cells and pure water free biomass can be obtained [64]. Table 15.1 shows the way in which these techniques work along with their advantages.

Table 15.1 Techniques involved in algae production.

Techniques	Advantages	Action
Filtration	Efficient and time-saving	Under pressure, large cells of more than 70 micrometer are filtered easily but small cells or cells less than 30 micrometers are filtered by using ultrafilters.
Flocculation	Efficient and time-saving	By using flocculent (chemicals, microorganisms and bioagents), cell size is aggravated and by increasing cell size, cells are easily obtained.

(Continued)

Table 15.1 Techniques involved in algae production. (*Continued*)

Techniques	Advantages	Action
Flotation	Cost-effective	Cells are obtained by trapping them via bubbling air.
Centrifugation	Fast and appropriate method for obtaining larger cells	Cells are obtained through sedimentation based on size, density and velocity of cells.
Sonication	Resists adsorption; efficient; cost-effective; no cell damage and takes up less space	Using acoustic forces, cells are pumped into resonator chamber and are separated.
Precipitation	Natural way of harvesting; efficient; effective and requires no chemical or energy	When circulation stops in pond algal cell settled down and are precipitated
Dewatering	Cost-effective; time-saving; does not require chemicals and doesn't harm cells	Water released from culture medium and algal cell containing biomass can easily be obtained.

15.3.2 Genetically Modified Organisms

To increase optimized production for novel forms of microorganisms, e.g., algae or bacteria, the recombinant DNA technique is used. In this technique, all types of microbes that undergo experimental modification are called genetically modified organisms. Novel and modified traits of microbes including algae are produced through this technique via gene transferring [67, 68]. Biotechnology researchers are trying to develop genetically modified organisms for the generation of biofuels [69].

15.3.3 Growth Control

Temperature of culture medium, irradiance, density of algal biomass, uniform mixing of nutrients and cells, concentration of nutrients and culture age are environmental conditions which could affect the growth of algae. All these environmental factors are examined and controlled by researchers. Nutrients which are required for algal production are sulfur, nitrogen and phosphorous. Iron and silica are trace elements and play a vital role in algal growth [66].

15.3.4 Production of Biofuels from Algae

After growing and harvesting, algae are extracted from broth culture by using alcohol or oil or other target products. Algal biomass is separated by different techniques used for the production of biofuels. Depending upon the type of algae and method of production, a substantial percentage of biomass is obtained, which could be lipid or oil. Commercial manufacture and solvent extracts are used to obtain lipid and oil from biomass. Biofuels obtained from algae are categorized as third generation biofuels. The carbohydrates obtained from algal biomass are used in the production of bioethanol, the oil obtained from biomass is used for the production of biodiesels, and the remaining biomass is used for fuel oil and methane production. The residual biomass is used in the production of nutraceuticals, therapeutics, dietary supplements as protein supplements, docosahexaenoic acid, eicosapentaenoic acid, fertilizers, biocontrol agents and animal feed.

15.3.5 Biochemical Conversion

Biochemical conversion requires the hydrolysis of cell wall microbes to ferment sugar, which is referred to as anaerobic digestion of sugar to bioethanol, biogas and biohydrogen. Biogas produced by acetogenesis involves the oxidation of all fermentable products into acetate, which are then converted to methane and carbon dioxide during methanogenesis [70]. Production of biogas could be influenced by C:N ration, pH, feeding rate, solid and temperature. Bioethanol is obtained from the fermentation of yeast of hydrolyzed carbohydrates [74]. Cyanobacterial glycogen was also reported to have a role in the production of bioethanol. For efficient production of bioethanol, pretreatments such as milling, liquefaction, hot H_2O wash, alginate extraction, and enzymatic hydrolysis are essential [75, 76]. Oxygen can suppress hydrogenase reaction by acting as photosynthetic byproduct.

15.3.6 Thermochemical Process

Thermochemical process requires the thermal breakdown of biomass. After the breakdown, the biomass converts into biofuels by gasification, pyrolysis and hydrothermal liquefaction [72]. Solid fuels, liquid fuels and gaseous fuels are obtained by pyrolysis of biomass. Algal pyrolysis is an important conversion process for the synthesis of bio-oils [73]. Algal pyrolysis is considered an efficient method for bio-oil production. Gasification involves the oxidation of biomass with controlled amount of oxygen, air or stream. Biocrude, char and gas are produced through the hydrothermal liquefaction of algal biomass [29].

15.3.7 Transesterification

Transesterification is a conversion process used for the synthesis of biodiesels and glycerol [34]. Transesterification process involves the reaction of triglycerides with alcohol in the presence of acidic or basic catalyst [71]. This process significantly reduces viscosity of algal oil and enhances fluidity in order to be applied on engine directly or mixed with petroleum diesel [70]. This process involves the extraction of algal lipids, which are then used for biodiesel production. The main fatty acids which are used for the synthesis of biodiesels are palmitic acid, oleic acid, stearic acid, linoleic acid and linolenic acid [77]. Diatoms, chlorophytes and cyanobacteria having high lipid content play a vital role in the synthesis of biodiesels [78].

15.4 Algal Biofuels from Macroalgae

Macroalgae seaweed contains no lignin and low cellulose content. They contain lipid which is used for the synthesis of biofuels. Macroalgal species *Ascophyllum nodosum* have high polyphenol content and play a vital role in biofuel production. *Ulva lactuca* have 5% sulfur content, leading to a significant level of hydrogen sulfide used for the synthesis of biomethane [39].

15.5 Algal Biofuels from Cyanobacteria and Microalgae

Cyanobacteria referred to as blue-green algae have recently gained great interest from researchers due to their potential for generation of chemical products and biofuels [58]. They are great platform and have possess potential for the production of biofuels due to their fast-growing ability and CO_2 fixation ability. Cyanobacteria and microalgae are considered the primary organisms for biofuel synthesis. Both of these do not require arable land to grow and fermentable sugar. Research on cyanobacteria has primarily cantered on their ability to produce a large amount of lipids, which are used for the synthesis of biodiesels [83]. Cyanobacteria have been engineered for the production of biofuels and related compounds [82]. Cyanobacterium *Synechococcus enlongatus* have been grown successfully by researchers for ethanol production by the addition of alcohol dehydrogenase and pyruvate decarboxylase redirection of carbon from pyruvate [79]. Cyanobacteria use for ethanol synthesis has been successfully improved [80, 81]. Cyanobacteria are also known as photoautotrophic prokaryotic organisms. They fix carbon dioxide and convert solar energy into chemical energy through photosynthesis. This photosynthetic ability of cyanobacteria is much higher than in higher plants. The use of cyanobacterium *Anabaena* sp. in carbon dioxide removal from flue gases has been increased. Cyanobacterial species *Synechocystis*, *Synechococcus*, and *Anabaena* have biomass doubling mass. Recently, cyanobacteria have become known as a novel cell factory for biofuels [60, 61]. They contain a large amount of lipids, carbohydrates, proteins and fatty acids. Carbohydrates used in bioethanol synthesis, lipids are converted to biodiesels, fatty acid form of butyrate, acetate and biodiesels obtained in fermentation, and the protein used in dietary supplements [88, 89].

Microalgae have been cultivated for many years on small scale for biofuel production, highly valuable products and animal feed. Second generation microalgal system has produced a high range of biofuels such as biodiesel, bioethanol, biohydrogen and biomethane. For high biodiesel production, increased lipid content is very important and microalgae possess a high lipid content. Highly efficient oil produced from lipid is used for the synthesis of biodiesel. The way in which microalgae are used for biodiesel production is the same as the synthesis of other biofuels. Biohydrogen production from microalgae has been used for 65 years and the algal species recognized for the first time for biohydrogen production was *Scenedesmus obliquus* [59]. Afterwards, cyanobacteria and green algae *Chlamydomonas reinhardtii* were also recognized for the synthesis of biohydrogen [86, 87].

It has been documented that by using sunlight microalgae could convert water to oxygen and hydrogen in two phases. Cyanobacteria also have the potential to produce hydrogen from water but uses it in an alternate way. Biogas production from biomass has gained importance all over the world. Biomethane is an important biogas which is obtained from microalgal biomass [65]. Microalgae have gained major attention as a point of interest by researchers for efficient production of biomass. Due to the abundance of lipid, protein and starch content in microalgae and absence of lignin (which is not fermented easily), microalgae are considered an ideal candidate for highly efficient biomethane production. Third generation biofuels are called advanced biofuels and microalgae have a tremendous variety for algal biomass production [84].

15.6 Types of Algal Biofuels

15.6.1 Hydrocarbons

Hydrocarbons, including gasoline, jet fuel and biodiesel, are known as fuels which do not contain oxygen but hydrogen and carbon. Hydrocarbon fuels derived from algal biomass or algal oil by thermochemical process are converted to hydrocarbons. Ethane, methane and propane also come under the category of hydrocarbon and are well known. Algal specie *Botryococcus braunii* is used for the synthesis of hydrocarbons [95].

15.6.2 Bioethanol

Bioethanol is the major fuel produced from cyanobacteria. Ethanol from cyanobacteria is synthesized by the expression of two enzymes; alcohol dehydrogenase and pyruvate decarboxylase. Ethanol synthesis from cyanobacteria has been increased by genetical approaches [80]. Additionally, algal biofuel from cyanobacteria in Nepal has produced ethanol by converting solar energy to chemical energy and fixing carbon dioxide.

15.6.3 Isobutanol

Isobutanol is an important biofuel which has been used widely in petrochemical industries. As compared to ethanol, isobutanol or butanol is considered an advanced biofuel due to its high density, less vapor pressure, low hygroscopicity and decreased corrosiveness [85]. Butanol is produced through the acetone butanol ethanol fermentation pathway by using *Clostridium acetobutylicum* or using keta acid pathway of *Escherichia coli* [93] in which sugar is used as raw material cyanobacterium *Synechococcus elongatus* PCC 7942, used for the synthesis of isobutanol from solar energy and carbon dioxide [93]; or by using isobutanol synthetic pathway in which ribulose 1,5, bisphosphonate oxygenase is used as key enzyme.

15.6.4 Isoprene

Isoprene is an important feedstock used in chemical industries. It is an advanced biofuel due to its greater energy storing capability. *Cyanobacterium synechocystis* is used for the synthesis of isoprene by using solar energy and carbon dioxide, which is cost-effective. Table 15.2 shows different types of microalgae used in the production of different biofuels.

Table 15.2 Different types of microalgae used in the production of different biofuels.

Microalgae	Type of biofuel
Chlorella sp. *Dunaliella tertiolecta* *Chlorella pyrenoidosa* *Oscillatoria* sp. *Nannochloropsis oculate* *Desmodesmus quadricaudatus* *Spirulina* sp. *Schizochytrium limacinum*	Biodiesel
Anabaena cylindrical *Chlamydomonas reinhardtii* *Mastigocladus laminosus*	Biohydrogen
Chlamydomonas reinhardtii *Chlorococcum* sp.	Bioethanol
Chlorella vulgaris	Ethanol
Chlorella vulgaris *Phaeodactylum tricornutum* *Chlorogloeopsis fritschii* *Porphyridium cruentum* *Nannochloropsis* sp. *Scenedesmus dimorphus* *Spirulina platensis* *Bacillariophyta* sp. *Tetraselmis* sp. *Cyanobacteria* *Desmodesmus* sp.	Bio-oil

(*Continued*)

Table 15.2 Different types of microalgae used in the production of different biofuels. (*Continued*)

Microalgae	Type of biofuel
Chlorella vulgaris *Tetraselmis* sp. *Nannochloropsis* sp. *Saccharina latissimi* *Nannochloropsis oculate* *Spirulina platensis* *Nannochloropsis gaditana*	Syngas
Emiliania huxleyi	Gas
Chlorella sp.	Oil/gas
Chlorella vulgaris *Euglena gracilis* *Spirulina* sp. *Ulva lactuca* *Arthrospira maxima* *Scenedesmus obliquus*	Methane
Botryococcus braunii	Biohydrocarbons
Cyanobacterium Synechococcus elongatus PCC 7942	Isobutanol
Cyanobacterium synechocystis	Isoprene
Microcystis aeruginosa *Chlorella protothecoides*	Oil
Chlorella vulgaris *Tetraselmis chuii* *Dunaliella tertiolecta* *Synechococcus* *Nannochloropsis* sp.	Char

15.6.5 Biodiesel

Biodiesel can be defined as mono-alkyl ester of long-chain fatty acids obtained from the oil of canola, soybean, sunflower, cotton seeds, animal feed and lipid derived from microalgae. Biodiesel is an important alternative fuel which can be synthesized from a variety of renewable resources. Usually, biodiesels are produced by mixing lipids with alcohol using basic catalyst as KOH through transesterification process. In this process, methane bonds with oil and forms mono-alkyl ester. In one study it was stated by researchers that lipase enzyme was used in the process of transesterification. Lipase enzyme is generally obtained from fungus and Gram-negative bacteria. In another process, ethanol was used instead of methanol in transesterification process because methanol is highly toxic and can burn skin on ignition [90–92].

15.6.6 Biohydrogen

Biohydrogen is a clear and good fuel obtained from algal biomass. Biohydrogen from microalgae obtained through the process of photobiological water splitting. Microalgal specie *Chlamydomonas reinhardtii* has been demonstrated to use for the synthesis of hydrogen. On sulpher deprivation stage, *Chlamydomonas reinhardtii* stop oxygen synthesis and increase hydrogen production. Hydrogen is used for the synthesis of power, electricity and heat used in transportation [55, 56].

15.6.7 Biomethane

The idea for the production of methane from algal biomass by carbohydrate fraction of cell was first proposed by Meier. Later on, this idea was further developed by Oswald and Golueke [94]. They introduced a conceptual technoeconomic engineering analysis of digesting microalgal biomass grown in large raceway ponds to produce methane gas. Biomethane is produced from "fresh" organic matter in landfills or biogas plants. Both methods are based on anaerobic digestion which is performed by the anaerobic microbes that thrive in absence of oxygen [96].

15.7 Biomass Supply

The extensive scope of biomass provenances can be utilized to deliver electricity generated from organic matter (bioenergy) in an assortment of structures. For example, a procedure to use deposits and energy yields may be used to produce power, heat, consolidated warmth and force and stream/strong/fluid biofuels. Bioenergy provides about 10% of the world's total primary energy supply today, a large portion of which is utilized in private sectors for heating purposes [32]. Conventional biofuels (fuelwood and charcoal, frequently utilized with low effectiveness) are created in nations in which it is acceptable that 95% of the national energy utilized depends on organic matter used to get energy. Conflictingly, in these nations, productive biomass use is becoming more significant at a time when less carbon is dispersed, and is becoming a sustainable segment of national energy policy frameworks. Truth be told, use of the current energy obtained from organic matter (bioenergy) appliances being developed particularly burn an organic matter with coal, covert the organic matter's carbonaceous gases (gasification) and utilize the organic-based fuel in vehicles (for the most part, bioethanol and biodiesel).

15.7.1 Biomass from Dedicated Energy Crops

Dedicated crop yields are primarily developed for energy; however, they may likewise have non-energy side effects. The perfect energy crop has productive sun-based energy transformation for bringing about exceptional returns (carbon fixation plants (C4 plant) effectively endure change under immense radiation and high warmth conditions), demands less agrochemical load, has less of level of H_2O prerequisite, and has lower moisture levels at harvest. At the same time, it is hard to find a yield that combines types of a particular models; although enduring C4 grasses, for example, miscanthus and switchgrass (*Panicum virgatum* L.), are especially encouraging in this regard [51].

Plants have a propensity for lasting development. Propensities admit the upsides regardless of foundation expenditures (just as arrived at this midpoint of over the pivot) and more noteworthy strength in dry spells. When ignition is the end use of organic material for bioenergy, output is very most likely the main point of elective harvests. Although as input limit the condition of usage, along with appropriateness as concern the profoundly huge cost of harvest (for example, bioethanol creation, diesel obtained from organic matter) condition along with appropriateness as concern the harvest endure profoundly huge. These corresponding financial restorations endure probably going to after time this significant commute into choosing the result as regard rivalry as countryside appliance among the energy obtained from organic materials (bioenergy) along with the manufacture of foodstuff, nutrients and grain. The correlative restoration (that is energy obtain from contrasted biomass and countryside appliance) will be affected by correlative outputs along with qualities, that endure issues through showcase powers and vend bends (for example endowments). The yield and estimation of side effects (e.g., feed) will likewise be noteworthy.

15.7.2 Biomass Debris and Waste

It has been observed that most organic biomass and waste most especially from horticulture or waste derived from agricultural farmland most especially from biomass debris could be redirected from a landfill and applied in energy recovery. A typical example of this is methane discharges are removed through the anaerobic breakdown of organic-based biomass in a site used in the eventual removal of the waste matter [33]. Unlike committed fields harvested for bioenergy, organic waste as well as build-ups that are created explicitly are being utilize as a vitality asset. They endure to the consequence as monetary action along with creation as the products into practically entire divisions of the financial system. These organic wastes that are being redirected away from landfills for energy recovery can likewise reduce a portion of the natural wastes related with landfill; for example, methane discharges removed through the anaerobic breakdown of organic-based biomass in a site used to dispose waste matter.

15.8 Organic Material-Based Energy: CO_2 Impartiality and Its Effects on Carbon Pools

Organic materials are utilized for providing energy to deal with "unbiased carbon" above its lifespan since ignition of the organic matter discharges a similar measure of CO_2 captured by the plant during its development. On the other hand, non-renewable energy sources discharge CO_2 that has been acquired and pent up for many years. Although energy obtained from biomass has a nearly closed carbon dioxide lifespan, there are GHG discharges throughout its life cycle, to a large extent from the creation stages: outside petroleum derivative sources of information are required as well as to deliver as well as collect the raw material used to provide energy, into preparing as well as taking care of organic matters, in biomass plant activity and to delivery of organic raw materials and fuel obtained from biomass. Furthermore, the accumulation of organic materials can cause changes in carbon storage above and below ground, and these changes are generally not taken into account in the GHG parity of bioenergy frameworks, with few exceptions [24, 26]. C is primarily stored in three distinct pools: planting (including rhizomes), trash, and land. When changing area usage, these specific capacity pools may change in anticipation of another balance being reached. For example, these additional uses as woodland deposits to obtain energy from biomass may result in a reduction of carbon stockpiling into backwoods garbage and land ponds, as the buildups endure to no, at this point left on the ground. This is a significant angle in light of the massive amounts of C in land natural issue: lands have approximately 50-300 t C/ha, compared to 2-20 t/ha in field as an alternative harvest organic material. These dirt C ponds are estimated to affect 2500Gt of carbon globally, as opposed to 560Gt C amount in vegetable agriculture and 760Gt in the air [38]. Due to the size of these dirt C ponds, even small increases rather than decreases in their capacity can be noteworthy on a global scale. These opportunities to remove carbon from the atmosphere into the land are primarily site-specific and highly dependent on past and present agricultural practices, the environment, and the characteristics of the land [38].

15.9 Non-CO_2 GHG Emissions in Bioenergy Systems

15.9.1 N_2O Emissions

The commitment to net GHG emissions of N_2O, which result from the use of nitrogen-rich compost and natural problem breakdown in soil, is an important factor in LCA considerations [48]. The supplement equalization of the dirt is impacted by the application of manure to horticulture land. Field emissions vary depending on the crop, culturing method, soil type, atmosphere, and the quantities of compost and fertilizer used [38]. These weaknesses into actual emissions continue to be magnified by N2O's significant potential to raise Earth's temperature, which is multiple times more significant than carbon dioxide [31]. In the view of preparation %, these impacts of Nitrous oxide, outflows survive particularly large as annual crops increase for the generation of biofuels, In comparison to lasting the vitality of crops. De-nitrification, the important process causing N2O production, is boosted under soggy land positions where O2 accessibility is lower, and this results in harvests developed in heavy precipitation situations instead below stream water systems having the most Nitrous oxide outflows [52]. Many life cycle assessment factors exclude N2O outflows because they use delinquency emanation factors provided by the IPCC, which are then measured by a few sources [97].

15.9.2 CH_4 Emanations

Methane is the third most significant greenhouse gas after carbon dioxide and nitrous oxide. It is released during the process of turning biomass into energy by burning petroleum derivatives in the absence of oxygen (anaerobic decomposition) as a natural feedstock, along with emissions that remove land-based environmental problems. Actually, increasing agricultural and lignocellulose yields may reduce the oxidation of CH4 in active areas and hence increase the concentration of methane in the atmosphere [49]. This decrease in the uptake (oxidizing) of CH4 by the land is caused by either the use of nitrogen compost or development; the outflow of

CH_4 from developed areas is correlated with this decrease in CH_4 uptake. This decline is very sensitive due to a number of position-specific factors, including soil warmth, earth moisture, along with these amounts and sort with regard to N compost. Estimated successful discharges can vary significantly depending on the outcome: Manure use-related CH_4 emissions can range from essentially nothing to on the order of 100g CH_4/kgN [28]. For instance, the conversion of local prairies and forests to managed pastures and developed crops reduces the oxidation of methane in the soil due to N preparation and soil unsettling influence. Ultimately, developed lands exhibit lower methane take-up percentages than those covered by local conditions [98]. However, Delucchi and Lipman found that under ideal conditions, an estimate of 10 gm CH_4/kg N for CH_4 take-up decrease (which compares to equivalent CH_4 emanation) is reasonable along with leading to a typically low commitment to life cycle GHG discharges of the bioenergy chain.

15.10 Microalgae for Biodiesel Production

Microalgae are unicellular or fundamentally multicellular prokaryotic or eukaryotic photosynthetic microorganisms that can grow quickly and endure harsh environmental conditions. Examples of eukaryotic microalgae include green growth (Chlorophyta) and diatoms, whereas prokaryotic microorganisms include cyanobacteria (Cyanophyceae) (Bacillariophyta). [45] gives a more comprehensive description of microalgae. Phases of the microalgae biodiesel value chain Microalgae may not seem to differ significantly from other biodiesel feedstocks at first glance, but they are microorganisms that thrive primarily in fluid environments and require particular development, collection, and handling techniques in order to be effectively converted into biodiesel. Every current method for making biodiesel from microalgae includes a unit for cell development, followed by a step where the cells are removed from the media where they were growing and a further step where the resultant lipids are extracted. Then, using current techniques and innovations that are used to produce various biofuels' feedstock, biodiesel or other biofuels are made. Recently, alternative methods for producing biofuel have come into focus in place of the trans esterification reaction, such as warm splitting (or pyrolysis), which involves the warm decay or cleavage of triglycerides and other natural mixtures introduced in the feedstock into simpler particles, such as alkanes, alkenes, aromatics, and carboxylic acids, among others [20].

15.10.1 Biodiesel Production

Biodiesel is a mixture of unsaturated fat alkyl esters made by transesterifying vegetable or animal fats (ester trade reaction). These lipid feedstock are composed of 90–98% (weight) triglycerides, a little amount of mono- and di-glycerides, free unsaturated fats (1–5%), and traces of water and phospholipids as well as phosphatides, carotenes, tocopherols, and sulfur mixtures [23]. A different advanced reaction is transesterification, which recalls three reversible steps for arranging in which triglycerides are converted to diglycerides, diglycerides are then converted to monoglycerides, and monoglycerides are then converted to esters (biodiesel) and glycerol (side-effect).

15.11 Futurity Progression in Bioenergy

15.11.1 Second Generation Biofuels

Biofuel is currently produced using carbohydrates and vegetable oils, which poses few problems. However, this raw material competes with food as their food source and ripe soil, which limits their potential accessibility. Additionally, per-harvest income and successful investment funds come from carbon dioxide emissions as well as fossil residues, and energy use is constrained by the large amount of energy required to improve. By using plant-based biomass materials, such as constructions rather than horticulture, ranger service and manufacturing, as well as committed biomass vitality of harvest yields, these limits may be partially overcome. When compared to oil-based fuels, the original biofuels—which are obtained by compressing grain and kernel—actually provide modest greenhouse gas reduction aids. However, due to their high land requirements and rather high costs, they may only be able to provide modest long-term non-renewable energy source uprooting. The main reason for the generally poor performance of grains and seeds is that the used division accounts for only a small portion of the above-ground biomass (for instance, the rapeseed kernel yield is 3.4 t/ha, but only 41% of the kernel is lubricant, reducing the 'successful' yield to 1.35 t/ha [51]). In addition to fuel obtained from biomass that is presently commercially available, sugarcane-derived bioethanol provides a significant rural area with a capability for greenhouse gas mitigation and, in turn, creates an appealing biofuel that reduces greenhouse gas emissions. The 'next' (often referred to as second) generation of biofuels, such as FT diesel produced by Fischer-Tropsch from plant-based biomass and bioethanol produced from lignocellulosic raw material, promise advantages over the first generation of biofuels in terms of long-term land use efficiency and ecological performance. The second generation of biofuel offers a wider range of applications and can be dynamically linked to food (plant-based biomass matters), change directions (thermochemical, decomposition at high temperatures, enzymes reaction, and so on), and finished products (gas or manufactured fluid biofuels); There are currently various second generation biofuel life cycle assays available [21, 43]. According to an ongoing life cycle assessment study focused on FT diesel fuel produced by Fischer-Tropsch creation along with wooded deposits through gasification followed by FT-amalgamation, this type of energy obtained from biomass could prevent 88% of greenhouse gas emissions when compared to a fossil fuel reference system [35].

15.11.2 Biorefinery

The term "biorefinery" is gaining traction among seasoned researchers; it encompasses a broad range of innovations capable of isolating biomass resources (woods, herbaceous plants, starch, etc.) within their own structural units (starches, amino acids, lipids), which may then be converted as value-added energy along with products. The International Energy Agency (IEA) Bioenergy Task 42 on Biorefineries recently planned the method to convert biomass into biofuel (biorefinery): "Biorefining: the sustainably preparation of plant-based biomass as in a range of pleasant items along with vitality." Accordingly, a biorefinery can be thought of as a facility that uses biomass change processes and equipment to distribute biofuels, force, and synthetic mixtures across all of the landscape plant materials. These biorefineries are designed to be analogous to the current oil processing facility, which produces a variety of fuels and products from oil. In contrast to oil, biomass is not uniform because the feedstocks can include grains, wood, grasses, organic wastes, etc. This biomass compositional variety is both a benefit and a drawback. The fact that biorefineries can produce a greater variety of product classes and rely on a wider range of raw materials than oil processing plants gives them some leeway. The requirement for a typically wider range of handling innovations is a downside [27]. Accordingly, biorefinery advancements should be uniform, modest, and appropriate for surrounding businesses. The term "biorefinery" refers to a shift away from traditional processing facilities that rely on the wasteful use of natural resources and substantial waste production. Throughout the full existence of the current corn wet processing factory, a case for how the biorefinery of the future can advance may be made [40].

15.12 Conclusion

Algae are an important and vital source of energy. The process of biofuel production from algal biomass includes the cultivation, harvesting, growth control, genetical engineering and separating techniques. Biochemical, chemical and transesterification processes are also used in the production of algal biomass from cyanobacteria, macro or microalgae. Algae containing high lipid content and no lignin are the best sources of third generation biofuels. Different species of algae used for the synthesis of biofuels were mentioned in the review. Biodiesel, biohydrogen, hydrocarbons, bioethanol, methane, gas, char, bio-oils, ethanol, isoprene and isobutanol are types of biofuels obtained from algal biomass of different algal species. After the production of biofuels, the remaining biomass is used as dietary supplement and feedstock. Algal-derived biofuels are renewable biofuels that are cost-effective and impactful. In this review, light was shed on different aspects of obtaining biofuels from algae. Increased production of biofuels will help save the environment by conserving natural resources.

References

1. Adapa PK, Bucko J, Tabil LG, Schoenau G and Sokhansanj S, Pelleting Characteristics of Fractionated Suncure and Dehydrated Alfalfa Grinds. ASAE/CSAE North-Central Intersectional Meeting, Saskatoon, Saskatchewan, Canada, September 27–28 2002.
2. Adetunji, C.O., Roli, O.I., Adetunji, J.B. Exopolysaccharides Derived from Beneficial Microorganisms: Antimicrobial, Food, and Health Benefits. In: Mishra, P., Mishra, R.R., Adetunji, C.O. (eds) *Innovations in Food Technology*. Springer, Singapore, 2020. https://doi.org/10.1007/978-981-15-6121-4_10
3. Adetunji, C O. Anani, Bio-fertilizer from Trichoderma: Boom for Agriculture Production and Management of Soil- and Root-Borne Plant Pathogens. Pages 245-256. *Innovations in Food Technology: Current Perspectives and Future Goals*. Editors: Pragya Mishra, Raghvendra Raman Mishra, Charles Oluwaseun Adetunji, 2020.
4. Adetunji C.O., Varma A. Biotechnological Application of *Trichoderma*: A Powerful Fungal Isolate with Diverse Potentials for the Attainment of Food

Safety, Management of Pest and Diseases, Healthy Planet, and Sustainable Agriculture. In: Manoharachary C., Singh H.B., Varma A. (eds) *Trichoderma: Agricultural Applications and Beyond. Soil Biology*, vol 61. Springer, Cham, 2020. https://doi.org/10.1007/978-3-030-54758-5_12

5. Olaniyan O.T., and Adetunji, C.O. Biochemical Role of Beneficial Microorganisms: An Overview on Recent Development in Environmental and Agro-Science. *Microbial rejuvenation of Polluted Environment*, 2021.
6. Adetunji, C.O., Akram, M., Imtiaz, A., Bertha, E.C., Sohail, A., Olaniyan, O,P., Zahid, R., J Adetunji, J.B., Enoyoze, G.E., Sarin, N.B. Genetically modified cassava; the last hope that could help to feed the world: recent advances. Edition title: *Genetically Modified Crops - Current Status, Prospects and Challenges*. Editors: Kishor, P. B. Kavi, Rajam, M. V., Pullaiah, T. (Eds.), 2020. https://www.springer.com/gp/book/9789811559310.
7. Inobeme, A., Jeevanandam, J., Adetunji, C.O., Anani, O.A., Thangadurai, D., Islam, S., Oyawoye, O.M., Oloke, J.K., Yerima, M.B., Olaniyan, O.T. Ecorestoration of soil treated with biosurfactant during greenhouse and field trials. Green Sustainable Process for Chemical and Environmental Engineering and Science. *Biosurfactants for the Bioremediation of Polluted Environments*, 2021, Pages 89-105, 2021. https://doi.org/10.1016/B978-0-12-822696-4.00010-3
8. Anani, O.A., Jeevanandam, J., Adetunji, C.O., Inobeme, A., Oloke, J.K., Yerima, M.B., Thangadurai, D., Islam, S., Oyawoye, O.M., Olaniyan, O.T. Application of biosurfactant as a noninvasive stimulant to enhance the degradation activities of indigenous hydrocarbon degraders in the soil. Green Sustainable Process for Chemical and Environmental Engineering and Science. *Biosurfactants for the Bioremediation of Polluted Environments*, 2021, Pages 69-87, 2021. https://doi.org/10.1016/B978-0-12-822696-4.00019-X
9. Adetunji, C.O., Inobeme, A., Osikemekha, A.A., Jeevanandam, J., Yerima, M.B., Thangadurai, D., Islam, S., Oyawoye, O.M., Oloke, J.K., Olaniyan, O.T. Isolation, screening, and characterization of biosurfactant-producing microorganism that can biodegrade heavily polluted soil using molecular techniques. Green Sustainable Process for Chemical and Environmental Engineering and Science. *Biosurfactants for the Bioremediation of Polluted Environments*, 2021, Pages 53-68, 2021. https://doi.org/10.1016/B978-0-12-822696-4.00016-4
10. Jeevanandam, J., Adetunji, C.O., Selvam, J.D., Anani, O.A., Inobeme, A., Islam, S., Thangadurai, D., Olaniyan, O.T. High industrial beneficial microorganisms for effective production of a high quantity of biosurfactant. In book: *Green Sustainable Process for Chemical and Environmental Engineering and Science*, 2021.
11. Adetunji, C.O., Jeevanandam, J., Anani, O.A., Inobeme, A., Thangadurai, D., Islam, S., Olaniyan, O.T. Strain improvement methodology and genetic engineering that could lead to an increase in the production of biosurfactants, 2021.

12. Adetunji, C.O., Jeevanandam, J., Inobeme, A., Olaniyan, O.T., Anani, O.A., Thangadurai, D., Islam, S. Application of biosurfactant for the production of adjuvant and their synergetic effects when combined with different agro-pesticides, 2021.
13. Adetunji, C.O., Olaniyan, O.T., Anani, O.A., Inobeme, A., Ukhurebor, K.E., Bodunrinde, K.E., Adetunji, J.B., Singh, K.R., Nayak, V., Palnam, W.D., and Singh, R.P. Bionanomaterials for green bionanotechnology. In book: *Bionanomaterials: Fundamentals and Biomedical Applications.* IOP Publishing, 2021.
14. Adetunji C.O., Inobeme A., Olaniyan O.T., Olisaka F.N., Bodunrinde R.E., Ahamed M.I. Microbial Desalination. In: Inamuddin, Khan A. (eds) *Sustainable Materials and Systems for Water Desalination.* Advances in Science, Technology & Innovation (IEREK Interdisciplinary Series for Sustainable Development). Springer, Cham, 2021. https://doi.org/10.1007/978-3-030-72873-1_13
15. Adetunji J.B., Adetunji C.O., Olaniyan O.T. African Walnuts: A Natural Depository of Nutritional and Bioactive Compounds Essential for Food and Nutritional Security in Africa. In: Babalola O.O. (eds) *Food Security and Safety.* Springer, Cham, 2021. https://doi.org/10.1007/978-3-030-50672-8_19
16. Adetunji, C.O., Olaniyan, O.T., Anani, O.A., Olisaka, F.N., Inobeme, A., Bodunrinde, R.E., Juliana Adetunji, J.B., Singh, K.R., Palnam, W.B., Singh. R.P. Current Scenario of Nanomaterials in the Environmental, Agricultural, and Biomedical Fields. Nanomaterials in Bionanotechnology. In book: *Nanomaterials in Bionanotechnology: Fundamentals and Applications.* 1. 6. CRC Press, 2021.
17. Adetunji C.O., Kremer, R.J., Makanjuola, R., BhallaSarin, N. Application of molecular biotechnology to manage biotic stress affecting crop enhancement and sustainable agriculture. *Advances in Agronomy,* 2021. https://www.sciencedirect.com/science/article/pii/S0065211321000304?dgcid=author
18. Adetunji, C.O., Olaniyan, O.T., Adetunji, J.B., Osemwegie, O., Ubi, B.E. African Mushrooms as Functional Foods and Nutraceuticals. In book: *Fermentation and Algal Biotechnologies for the Food, Beverage and Other Bioproduct Industries.* 1st Edition. 2022. CRC Press. p 19, 2022.
19. Adetunji, C.O., Ukhurebor, K.E., Olaniyan, O.T., Ubi, B.E., Oloke, J.K., Dauda, W.T., Hefft, D.I. Recent Advances in Molecular Techniques for the Enhancement of Crop Production. In book: *Agricultural Biotechnology, Biodiversity and Bioresources Conservation and Utilization.* 1st Edition. 2022. CRC Press. p 20, 2022.
20. Adetunji C.O., Inobeme A., Olaniyan O.T., Olisaka F.N., Bodunrinde R.E., Ahamed M.I. Microbial Desalination. In: Inamuddin, Khan A. (eds) Sustainable Materials and Systems for Water Desalination. Advances in Science, Technology & Innovation (IEREK Interdisciplinary Series

for Sustainable Development). Springer, Cham, 2021. https://doi.org/10.1007/978-3-030-72873-1_13

21. Adetunji J.B., Adetunji C.O., Olaniyan O.T. African Walnuts: A Natural Depository of Nutritional and Bioactive Compounds Essential for Food and Nutritional Security in Africa. In: Babalola O.O. (eds) *Food Security and Safety*. Springer, Cham, 2021. https://doi.org/10.1007/978-3-030-50672-8_19
22. Bahadur NP, Boocock DGB, Konar SK. Liquid hydrocarbons from catalytic pyrolysis of sewage sludge lipid and canola oil: evaluation of fuel properties. Energy Fuels 9:248–56, 1995.
23. Baitz M, Deimling S, Binder M. Vergleichende Ökobilanz von SunDiesel (Choren-Verfahren) und konventionellem Dieselkraftstoff. IKP Universität Stuttgart, Institut für Kunststoffprüfung und Kunststoffkunde, PE Europe GmbH, Stuttgart, DE, http://www.choren.com/de/choren industries/informationen presse/info downloads/ (in German); 2004.
24. Bapat DW, Kulkarni SV and Bhandarkar VP, Design and operating experience on fl uidized bed boiler burning biomass fuels with high alkali ash, in *Proceedings of the 14th International Conference on Fluidized Bed Combustion*, ed by Preto FDS. ASME Publishers, New York, USA, pp. 165–174 1997.
25. Bozbas K. Biodiesel as an alternative motor fuel: production and policies in the European Union. *Renewable and Sustainable Energy Reviews* 12:542–52, 2008.
26. Bradley D. GHG Balances of forest sequestration and a bioenergy system, Case study for IEA Bioenergy Task 38 on GHG balance of biomass and bioenergy system; full report available at: ttp://www.ieabioenergy-task38.org/projects/task38casestudies/index1.htm; 2004.
27. Briggs JL, Maier DE, Watkins BA and Behnke KC, Effects of ingredients and processing parameters on pellet quality. *Poult Sci* 78:1464–1471 1999.
28. Cowie AL. Greenhouse gas balance of bioenergy systems based on integrated plantation forestry in North East New South Wales, Australia, Case study for IEA Bioenergy Task 38 on GHG balance of biomass and bioenergy system; full report available at: http://www.ieabioenergy-task38.org/ projects/task-38casestudies/index1.htm; 2004.
29. Dale BE, Kim S. Biomass refining global impact—the biobased economy of the 21st century. In: Kamm B, Gruber PR, Kamm M, editors. *Biorefineries—industrial processes and products (Status Quo Future Directions)*, vol. I. Wiley-VCH; 2006.
30. Delucchi MA, Lipman T. A Lifecycle Emissions Model (LEM): Lifecycle Emissions From Transportation Fuels, Motor Vehicles, Transportation Modes, Electricity Use, Heating and Cooking Fuels, and Materials. Appendix C: Emissions related to cultivation and fertilizer use. Davis, CA, USA: Institute of Transportation Studies, University of California, 2003; website: http://www.its.ucdavis.edu/publications/2003/UCD-ITS-RR-03-17C.pdf; 2003
31. Dewulf J, Van Langenhove H. Renewables-based technology: sustainability assessment. John Wiley & Sons, Ltd; 2006.

32. Farrell AE, Plevin RJ, Turner BT, Jones AD, O'Hare M, Kammen DM. Ethanol can contribute to energy and environmental goals. *Science* 311:506–8, 2006.
33. Forster P, Ramaswamy V, Artaxo P, Berntsen T, Betts R, Fahey DW, *et al.* Changes in atmospheric constituents and in radiative forcing. In: Solomon S, Qin D, Manning M, Chen Z, Marquis M, Averyt KB, Tignor M, Miller HL, editors. *Climate change 2007: the physical science basis. Contribution of working group I to the fourth assessment report of the intergovernmental panel on climate change.* Cambridge University Press, Cambridge, United Kingdom and New York, NY, USA; 2007.
34. GBEP. Global Bioenergy Partnership, A review of the current state of bioenergy development in G8+ 5 countries, GBEP Secretariat, Food and Agriculture Organization of the United Nations (FAO), Rome 2007; website: ftp://ftp.fao.org/docrep/fao/010/a1348e/a1348e00.pdf; 2007.
35. Hoogwijk M, Faaij A, Broek R, van den, Berndes G, Gielen D, *et al.* Exploration of the ranges of the global potential of biomass for energy. *Biomass Bioenergy* 25(2):119–33, 2003.
36. IEA (International Energy Agency) Bioenergy Task 42 on Biorefineries, Minutes of the third Task meeting, Copenhagen, Denmark, 25 and 26 March 2008, www.biorefinery.nl\IEABioenergy-Task42.
37. Jungmeier G, Lingitz A, Spitzer J, Hofbauer H, Fürnsinn S.Wood to biofuels: feasibility study for a biofuel plant in the Austrian province of Styria. In: *Proceedings of the 15th European biomass conference and exhibition—from research to market deployment*; 2007.
38. Kaliyan N and Morey RV, Natural binders and solid bridge type binding mechanisms in briquettes and pellets made from corn stover and switchgrass. *Bioresource Technol* 101:1082–1090 2010.
39. Kim S, Dale BE. Life cycle assessment of various cropping systems utilized for producing biofuels: bioethanol and biodiesel. *Biomass Bioenergy* 29:426–39, 2005.
40. Lal R. Promise and limitations of soils to minimize climate change. J Soil Water Conserv 2008;63:113–8 Larson E. A reviewof LCA studies on liquid biofuels for the transport sector, Scientific and technical advisory panel of the global environment facility (STAP)workshop on liquid biofuels, 29 August to 1 September 2005, New Delhi, India; 2005.
41. Sokhansanj S and Fenton J, Cost benefit of biomass supply and preprocessing: BIOCAP (Canada) research integration program synthesis paper. (Online). Available at: http://www.biocap.ca/rif/reprt/Sokhansanj_S.pdf (June 22, 2011)
42. Yenikaya C, Yaman H, Atar N, Eredogan Y, Colak F. Biomass resources and decolorization of acidic dyes from aqueous solutions by biomass biosorption. *Energy Educ Sci Technol Part A* 24: 1–13, 2009.
43. Beer, L. L.; Boyd, E. S.; Peters, J. W.; Posewitz, M. C. Engineering algae for biohydrogen and biofuel production. *Curr. Opin. Biotechnol.* 20, 264–271, 2009.

44. Mullner, K.; Happe, T. Biofuel from algae-photobiological hydrogen production and CO2-fixation. *Int. J. Energy Technol. Policy* 5, 290–295, 2007.
45. Kan A. neral characteristics of waste management: a review. *Energy Educ Sci Technol Part A* 23:55–69, 2009.
46. Evangelista, V., Barsanti, L., Frassanito, A. M., Passarelli, V., Gualieri, P. Algal toxins: Nature, occurrence, effect and detection. In *Proceedings of the NATO Advanced Study Institute on Sensor Systems for Biological Threats*; The Algal Toxins Case: Pisa, Italy, 2008.
47. Graham, L. E.; Graham, J. E.; Wilcox, L. W. Algae, 2nd ed.; Benjamin-Cummings Publishing: Menlo Park, CA, 2008.
48. Li, X.; Xu, H.; Wu, Q. Large-scale biodiesel production from microalga Chlorella protothecoides through heterotrophic cultivation in bioreactors. *Biotechnol. Bioeng.* 98, 764–771, 2007.
49. Briggs, M. Widescale Biodiesel Production from Algae; University of New Hampshire Biodiesel Group, 2004.
50. Gonzalez-Fernández, C.; Molinuevo-Salces, B.; Garcíá-Gonzalez,, M. C. Nitrogen transformations under different conditions in open ponds by means of microalgae–bacteria consortium treating pig slurry. *Bioresour. Technol.* 102, 960–966, 2011.
51. Venkata Mohan, S.; Bhaskar, Y. B.; Krishna, T. M.; Chandrasekhara Rao, N.; Lalit Babu, V.; Sarma, P. N. Biohydrogen production from chemical wastewater as substrate by selectively enriched anaerobic mixed consortia: Influence of fermentation pH and substrate composition. *Int. J. Hydrogen Energy* 32, 2286–2295, 2007.
52. Butterfi, B.A.; Jones, J. Harvesting of algae grown in agricultural wastewaters. Trans. Geophys. Union. 50, 612, 1969.
53. Chen, C.-L.; Chang, J.-S.; Lee, D.-J. Dewatering and drying methods for Microalgae. *Dry. Technol.* 33, 443–454, 2015. [CrossRef].
54. Sukenik, A.; Levy, R. S.; Levy, Y.; Falkowski, P. G.; Dubinsky, Z. Optimizing algal biomass production in an outdoor pond: A simulation model. *J. Appl. Phycol.* 3, 191–201, 1991.
55. Centers for Disease Control and Prevention (CDC). National Center for Environmental Health, Investigation of Human Health Effects Associated with Potential Exposure to Genetically Modified Corn; A Report to the U.S. Food and Drug Administration from the Centers for Disease Control and Prevention: Atlanta, GA, 2001.
56. U.S. Department of Energy (DOE), Office of Science, Genomics: GTL Systems Biology for Energy and Environment, Cellulosic Ethanol: Fuel Ethanol Production, 2008. http:// genomicsgtl.energy.gov (accessed).
57. Adeleke, D. A., Olajide, P. A., Omowumi, O. S., Okunlola, D. D., Taiwo, A. M., & Adetuyi, B. O. Effect of Monosodium Glutamate on the Body System. *World News of Natural Sciences*, *44*, 1-23, 2022.
58. Adetuyi B.O, Oluwole E.O Dairo J.O Chemoprotective Potential of Ethanol Extract of *Ganoderma Lucidum* on Liver and Kidney Parameters in

Plasmodium Beghei-Induced Mice, *International Journal of Chemistry and Chemical Processes* (IJCC). Vol 1(8).:29-36, 2015.
59. Adetuyi B.O, Oluwole E.O, Dairo J.O Biochemical effects of shea butter and groundnut oils on white albino rats, *International Journal of Chemistry and Chemical Processes* (IJCC). Vol 1(8).:1-17, 2015.
60. Adetuyi BO, Dairo JO, Didunyemi MO Anti-Hyperglycemic Potency of *Jatropha Gossypiifolia* in Alloxan Induced Diabetes. *Biochem Pharmacol (Los Angel)* 4(5):193, 2015.
61. Adetuyi, B. O., Omolabi, F. K., Olajide, P. A., Oloke, J. K. Pharmacological, Biochemical and Therapeutic Potential of Milk Thistle (Silymarin): A Review *World News of Natural Sciences* 37:75-91, 2021.
62. Adetuyi, B. O., & Farombi, E. O. 6-Gingerol, an active constituent of ginger, attenuates lipopolysaccharide-induced oxidation, inflammation, cognitive deficits, neuroplasticity, and amyloidogenesis in rat. *Journal of Food Biochemistry, 45*(4), e13660, 2021.
63. Adetuyi, B. O., Adebayo, P. F., Olajide, P. A., Atanda, O. O., Oloke, J. K. . Involvement of Free Radicals in the Ageing of Cutaneous Membrane. *World News of Natural Sciences* 43, 11-37, 2022.
64. Adetuyi, B. O., Adebisi, O. A., Adetuyi, O. A., Ogunlana, O. O., Toloyai, P-B., Egbuna, C., Uche, C. Z., Khan, J., Adumanya, O. C., Patrick-Iwuanyanwu, K.C., Ficus exasperata Attenuates Acetaminophen-Induced Hepatic Damage via NF-κB Signaling Mechanism in Experimental Rat Model BioMed Research International, 2022, 2022.
65. Adetuyi, B. O., Adebisi, O. A., Awoyelu, E. H., Adetuyi, O. A., Ogunlana, O. O. Phytochemical and Toxicological effect of Ethanol extract of *Heliotropium indicum* on Liver of Male Albino Rats. *Letters in Applied NanoBioscience (LIANB)* 10(2):2085-2095, 2020. https://doi.org/10.33263/LIANBS102.20852095
66. Adetuyi, B. O., Odine, G. O., Olajide, P. A., Adetuyi, O. A., Atanda, O. O., Oloke J. K. Nutraceuticals: role in metabolic disease, prevention and treatment, *World News of Natural Sciences* 42, 1-27, 2022.
67. Adetuyi, B. O., Ogundipe, A. E., Ogunlana, O. O., Egbuna, C., Estella, O. U., Mishra, A. P., Akram, M., & Achar, R. R. Banana Peel as a Source of Nutraceuticals. Food and Agricultural Byproducts as Important Source of Valuable Nutraceuticals. 1st ed. 2022. Springer, Berlin. 400 p. 2022. ISBN 978-3-030-98759-6
68. Adetuyi, B. O., Okeowo, T. O., Adetuyi, O. A., Adebisi, O. A., Ogunlana, O., Oretade, O. J., Marraiki, N., Beshbishy, A. M., Welson, N. N., & Batiha, G. E Ganoderma lucidum from red mushroom attenuates formaldehyde- induced liver damage in experimental male rat model. *Biology* 2020, *9*(10), 313, 2020. https://doi.org/10.3390/biology9100313
69. Adetuyi, B. O., Olajide, P. A., Awoyelu, E. H., Adetuyi, O. A., Adebisi, O. A., Oloke, J. K., Epidemiology and Therapeutic measure for COVID-19; A review. *African Journal of Reproductive Health June 2020* (Special Edition on

COVID-19); 24 (2):142, 2020. https://www.ajrh.info/index.php/ajrh/article/view/2300

70. Adetuyi, B. O., Olajide, P. A., Omowumi, O. S., Odine, G. O., Okunlola, D. D., Taiwo, A. M., & Opayinka, O. D. Blockage of Alzheimer's gene: Breakthrough effect of Apolipoprotein E4. *African Journal of Advanced Pure and Applied Sciences (AJAPAS)*, 26-33, 2022.
71. Adetuyi, B. O., Toloyai, P. Y., Ojugbeli, E. T., Oyebanjo, O. T., Adetuyi, O. A., Uche, C. Z., Olisah, M. C., Adumanya, O. C., Jude, C., Chikwendu, J. K., Akram, M., Awuchi, C.G., Egbuna, C. Neurorestorative Roles of Microgliosis and Astrogliosis in Neuroinflammation and Neurodegeneration. *Scicom Journal of Medical and Applied Medical Sciences* 1(1):1-5, 2021.
72. Adetuyi, B.O., Olajide, P. A., Oluwatosin, A., Oloke, J. K. Preventive Phytochemicals of Cancer as Speed Breakers in Inflammatory Signaling. *Research Journal of Life Sciences, Bioinformatics, Pharmaceutical and Chemical Sciences* 8 (1) 30-61, 2022.
73. Adewale, G. G., Olajide, P. A., Omowumi, O. S., Okunlola, D. D., Taiwo, A. M., & Adetuyi, B. O. Toxicological Significance of the Occurrence of Selenium in Foods. *World News of Natural Sciences, 44*, 63-88, 2022.
74. Awoyelu, E. H., Oladipo, E. K., Adetuyi, B. O., Senbadejo, T. Y., Oyawoye, O. M., Oloke, J. K. Phyloevolutionary analysis of SARS-CoV-2 in Nigeria. *New Microbes and New Infections* Vol 36. 100717, 2020. https://doi.org/10.1016/j.nmni.2020.100717
75. Batiha, G. B., Awad, D. A., Algamma, A. M., Nyamota, R., Wahed, M. I., Shah, M. A., Amin, M. N., Adetuyi, B. O., Hetta, H. F., Cruz-Marins, N., Koirala, N., Ghosh, A., & Sabatier, J. Diary-derived and Egg White Proteins in Enhancing Immune System against COVID-19 *Frontiers in Nutritionr. (Nutritional Immunology)* 8:629440, 2021.
76. Didunyemi, M. O., Adetuyi, B. O., & Oyewale, I. A. Inhibition of lipid peroxidation and *in-vitro* antioxidant capacity of aqueous, acetone and methanol leaf extracts of green and red Acalypha wilkesiana Muell Arg. *Int J Biol Med Res.*11(3):7089-7094, 2020.
77. Didunyemi, M. O., Adetuyi, B. O., Oyebanjo, O. O. Morinda lucida attenuates acetaminophen-induced oxidative damage and hepatotoxicity in rats, *Journal of Biomedical sciences Vol 8; No*, 2019. https://www.jbiomeds.com/biomedical-sciences/morinda-lucida-attenuates-acetaminopheninduced-oxidative-damage-and-hepatotoxicity-in-rats.php?aid=24482
78. Farombi, E. O., Abolaji, A. O., Adetuyi, B. O., Awosanya, O., & Fabusoro, M. Neuroprotective role of 6-Gingerol-rich fraction of Zingiber officinale (Ginger) against acrylonitrile-induced neurotoxicity in male Wistar rats. *Journal of Basic and Clinical Physiology and Pharmacology, 30*(3), 2019.
79. James-Okoro, P. O., Iheagwam, F. N., Sholeye, M. I., Umoren, I. A., Adetuyi, B. O., Ogundipe, A. E., Braimah, A. A., Adekunbi, T. S., Ogunlana, O. E., & Ogunlana, O. O. Phytochemical and *in vitro* antioxidant assessment of Yoyo bitters *World News of Natural Sciences* 37:1-17, 2021.

80. Nazir, A., Itrat, N., Shahid, A., Mushtaq, Z., Abdulrahman, S. A., Egbuna, C., Adetuyi, B. O., Khan, J., Uche, C. Z, Toloyai, P. Y. Orange Peel as a Source of Nutraceuticals. Food and Agricultural Byproducts as Important Source of Valuable Nutraceuticals.1st ed. 2022 2022. Springer, Berlin. 400 pp. 2022.
81. Ogunlana, O. O., Adetuyi, B. O., Adekunbi, T. S., Adegboye, B. E., Iheagwam, F. N., Ogunlana, O. E. Ruzu bitters ameliorates high–fat diet induced non-alcoholic fatty liver disease in male Wistar rats. *Journal of Pharmacy and Pharmacognosy Research* 9(3), 251-26, 2021.
82. Ogunlana, O. O., Adetuyi, B. O., Rotimi, M., Esalomi, I., Adeyemi A., Akinyemi, J., Ogunlana, O., Adetuyi, O., Adebisi, O., Okpata, E., Baty, R., & Batiha, G. Hypoglycemic Activities of Ethanol Seed Extract of *Hunteria umbellate* (Hallier F.) on Streptozotocin-induced Diabetic Rats. *Clinical Phytoscience,* 7 (1), 1-9, 2021.
83. Ogunlana, O. O., Ogunlana, O. E., Adekunbi, T. S., Adetuyi, B. O., Adegboye, B. E., & Iheagwam, F. N. Anti-inflammatory Mechanism of Ruzu Bitters on Diet-Induced Nonalcoholic Fatty Liver Disease in Male Wistar Rats. *Evidence-Based Complementary and Alternative Medicine, 2020,* 2020. https://doi.org/10.1155/2020/5246725
84. Ogunlana, O. O., Ogunlana, O. E., Popoola, J. O., Adetuyi, B. O., Adekunbi, T. S., David, O. L., Adeleye, O. J., Udeogu, S. A., & Adeyemi, A. O. Twigs of Andrographis paniculata (Burn. F) Nees attenuates Carbon Tetrachloride (CCl4) Induced Liver Damage in Wistar Albino Rats. *RASAYAN Journal of Chemistry* 14(4): 2598-2603, 2021.
85. Ogunlana, O. O., Adetuyi, B. O., Esalomi, E. F., Rotimi, M. I., Popoola, J. O., Ogunlana, O. E., & Adetuyi, O. A. Antidiabetic and Antioxidant Activities of the Twigs of Andrograhis paniculata on Streptozotocin-Induced Diabetic Male Rats. *BioChem* 1(3):238-249, 2021.
86. Olajide, P. A., Adetuyi, O. A., Omowumi, O. S. & Adetuyi, B. O. Anticancer and Antioxidant Phytochemicals as Speed Breakers in Inflammatory Signaling. *World News of Natural Sciences* 44, 231-259, 2022.
87. Olajide, P. A., Omowumi, O. S., Okunlola, D. D., & Adetuyi, B. O. Deadly Pandemia: Monkeypox Disease, a Case Study. *African Journal of Advanced Pure and Applied Sciences (AJAPAS)*, 34-37, 2022.
88. Olajide, P. A., Omowumi, S. O., & Odine, G. O. Pathogenesis of Reactive Oxygen Species: A Review. *World News of Natural Sciences* 44:150-164.
89. Teixeira, A.C.R.; Sodré, J.R.; Guarieiro, L.L.N.; Vieira, E.D.; de Medeiros, F.F.; Alves, C.T. A Review on Second and Third Generation Bioethanol Production; SAE International: Warrendale, PA, USA, 2016; pp. 1–7, 2022.
90. Lee, J.H.; Kim, S.B.; Kang, S.W.; Song, Y.S.; Park, C.; Han, S.O.; Kim, S.W. Biodiesel production by a mixture of Candida rugosa and Rhizopus oryzae lipases using a supercritical carbon dioxide process. *Bioresour. Technol.* 102, 2105–2108, 2011. [CrossRef] [PubMed]

91. Bajhaiya, A.K.; Mandotra, S.K.; Suseela, M.R.; Toppo, K.; Ranade, S. Algal Biodiesel: The next generation biofuel for India. *Asian J. Exp. Biol. Sci* 1, 728–739, 2010.
92. Deng, M.-D., and Coleman, J. R. Ethanol synthesis by genetic engineering in cyanobacteria. *Appl. Environ. Microbiol.* 65, 523–528, 1999.
93. Dexter, J., and Fu, P. Metabolic engineering of cyanobacteria for ethanol production. *Energy Environ. Sci.* 2, 857, 2009.
94. Gao, Z. X., Zhao, H., Li, Z. M., Tan, X. M., and Lu, X. F. Photosynthetic production of ethanol from carbon dioxide in genetically engineered cyanobacteria. *Energy Environ. Sci.* 5, 9857–9865, 2012.
95. Machado, I. M., and Atsumi, S. Cyanobacterial biofuel production. *J. Biotechnol.* 162, 50–56, 2012.
96. Pate, R., Klise, G., and Wu, B. Resource demand implications for US algae biofuels production scale-up. *Appl. Energy* 88, 3377–3388, 2011.
97. IPCC, 2006 IPCC Guidelines for National Greenhouse Gas Inventories, volume 1, 2006. https://www.ipcc-nggip.iges.or.jp/public/2006gl/
98. Mosier, A.R., Duxbury, J.M., Freney, J.R., Heinemeyer, O. & Minami, K., Nitrous oxide emissions from agricultural fields: As-sessment, measurement and mitigation. *Plant and Soil* 181:95–108, 1998.

Index

Printed and bound by CPI Group (UK) Ltd, Croydon, CR0 4YY

17/09/2026

14964274-0006